Precise Positioning with Smartphones

Precise Positioning with Smartphones

Editors

Yang Gao
Jacek Paziewski
Michael Fu
Augusto Mazzoni

MDPI

Basel • Beijing • Wuhan • Barcelona • Belgrade • Novi Sad • Cluj • Manchester

Editors

Yang Gao
Department of Geomatics Engineering, Schulich School of Engineering
University of Calgary
Calgary, AB, Canada

Jacek Paziewski
The Faculty of Geoengineering, Department of Geodesy
University of Warmia and Mazury in Olsztyn (UWM)
Olsztyn, Poland

Michael Fu
Google Android Location Team
Mountain View, CA, USA

Augusto Mazzoni
Geodesy and Geomatics Division, DICEA
Sapienza University of Rome
Rome, Italy

Editorial Office
MDPI
St. Alban-Anlage 66
4052 Basel, Switzerland

This is a reprint of articles from the Special Issue published online in the open access journal *Sensors* (ISSN 1424-8220) (available at: https://www.mdpi.com/journal/sensors/special_issues/precisepositioning_sensors).

For citation purposes, cite each article independently as indicated on the article page online and as indicated below:

Lastname, A.A.; Lastname, B.B. Article Title. *Journal Name* **Year**, *Volume Number*, Page Range.

ISBN 978-3-0365-8818-6 (Hbk)
ISBN 978-3-0365-8819-3 (PDF)
doi.org/10.3390/books978-3-0365-8819-3

Cover image courtesy of Michael Fu

Contents

About the Editors

Yang Gao

Dr. Yang Gao is a Professor at the Department of Geomatics Engineering at the University of Calgary. His research expertise includes both theoretical aspects and practical applications of satellite-based positioning and navigation systems. His current research focuses on high-precision GNSS positioning and multi-sensor integrated navigation systems. Dr. Gao has received many honors and awards for his significant contributions to the field, and is a Fellow of the US Institute of Navigation (ION), the UK Royal Institute of Navigation (RIN), the Engineering Institute of Canada (EIC), and the International Association of Geodesy (IAG).

Jacek Paziewski

Dr. Jacek Paziewski is an Associate Professor at the Department of Geodesy at the University of Warmia and Mazury in Olsztyn, Poland. He received his doctorate and habilitation degree in Satellite Geodesy in 2012 and 2018, respectively. His research interests cover precise GNSS positioning algorithm development, atmosphere modeling with GNSS measurements, the application of high-rate GNSS observations, and positioning with smartphones and low-cost receivers. He has been the investigator or principal investigator in several research projects funded by national and international agencies (e.g. European Space Agency, National Science Centre). He is an active member of the International Association of Geodesy (IAG), the European Geosciences Union (EGU), and the International Federation of Surveyors (FIG). He is also a member of the editorial boards of the *Journal of Geodesy*, *Journal of Surveying Engineering*, *Measurement Science and Technology*, *Journal of Geodetic Science*, and *Advances in Geodesy and Geoinformation*, and a reviewer for several other journals. His scientific contributions have been recognized through awards, including a scholarship for outstanding young researchers funded by the Polish Ministry of Science and Higher Education (2018–2020). Since 2020, he has been a member of the Committee on Geodesy of the Polish Academy of Science and chair of the IAG's Working Group "Reliability of Low-Cost & Android GNSS in Navigation and Geosciences".

Michael Fu

Dr. Michael Fu is a senior software engineer in the Google Android Context team, where he works on providing accurate positioning for worldwide Android users. He is a core member in the development of Android raw GNSS APIs, Google GNSS positioning algorithms, and Google Smartphone Decimeter Challenges. He served as a session chair for smartphone precision positioning at ION GNSS+ 2021, 2022, and 2023. He is a member of ION and the ION/IAG Working Group 4.1.5 "Positioning and Navigation in Asian Urban". Prior to Google, he received a Ph.D. degree in Computer Engineering from Texas A&M University in 2019, and a bachelor degree in Automation from Nanjing University in 2013.

Augusto Mazzoni

Augusto Mazzoni has been Associate Professor of Positioning and Geomatics at Sapienza University of Rome since 2018. His recent investigation topics include the exploitation of Android GNSS raw measurements, GNSS-based PVT estimation, high-rate GNSS for seismology, and ground and infrastructure deformation control. He was a recipient of the following awards: Best National PhD Thesis in Geomatics in 2009; DLR Special Topic Prize and Audience Award of the European Satellite Navigation Competition in 2010; Success Story of the European Satellite Navigation Competition in 2012; and ESA Certificate for Galileo In-Orbit Validation in 2014. He is a cofounder of Kuaternion srl (Sapienza University of Rome Start-Up Company) and co-inventor of the patent VADASE (2010).

sensors

MDPI

Article

GNSS Observation Generation from Smartphone Android Location API: Performance of Existing Apps, Issues and Improvement

Farzaneh Zangenehnejad *, Yang Jiang and Yang Gao

Department of Geomatics Engineering, Schulich School of Engineering, University of Calgary, Calgary, AB T2N 1N4, Canada
* Correspondence: farzaneh.zangenehnej@ucalgary.ca

Abstract: Precise position information available from smartphones can play an important role in developing new location-based service (LBS) applications. Starting from 2016, and after the release of Nougat version (Version 7) by Google, developers have had access to the GNSS raw measurements through the new application programming interface (API), namely android.location (API level 24). However, the new API does not provide the typical GNSS observations directly (e.g., pseudorange, carrier-phase and Doppler observations) which have to be generated by the users themselves. Although several Apps have been developed for the GNSS observations generation, various data analyses indicate quality concerns, from biases to observation inconsistency in the generated GNSS observations output from those Apps. The quality concerns would subsequently affect GNSS data processing such as cycle slip detection, code smoothing and ultimately positioning performance. In this study, we first investigate algorithms for GNSS observations generation from the android.location API output. We then evaluate the performances of two widely used Apps (Geo++RINEX logger and GnssLogger Apps), as well as our newly developed one (namely UofC CSV2RINEX tool) which converts the CSV file to a Receiver INdependent Exchange (RINEX) file. Positioning performance analysis is also provided which indicates improved positioning accuracy using our newly developed tool. Future work finding out the potential reasons for the identified misbehavior in the generated GNSS observations is recommended; it will require a joint effort with the App developers.

Keywords: smartphone positioning; precise point positioning (PPP); android location API; smartphone GNSS logging apps; GnssLogger app; Geo++ RINEX logger app; UofC CSV2RINEX tool

Citation: Zangenehnejad, F.; Jiang, Y.; Gao, Y. GNSS Observation Generation from Smartphone Android Location API: Performance of Existing Apps, Issues and Improvement. *Sensors* **2023**, *23*, 777. https://doi.org/10.3390/s23020777

Academic Editor: Maorong Ge

Received: 23 December 2022
Revised: 5 January 2023
Accepted: 5 January 2023
Published: 10 January 2023

1. Introduction

Precise position information available from smartphones is of great importance when enabling many smartphone-based location-based service (LBS) applications. Since the GNSS observations, including carrier-phase, became available to users from smart devices running Android Nougat (version 7.0) in 2016, many methods and algorithms have been developed to enable precise positioning using these mass-market devices, such as analysis of GNSS smartphone observations [1–5], PPP smartphone positioning [6–10], real-time kinematic (RTK) smartphone positioning [11–15] and GNSS/INS integration using smartphone observations [16–18]. A comprehensive review of the recent advances and research done in the field of GNSS smartphone positioning, including those published up until 2021, can be found in [19,20], while some more recent contributions in 2022 are provided in the following. Bahadur [21] investigated the real-time standalone positioning accuracy employing the single-frequency code observations form the three smartphones, Xiaomi Mi8, Google Pixel 4 and Pixel 4 XL, in the kinematic mode. The study addressed two issues: (1) comparing the ultra-rapid and IGS real-time service (IGS-RTS) and (2) investigating the effect of an improved weighting model, utilizing the variance component estimation method, on the positioning accuracy. The experimental tests indicated that the use of IGS-RTS products augments had better performance compared with the ultra-rapid products. Moreover, introducing the actual stochastic characteristics of multi-GNSS observations improved the smartphone positioning performance by 11.0% on average. Li et al. [22] proposed a combined elevation angle and carrier-to-noise density ratio (C/N0) weighting method for smartphone-based GNSS PPP by normalizing the C/N0-derived variances to the scale of the elevation-angle-derived variances. The results indicated an improvement in the three-dimensional positioning accuracy by 22.7% and 24.2% in an open-sky area, and by 52.0% and 26.0% in a constrained visibility area, compared with the elevation-angle-only and C/N0-only weighting methods, respectively. Li and Cai [23] proposed a mixed single- and dual-frequency quad-constellation PPP model to improve the smartphone positioning performance by taking advantage of all available GNSS observations. The effectiveness of the proposed model has then been investigated using both static and kinematic tests. Based on the results, the mixed frequency model could effectively improve the positioning performance compared to the traditional dual frequency PPP and the single-frequency PPP. Li et al. [24] proposed a real-time PPP algorithm for land vehicle navigation with smart devices. The smartphones were placed on the roof and the dashboard. The positioning accuracy of vehicle-roof mode was in the order of 1.0 m for the horizontal component and 1.5 m for the vertical component, while the positioning accuracy of the dashboard test were about 1.0–1.5 m in the horizontal direction and 1–2 m in the vertical direction. Li et al. [25] proposed an uncombined PPP-RTK model to achieve rapid integer ambiguity resolution (IAR) with the regional atmospheric augmentation with an external low-cost helical antenna. The results indicated that PPP rapid ambiguity resolution could be achieved using the smart devices' GNSS raw observations with a low-cost helical antenna. The method, therefore, has the potential to provide high-precision positioning services and can be widely used in massive market applications because of the advantages of low weight, low-power consumption and portability. Xu et al. [26] investigated the performance of ionospheric total electron content (TEC) determined by GNSS dual-frequency measurements derived from the Xiaomi Mi8, as an example. In this contribution, the ionospheric observable was retrieved from the code and carrier-phase measurements using the carrier-to-code leveling technique and a new carrier-to-noise weighting strategy instead of an elevation weighting strategy. The slant TEC derived from the Xiaomi Mi8 was then compared to the slant TEC derived from a geodetic receiver as the reference. According to the results, applying smart device-level GNSS observations in ionospheric studies is feasible. Zhu et al. [27] proposed an inertial measurement unit (IMU)-aided uncombined PPP coupled mathematical model, suitable for smartphone positioning. The proposed PPP/INS-coupled model integrated the dual-frequency GNSS observations and IMU data from smartphones with C/N0-dependent stochastic model and robust Kalman filter (RKF) model to improve the positioning performance under GNSS-degraded environments. Experimental results indicated that the proposed PPP/INS method could effectively improve the smartphone positioning performance compared with the conventional smartphone PPP method. Yi et al. [28] presented a novel sensor fusion technique using PPP and the inertial sensors in smartphones, combined with a single- and dual-frequency (SFDF) optimisation scheme for smartphones. Using several vehicle experiments, a significant improvement in the final solutions has been confirmed in the case of multi-GNSS PPP/IMU integration, providing consistent horizontal positioning accuracy of <2 m RMS in real-world driving scenarios.

Among the contributions in the field of differential/relative positioning, we can also refer to the following. Bakuła et al. [29] investigated the effect of L1 and L5 frequencies on the positioning accuracy of the pseudo-range differential GNSS (DGPS) using data from two Huawei P30 Pro devices. The results showed a better positioning accuracy employing the P(L5) code compared to the P(L1) code. Li et al. [30] proposed a combined RTK/fifth generation (5G) mobile communication technology positioning model by combining global positioning system-RTK with 5G time-of-arrival observations to improve the positioning accuracy under medium and long baselines. The results indicated that good positioning results could be achieved in the case of combined RTK/5G positioning model, even while some satellites are occluded. Benvenuto et al. [31] presented a method for mitigating the multipath effect in order to improve the accuracy and robustness of GNSS RTK positioning from Android smartphones. The main idea is to weigh GNSS observations of each piece of satellite data considering the proposed parameter MDP (multipath detection parameter) and signal noise ratio (SNR) values. It assigns lower weights to the unreliable observations potentially affected by the multipath error (and vice versa). Li et al. [32] proposed a single-frequency RTK robust adaptive Kalman filtering algorithm applied to smartphone GNSS positioning. It is based on the quartile method to dynamically determine the threshold value and eliminate the gross error of observation. The effectiveness of the proposed quartile robust RTK algorithm has been validated using the simulated and real dynamic experiments. According to the results, the proposed method could significantly eliminate large gross errors, and reasonably allocate weights to different observations according to the innovation vector. As a result, the overall positioning accuracies have been improved. Liu et al. [33] proposed a method to detect and correct the non-line-of-sight (NLOS) signals, which is an important issue in urban environments. This method is based on a convolutional neural network constructed using the original observations of smartphones providing the detection accuracy of more than 95%. The detected NLOS signals were decomposed using the variational mode decomposition method to eliminate the NLOS part and improve the data quality. They then evaluated the efficiency of the proposed method in both static and dynamic modes in an urban environment using the RTK method. The results confirmed an improvement in the RTK positioning accuracy in both static and dynamic tests using the proposed method. Various researchers have also been conducting investigations into the feasibility of ambiguity resolution with a smartphone receiver. For example, Miao et al. [34] first investigated the quality of L5/E5a/B2a signals, their superiorities in IAR and precise positioning with respect to the L1/E1/B1 signals from GPS, QZSS, Galileo and Beidou−2/3 satellites. The authors then proposed a new weighting model that takes into account the variation range of C/N0, providing a better weighting model than the traditional weighting model, thus improving the positioning accuracy. The results indicated that the L5/E5a/B2a signals could generally obtain higher IAR fix-rate and positioning accuracies than the L1/E1/B1 signals. Yong et al. [35] compared the best integer equivariant (BIE) estimator to the integer least squares (ILS) and float contenders using GNSS data collected by Google Pixel 4 smartphones for short-baseline RTK positioning. The results indicated that the BIE estimator will always give a better RTK positioning performance than that of the ILS and float solutions while using both single- and dual-frequency smartphone measurements for the combination of GPS + Galileo + QZSS + BDS. Li et al. [36] investigated the effect factors for integer property of phase ambiguities, data quality, IAR efficiency and positioning accuracy for the smartphone. The results indicated that the smartphone brands, operating systems and smartphone attitudes would affect the integer property of phase ambiguities and data quality. The kinematic positioning results showed the meter-level accuracy with an embedded antenna, and the centimeter to decimeter-level accuracy with the external antenna. Finally, we should note that although the results of current studies are promising, we still need more effort to achieve satisfying accuracy for many location-based services.

In 2021 and 2022, the Android GPS team of Google hosted two Google smartphone decimeter challenges (GSDC), where various smartphone GNSS datasets of real vehicular applications were used to determine smartphone positioning accuracies [37]. It was revealed that meter-level accuracy is generally achieved by the leading participants, which is still not enough to enable smartphone precise positioning. Several challenges must be taken into account in order to further improve smartphone positioning such as: carefully analyzing the smartphone GNSS observations, investigating the environment effect and smartphone holding modes and improving positioning algorithm and implementation.

It is obvious that the quality of the GNSS observations plays an important role in the final positioning performance. Currently the GNSS pseudorange, carrier-phase and Doppler observations are not directly accessible from the API 24 implemented in Android 7 or higher. They have to be generated by smartphone users from the GNSS raw information accessible through Google's application programming interface (API), namely android.location (API level 24). Various Apps for smartphone GNSS data logging and GNSS observations generation have been developed and two of them are widely used as described in the following. In 2016, Google released an open source application, namely GnssLogger App, which logs the raw measurements of the GnssClock and GnssMeasurement classes from the android.location API. This information can then be used to generate the GPS time, pseudorange, carrier-phase and Doppler observations, which, however, must be done by the users themselves. Later In 2017, the Geo++ GmbH Company released an open-source application, namely Geo++ RINEX Logger App, capable of providing the GNSS pseudorange, carrier-phase and Doppler observations in the Receiver INdependent Exchange (RINEX) format [38]. The GnssLogger App in its updated version (v3.0.0.1) can now provide not only the android.location API raw GNSS measurements in CSV format, but also GNSS observations in the RINEX format. Generating the carrier-phase and Doppler observations is straightforward (refer to Equations (3) and (4) in Section 2). However, generating the pseudorange observations might be challenging. Some further information can be found in Section 2.

Although GnssLogger and GEO++ RINEX Logger are widely used, the quality of GNSS observations output from them was found inconsistent in different aspects. In a previous work based on dataset from the Xiaomi Mi8, Google Pixel 5 and Samsung S20 smartphones, the first-order differences of pseudorange and carrier-phase observations were found not following the same trend of the Doppler observations for all three smartphones. The results also showed that the Doppler observations from the Xiaomi Mi8 and Samsung S20 smart devices were biased with respect to the pseudorange and carrier-phase observations [39]. The data logging in the RINEX format (i.e., generating the typical GNSS observations from the Android location API) was likely the cause of such observations' misbehavior. This is due to the fact that each logging App implements its own GNSS observation conversion algorithm and uses different parameter settings, thresholds and float computing accuracies. The quality concerns in the generated GNSS observations would affect GNSS data processing such as cycle slip detection, code smoothing and ultimately positioning performance. Since the data logging and GNSS observations generation are a critical part of smartphone positioning algorithm development, they should be carefully evaluated for the purpose of precise position determination. A comparison between different Apps should therefore be made to assess their consistency.

The aim of this paper is to investigate the quality of smartphone GNSS observations in RINEX format from existing smartphone GNSS logging Apps and improvement in smartphone GNSS observation generation with a focus on the following aspects:

- To provide a performance evaluation of RINEX outputs from two widely used smartphone GNSS data loggers, namely the GnssLogger App, and the Geo++ RINEX App, and also compare to our newly developed software (UofC CSV2RINEX tool). It gives the reader a great insight into the potential issues in the GNSS observations such as their inconsistency and bias issues in the smartphone pseudorange, carrier-phase and Doppler observations;
- To introduce our newly developed software (UofC CSV2RINEX tool) available at https://github.com/FarzanehZangeneh/csv2rinex, which provides improved performance. Such a tool is of value to researchers and engineers who are developing precise positioning algorithms and products with smartphones GNSS observations;
- To investigate the positioning performance of the three RINEX files in the post-processed mode using a real kinematic dataset.

The paper is organized as follows. How to convert the Android raw location-related measurements to the typical GNSS observations (e.g., pseudorange, carrier-phase and Doppler) is first explained in the next section. In the subsequent section, the employed mathematical model, which is the uncombined precise point positioning (UPPP) model, is described in detail. In the numerical results section, the quality of generated GNSS observations from different smart devices and using different loggers is assessed. In this section, the inconsistency between the pseudorange, carrier-phase and Doppler measurements reported also in Zangenehnejad et al. [39] is thoroughly investigated. The presence of some carrier-phase observations without changes over time and its possible reason will also be addressed in this part. Finally, the positioning performance of the three RINEX files (RINEX outputs from GnssLogger, Geo++ RINEX logger and UofC CSV2RINEX convertor) is investigated using the GNSS observations from the Xiaomi Mi8 in kinematic mode. We then draw some conclusions in the last section.

2. Access to Android Raw GNSS Measurements and GNSS Observation Generation

Since releasing the Nougat version of the Android system (Version 7) in 2016, the users have access to the raw GNSS measurements through the new location API consisting of two classes, GNSSMeasurement class and GNSSClock class. However, the users still need to extract the typical GNSS observations, such as pseudorange, carrier-phase and Doppler observations, from the raw data logged in the two classes. A list of raw measurements of Android 7 Location API in GNSSClock and GNSSMeasurement classes can be found in [20,40]. How to convert the raw measurements logged through the android.location API to the GNSS observations (i.e., pseudorange, carrier-phase and Doppler observations) has been provided in details in the white paper published by the European GNSS Agency's (GSA) [40]. Table 1 gives a list of available GNSS logger Android applications and their output formats.

Table 1. Available GNSS logger Android applications [20].

App	Developer	Output Format
GnssLogger	Google	CSV, NMEA and RINEX
Geo++ RINEX Logger	Geo++ GmbH Company	RINEX
rinexON	FLAMINGO	NMEA, RINEX
GalileoPVT	European Space Agency (ESA)	CSV and NMEA
G-RitZ logger	Ritsumeikan University	NMEA, RINEX
GNSS/IMU Android Logger	Universität der Bundeswehr München	CSV, RINEX and IMU data

A brief explanation about how to generate the pseudorange, carrier-phase and Doppler observations is provided in continue.

2.1. *Pseudorange Observation Generation*

The pseudorange observation is the travelling time of the signal to propagate from the satellite to the receiver (here smartphone). It is of the form [24]:

$$P = (t_{Rx} - t_{Tx}) \times 10^{-9} \times c, \tag{1}$$

where P is the pseudorange observation in meter, t_{Rx} is the received time (measurement time) in nanosecond, $t_{Tx} = ReceivedSvTimeNanos$ [ns] is the received GNSS satellite time at the measurement time in nanosecond reported in the CSV file (one of the variables in the GNSSMeasurement class) and $c = 299792458.0$ [m/s] is the speed of light. The measurement time t_{Rx} in GNSS time system in nanosecond is as follows:

$$t_{Rx\ GNSS} = TimeNanos + TimeOffsetNanos - (FullBiasNanos(1) + BiasNanos(1)), \tag{2}$$

where $TimeNanos$ is the GNSS receiver's internal hardware clock value, $TimeOffsetNanos$ is the time offset at which the measurement was taken, $FullBiasNanos$ is the difference between $TimeNanos$ inside the GPS receiver and the true GPS time since 6 January 1980 and $BiasNanos$ is the clock's sub-nanosecond bias. All of these variables can be found either in the GNSSMeasurement class or in the GNSSClock class. They all are reported in nanosecond. It should be noted that only the first value of $FullBiasNanos$ and $BiasNanos$

must be used to compute all the received times (i.e., $FullBiasNanos(1)$ and $BiasNanos(1)$) as long as there is no discontinuity in the internal received time. $t_{Rx\ GNSS}$ and t_{Tx} must also be in the same time system for all GNSS systems which is not the case here as $t_{Rx\ GNSS}$ is in the GNSS reference system while t_{Tx} is given for each GNSS system. Therefore, one must convert to other one (i.e., same GNSS time system). How to do this can be found in [20,40].

2.2. Carrier-Phase Observation Generation

The carrier-phase observation in cycle can be obtained as:

$$\varphi = AccumulatedDeltaRangeMeters/\lambda, \tag{3}$$

where $AccumulatedDeltaRangeMeters$ is the accumulated delta range (ADR) since the last channel reset which is one of the variables from GNSSMeasurement class within the Android API package "location". λ also denotes the signal's wavelength. It should also be noted that it is better to use only valid measurements for the carrier phase observation calculation. Validity of the carrier measurements can be checked using the $AccumulatedDeltaRangeState$ variable.

2.3. Doppler Observation Generation

The Doppler shift causing from the satellite movement can be obtained as follows:

$$dopplershift = -PseudorangeRatemetersperSecond/\lambda, \tag{4}$$

where $PseudorangeRatemetersperSecond$ is the pseudorange rate at the timestamp in m/s and can be found as one of the variables in GNSSMeasurement class.

Computing the carrier-phase and Doppler observations are straightforward and we do not face any numerical problems while converting them. However, we might have some numerical issues/errors while generating the pseudorange observations. This is due to the fact that each logging App implements its own GNSS observation conversion algorithm and uses different parameter settings, thresholds and float computing accuracies. Such an issue will affect the quality of the generated observations saved into the RINEX file. In this contribution, we have also developed our in-house convertor in C++, namely UofC CSV2RINEX, to convert a CSV file into a RINEX file. In Section 4, three RINEX files coming from GnssLogger App, Geo++ RINEX logger and our converted RINEX using UofC CSV2RINEX are thoroughly investigated and compared from different aspects.

3. Precise Positioning Using Uncombined PPP (UPPP) Algorithm

With GNSS observations of pseudorange, carrier-phase and Doppler, the uncombined PPP (UPPP) model can be employed for precise positioning. The undifferenced GNSS pseudorange and carrier-phase observations for the satellite s and the receiver r on frequency j are as follows [41]:

$$\begin{aligned} E\left(\Phi^{s}_{r,j}\right) &= \rho^{s}_{r} + T^{s}_{r} + cdt_{r} - cdt^{s} - \gamma_{j} I^{s}_{r,1} + \lambda_{j} N^{s}_{r,j} + B_{r,j} - B^{s}_{j} \\ E\left(\mathrm{P}^{s}_{r,j}\right) &= \rho^{s}_{r} + T^{s}_{r} + cdt_{r} - cdt^{s} + \gamma_{j} I^{s}_{r,1} + b_{r,j} - b^{s}_{j} \end{aligned}, \tag{5}$$

where E is the mathematical expectation operator, P_j and $\Phi_j = \lambda_j \varphi_j$ denote the pseudorange and carrier-phase observations on the frequency j in meters, ρ is the geometric range between satellite and receiver as a function of the satellite and the receiver coordinates, T is the tropospheric delay (m) which can be spilt into dry and wet parts, c is the vacuum speed of light (m/s), dt_r and dt^s are the receiver and satellite clock errors (s), respectively, $I^{s}_{r,1}$ is the first-order slant ionospheric delay on frequency L1 (m), $\gamma_j = f_1^2 / f_j^2$ is the frequency-dependent multiplier factor (in the case of L1 frequency $\gamma_j = 1$), f_f is the corresponding frequency, λ_j is the corresponding carrier-phase wavelength (m), $N^{s}_{r,j}$ denotes the integer carrier-phase ambiguity term in cycle, $b_{r,j}$ and $B_{r,j}$ denote the frequency-dependent receiver pseudorange and carrier-phase hardware delays (biases), respectively, and b_f^s and B_f^s are the frequency-dependent satellite pseudorange and carrier-phase hardware delays (biases), respectively.

The precise satellite clock errors provided by International GNSS Service (IGS) are based on the ionosphere-free (*IF*) linear combination of code observations on L1 and L2 frequencies, i.e., P1 and P2, as follows [42]:

$$dt^{s,IF} = cdt^s + b_{IF(1,2)}^s, \tag{6}$$

where $b_{IF(1,2)}^s = \alpha_{IF}^{L1,L2} b_1^s + \beta_{IF}^{L1,L2} b_2^s$ is the satellite ionosphere-free code bias in which b_1^s and b_2^s are the satellite pseudorange hardware delays for P1 and P2, respectively. The coefficients $\alpha_{IF}^{L1,L2}$ and $\beta_{IF}^{L1,L2}$ are also of the following form:

$$\alpha_{IF}^{L1,L2} = f_1^2 \Big/ \left(f_1^2 - f_2^2\right) \; and \; \beta_{IF}^{L1,L2} = -f_2^2 \Big/ \left(f_1^2 - f_2^2\right), \tag{7}$$

The uncombined PPP model for L1 and L5 frequencies can then be rewritten as follows:

$$\begin{aligned} E\left(P_{r,1}^s\right) &= \rho_r^s + T_r^s + \left(cdt_r + b_{r,1}\right) - cdt^{s,IF} + I_{r,1}^s + (b_{IF(1,2)}^s - b_1^s) \\ E\left(\Phi_{r,1}^s\right) &= \rho_r^s + T_r^s + \left(cdt_r + b_{r,1}\right) - cdt^{s,IF} - I_{r,1}^s + [\lambda_1 N_{r,1}^s - b_{r,1} + B_{r,1} - B_1^s + b_{IF(1,2)}^s] \\ E\left(P_{r,3}^s\right) &= \rho_r^s + T_r^s + \left(cdt_r + b_{r,1}\right) - cdt^{s,IF} + \gamma_3 I_{r,1}^s + b_{r,3} - b_{r,1} + (b_{IF(1,2)}^s - b_3^s) \\ E\left(\Phi_{r,3}^s\right) &= \rho_r^s + T_r^s + \left(cdt_r + b_{r,1}\right) - cdt^{s,IF} - \gamma_3 I_{r,1}^s + [\lambda_3 N_{r,3}^s + B_{r,3} - B_3^s + b_{IF(1,2)}^s - b_{r,1}] \end{aligned}, \tag{8}$$

By introducing $\widetilde{cdt}_r = cdt_r + b_{r,1}$, $\lambda_1 \widetilde{N}_{r,1}^s = \lambda_1 N_{r,1}^s + B_{r,1} - B_1^s + b_{IF(1,2)}^s - b_{r,1}$ and $\lambda_3 \widetilde{N}_{r,3}^s = \lambda_3 N_{r,3}^s + B_{r,3} - B_3^s + b_{IF(1,2)}^s - b_{r,1}$, have:

$$\begin{cases} E\left(P_{r,1}^s - \frac{1}{\gamma_2 - 1} DCB_{1,2}^s\right) = \rho_r^s + T_r^s + \widetilde{cdt}_r - cdt^{s,IF} + I_{r,1}^s \\ E\left(\Phi_{r,1}^s\right) = \rho_r^s + T_r^s + \widetilde{cdt}_r - cdt^{s,IF} - I_{r,1}^s + \lambda_1 \widetilde{N}_{r,1}^s \\ E\left(P_{r,3}^s - \frac{\gamma_2}{\gamma_2 - 1} DCB_{1,2}^s + \frac{1}{\gamma_2 - 1} DCB_{2,3}^s\right) = \rho_r^s + T_r^s + \widetilde{cdt}_r - cdt^{s,IF} + \gamma_3 I_{r,1}^s - DCB_{1,3}^r \\ E\left(\Phi_{r,3}^s\right) = \rho_r^s + T_r^s + \widetilde{cdt}_r - cdt^{s,IF} - \gamma_3 I_{r,1}^s + \lambda_3 \widetilde{N}_{r,3}^s \end{cases}, \tag{9}$$

where $DCB_{1,3}^r = b_{r,1} - b_{r,3}$, $b_{IF(1,2)}^s - b_1^s = \frac{1}{\gamma_2 - 1} DCB_{1,2}^s$, $b_{IF(1,2)}^s - b_3^s = \frac{\gamma_2}{\gamma_2 - 1} DCB_{1,2}^s - \frac{1}{\gamma_2 - 1} DCB_{2,3}^s$ with $DCB_{1,2}^s = b_1^s - b_2^s$, $DCB_{2,3}^s = b_2^s - b_3^s$ which are the satellite differential code biases (DCB) available from the IGS. The unknowns here are the receiver position, the receiver clock error $\widetilde{cdt}_r$, the real-valued carrier-phase ambiguity terms $\lambda_1 \widetilde{N}_{r,1}^s$ and $\lambda_3 \widetilde{N}_{r,3}^s$, the slant ionospheric delay $I_{r,1}^s$, the tropospheric delay and $DCB_{1,3}^r$. The ionospheric delay can be also modeled by the global ionospheric maps (GIM) or the empirical models, i.e., ionosphere-corrected.

4. Quality Analysis of GNSS Observations from Different Logging Apps and Improvement

This section consists of two parts. First, the quality of GNSS measurements saved into the RINEX files obtained from the two widely used logging Apps, namely GnssLogger and Geo++ RINEX logger, is assessed from different aspects. In this section, we also assess our newly developed in-house software (UofC CSV2RINEX) for converting a CSV file into a RINEX file which provides improved GNSS observations. Second, the positioning performance of the three RINEX files is investigated in kinematic mode.

4.1. Analysis of GNSS Observations from Different Logging Apps

In this section, we first thoroughly investigate the inconsistency between the pseudorange, carrier-phase and Doppler observations in Section 4.1.1. Such misbehavior has been recently reported in [39]. Another issue which will be covered in Section 4.1.2 is the presence of some carrier-phase observations without changes over time. The results of these two subsections clearly indicate the importance of analyzing GNSS logger outputs before using them.

4.1.1. Inconsistency between Pseudorange, Carrier-Phase and Doppler Observations

GnssLogger app and Geo++ RINEX both are capable of providing GNSS observables in RINEX format. Having RINEX format available allows the users to post-process the logged data, improving the accuracy. However, different logging apps have different performance which affects the positioning results as well. In this section, we first investigate outputs of the three RINEX files, (1) RINEX file saved by GNSSLogger App, (2) RINEX file logged by Geo++ RINEX Logger App and (3) RINEX file generated by our convertor toolbox which converts the CSV file to the RINEX format.

To this end, the Xiaomi Mi8, Samsung S20 and Google Pixel 5 smartphones were put on the top of the geodetic pillars with known coordinates on the rooftop of the Civil Engineering building, University of Calgary, Calgary, Canada. The first two devices used the Broadcom chipset, while the last one used the Qualcomm chipset. All three devices were dual-frequency smartphones supporting L5/E5a frequencies for GPS and Galileo, respectively. The dataset was collected on 23 November 2022 with a sampling interval of 1 sec for about 1:30 h. For further investigation and presenting results, GPS PRN 01, Galileo PRN 31 and GLONASS PRN 17 were selected. The reason for selecting these PRNs is their better availability and continuity during the observation period. We should also mention that the same results were observed for other PRNs. To have a better view, the first 900 epochs (15 min) were used for plotting the figures. Table 2 also provides a brief summary of the experiment.

Table 2. GNSS data information.

Devices	Xiaomi Mi8, Google Pixel 5 and Samsung S20
PRNs	PRN 01 (GPS), 17 (GLONASS), 31 (Galileo)
Mode	Static
App logger	Geo++ RINEX logger (v2.1.6), GnssLogger (v3.0.5.6)
Date	23 November, 2022
Duration	~1 h 30 min
Sampling interval	1 sec

Figure 1 provides the C/N0 measurements of the selected PRNs on the L1 frequency for the Xiaomi Mi8, Google Pixel 5 and Samsung S20, from left to right, respectively. The plot reveals that the three smartphones did not have similar performances in terms of their C/N0 records, even though the data have been collected in the same environment at the same time. As can be seen, the C/N0 records of the Xiaomi Mi8 and Samsung S20 are smoother than the ones recorded by the Google Pixel 5.

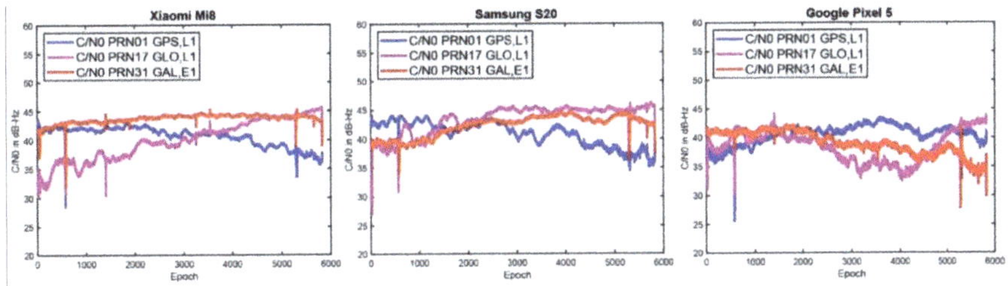

Figure 1. C/N0 measurements for selected PRNs and three smartphone (Xiaomi Mi8, Google Pixel 5 and Samsung S20).

To investigate the quality of GNSS observations from the three RINEX files, different indicators are selected and reported in Table 3. They are as follows:

- First-order differentiation of pseudorange and carrier-phase versus Doppler observations: The first-order differentiation of GNSS pseudorange and carrier-phase observations are obtained by calculating differences between adjacent elements of GNSS pseudorange and carrier-phase observations divided by the sampling interval (i.e., $diff(\mathrm{P}^{s}_{r,j})/T$ and $diff(\Phi^{s}_{r,j})/T$ where T is the sampling interval which is 1 s here). They then compare to the Doppler observations ($-\lambda_j D^{s}_{r,j}$). The first-order differences of pseudorange and carrier-phase observations have to follow the same trend of the Doppler observations in theory;
- Geometry-free combination (Code-minus-phase: CMP): It cancel the geometric part of the measurement (i.e., geometric range, receiver and satellite clock and tropospheric delay), leaving ambiguity, ionosphere term, multipath and noise. This combination can also be used to detect cycle-slips in the carrier-phase observations as a cycle-slip appears as a jump in the CMP plot;
- Carrier-phase predicted error: The predicted carrier-phase in cycle can be obtained using the discrete Doppler measurements as $\hat{\varphi}^{s}_{r,j}(k+1) = \varphi^{s}_{r,j}(k) + \frac{D^{s}_{r,j}(k+1)+D^{s}_{r,j}(k)}{2}T$. The carrier-phase predicted error is then estimated as $\hat{\Phi}^{s}_{r,j} - \Phi^{s}_{r,j} = \lambda_j\hat{\varphi}^{s}_{r,j} - \lambda_j\varphi^{s}_{r,j}$ in meters.

Table 3. Different indictors used to analyze raw GNSS observations.

Indicator	Formula
First-order differentiation of pseudorange and phase versus Doppler observations	$\begin{cases} \mathrm{diff}(\mathrm{P}^{s}_{r,j})/\mathrm{T} \\ \mathrm{diff}(\Phi^{s}_{r,j})/\mathrm{T} \\ -\lambda_j D^{s}_{r,j} \end{cases}$
Geometry-free (Code minus phase: CMP)	$\mathrm{P}^{s}_{r,j} - \Phi^{s}_{r,j}$
Carrier-phase predicted error	$\hat{\Phi}^{s}_{r,j} - \Phi^{s}_{r,j}$

Let us start with the first indicator. Figure 2 shows the first-order differentiation of GPS pseudorange and carrier-phase observations, as well as the Doppler observations for PRN 01 on the L1 frequency from the three RINEX files for the Xiaomi Mi8. To have a better view, the difference between Doppler observations and the first-order differentiation of the pseudorange (i.e., $-\lambda_j D^{s}_{r,j} - diff(\mathrm{P}^{s}_{r,j})/T$) and the difference between Doppler observations and the first-order differentiation of the carrier-phase observations, (i.e., $-\lambda_j D^{s}_{r,j} - diff(\Phi^{s}_{r,j})/T$) are also depicted in the right panel of this figure. In some graphs, the red line cannot be seen at this zoom setting since it is under the green one (Doppler). A few observations can be highlighted from the Figure 1. The Doppler observations of the three RINEX files are the same. As mentioned, the Doppler shift was obtained as $dpplershift = -PseudorangeRatemetersperSecond/\lambda$ (see Equation (4)), showing that generating the Doppler observation was straightforward and without any complication.

(2) Shown in the top panel of Figure 2 is related to the GnssLogger RINEX file. As observed, there was an offset between the Doppler and the pseudorange observations. Such an offset could have been caused during the pseudorange generation from the raw measurements in the Android API "location"-related classes. An offset was probably applied to the pseudorange observations. Applying such an offset would not be affected the solution as it could be lumped into the receiver clock bias and the real-valued ambiguities. (3) Shown in the middle panel of Figure 2 is related to the Geo++ RINEX logger output. Similar offset could be observed here not only for the pseudorange, but also for the carrier-phase observations. The carrier-phase observations followed the pseudorange observations behavior in terms of the anomalies, spikes and jumps. This shows that what happened to the pseudorange observations during their generation procedure also happened to the carrier-phase observations. (4) Shown in the bottom panel of Figure 2 is related to the converted RINEX file from our developed convertor following the equations in [40]. Unlike the other two RINEX files, there was no offset between the pseudorange and carrier-phase and Doppler observations. The first-order differences of pseudorange and carrier-phase observations also followed the same trend of the Doppler observations. (5) The Doppler observations can be employed for cycle slip detection and/or code smoothing. Considering the possible biases and anomalies in the data, the Doppler observations have to be carefully analyzed before use. The other two indicators, the CMP combination and the carrier-phase predicted error, are then utilized to further investigate the effect of possible biases and anomalies in the data. Before that, the same plots for the GLONASS PRN 17 and Galileo PRN 31 on the first frequency are shown in Figures 3 and 4, respectively. Shown in the top, middle and bottom panels of Figures 3 and 4 are related to the GnssLogger RINEX file, the Geo++ RINEX logger output and the converted RINEX file from our developed convertor, respectively. The same conclusions hold for the GLONASS and Galileo constellations. Therefore, we only present the GPS results in continue.

Figure 5 represents the CMP and the carrier-phase predicted error for GPS PRNs 01 from the three RINEX files for the Xiaomi Mi8 on the L1 frequency. The main purpose of this plot is to evaluate how the possible anomalies/jumps or offsets affected the CMP and the carrier-phase predicted error. The top, middle and bottom panels of Figure 5 include the CMP plot computed by using the GnssLogger RINEX, the Geo++ RINEX logger output and the converted RINEX, respectively. There are two important points about this figure that needed to be expressed. (1) First, let us look at the plot of the CMP values obtained from the GnssLogger RINEX (the top panel). The CMP does not include the geometric part while it includes the carrier-phase ambiguity, twice the ionospheric error, pseudorange and carrier-phase noise and multipath. Therefore, such behavior is not expected from the CMP values for the GnssLogger RINEX file, as it must be a constant value with a reasonable noise level as long as there is no cycle-slip in the data. It shows that the pseudorange and carrier-phase observations are not consistent (i.e., they are divergent, see the smaller panel in Figure 5 (top-left) in which the pseudorange and carrier-phase observations shifted to zero to have a better view). This indicates that the CMP combination cannot be employed to detect the possible cycle-slips in this case. We should note that this slop is the same for all PRNs. Therefore, in the case of using single-difference between satellites or double-difference observations, this issue would not affect the positioning results as it is removed through the differencing procedure. (2) As can also be observed in this figure, the carrier-phase predicted error obtained from the Geo++ RINEX logger, depicted in the middle panel, does not have an expected behavior for any reason. This also indicates that the Doppler observations cannot be employed here to detect the possible cycle-slips in the data. It is not clear to us why these plots are like that, as these Apps have not disclosed their internal algorithms.

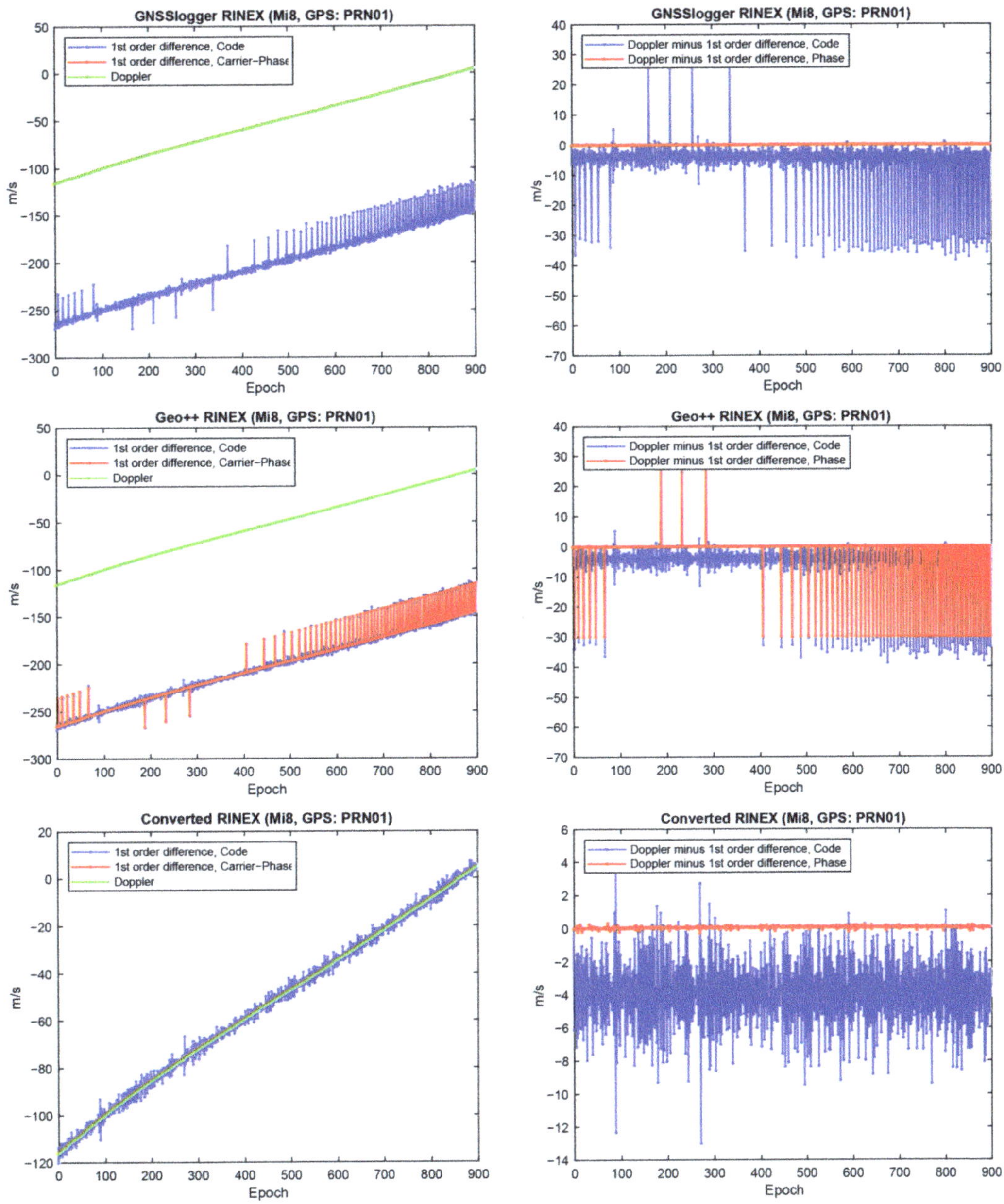

Figure 2. First-order differentiation of GPS pseudorange and carrier-phase observations as well as Doppler observations on L1 frequency for PRN 01 (Xiaomi Mi8).

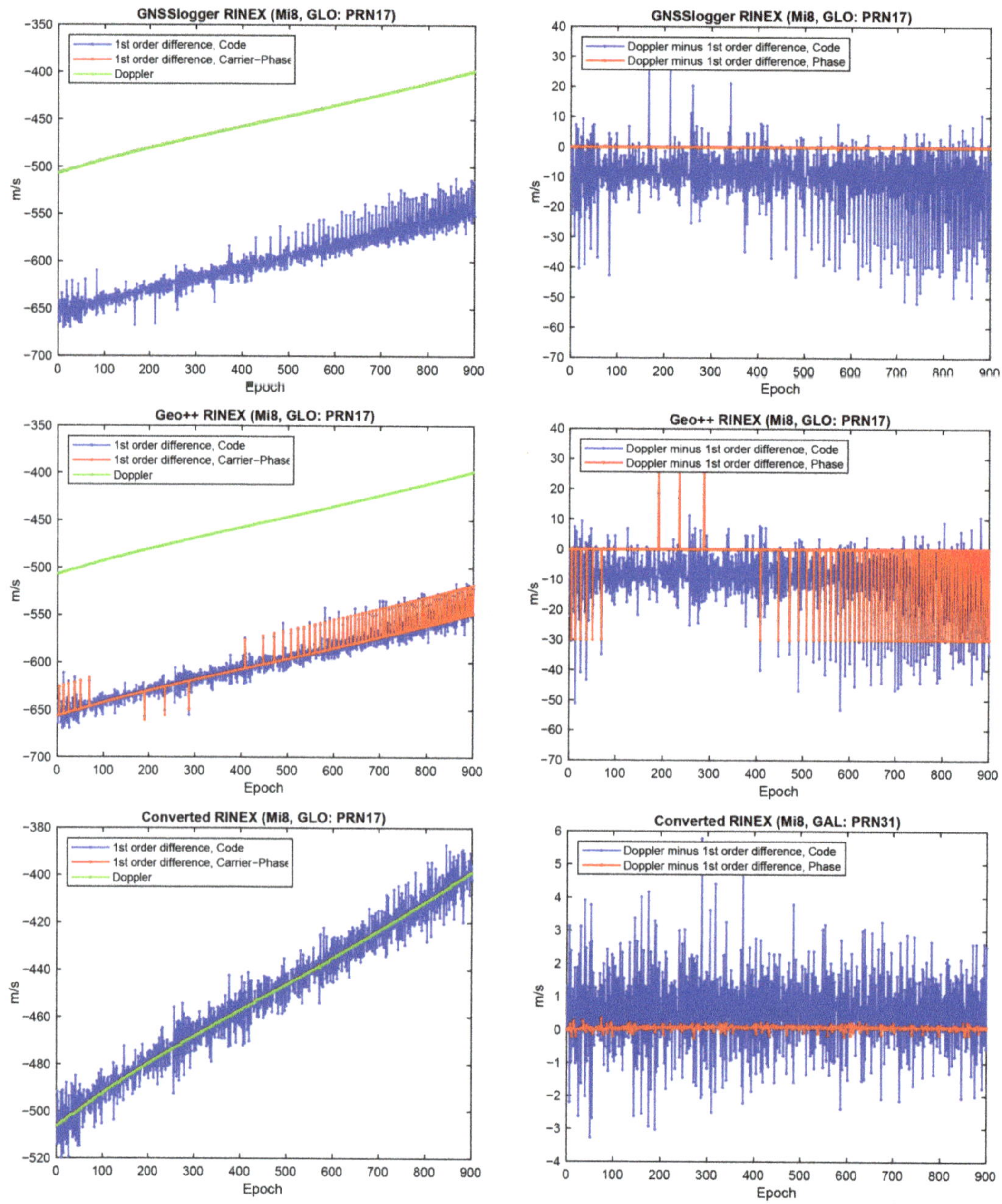

Figure 3. First-order differentiation of GLONASS pseudorange and carrier-phase observations as well as Doppler observations on L1 frequency for PRN 17 (Xiaomi Mi8).

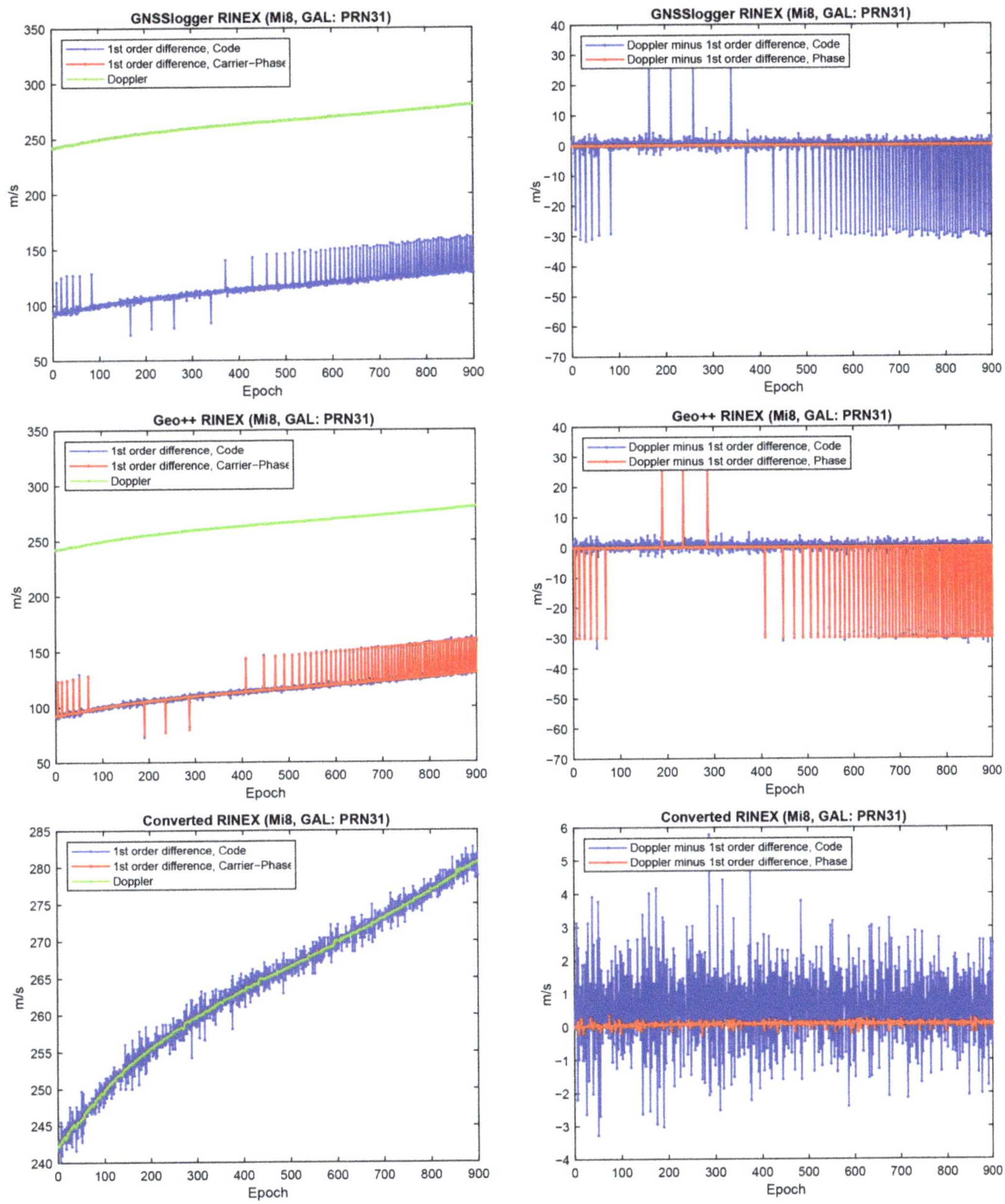

Figure 4. First-order differentiation of Galileo pseudorange and carrier-phase observations as well as Doppler observations on L1 frequency for PRN 31 (Xiaomi Mi8).

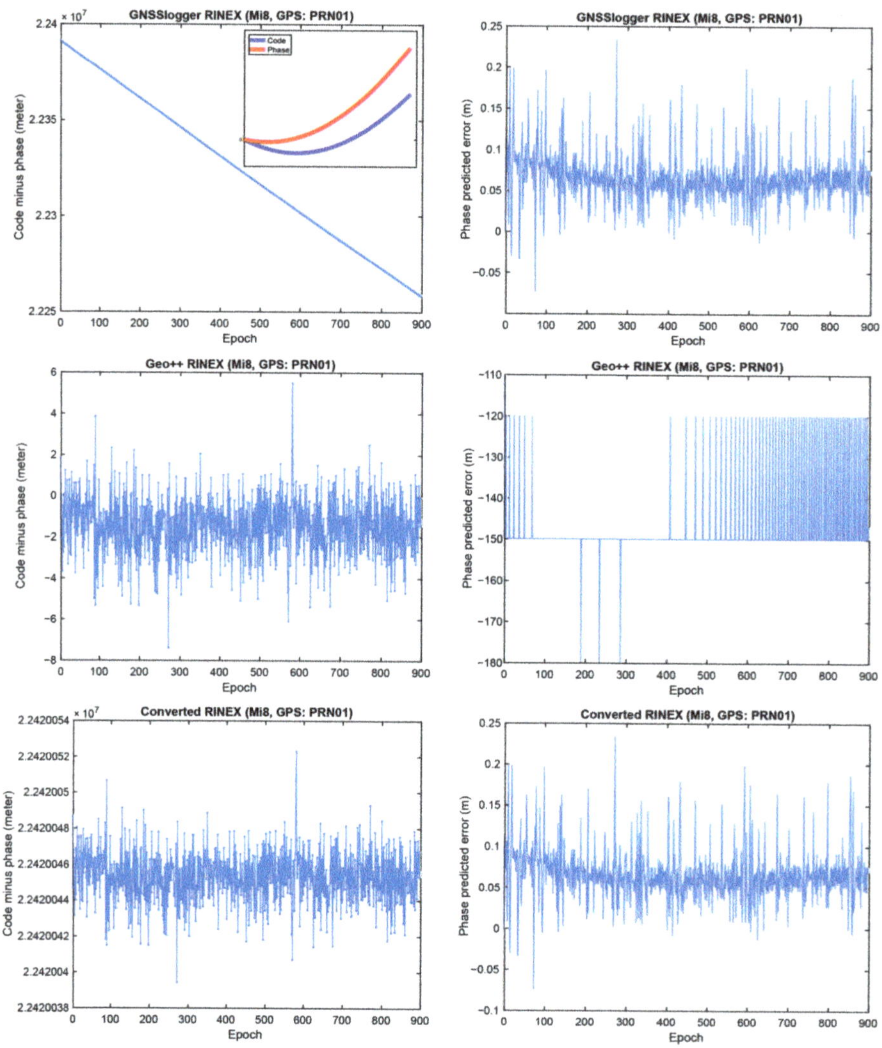

Figure 5. CMP (**left**) and carrier-phase predicted error (**right**) for GPS PRNs 01 (Xiaomi Mi8).

Finally, Table 4 provides a summary of consistency or inconsistency between different GNSS observations in the three RINEX files. The highlighted cells in this table indicate that the CMP combination cannot be implemented in the cycle-slip detection procedure while using the RINEX by GNSSLogger App. In addition, the Doppler observations cannot be employed for the cycle-slip detection while using Geo++ RINEX logger output.

Finally, the same plots for the Samsung S20 and the Google Pixel 5 are given in Figures 6–9. Due to the page limitation, the plots are only provided for the GPS PRN 01. They support the similar conclusion as before.

Table 4. Consistency between different GNSS observations in three RINEX files.

Combination	GnssLogger	Geo++ RINEX Logger	UofC CSV2RINEX
Code & Phase	**No (Attention required!)**	Yes	Yes
Code & Doppler	No	No	Yes
Phase & Doppler	Yes	**No (Attention required!)**	Yes

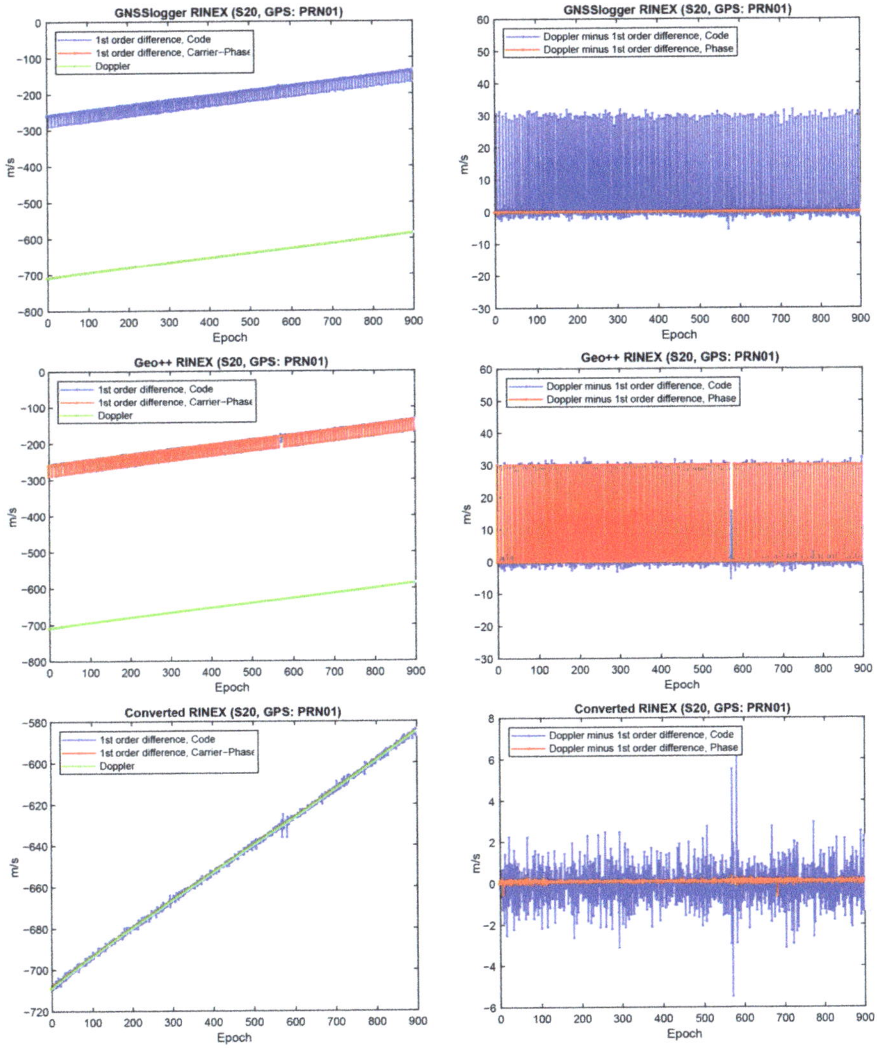

Figure 6. First-order differentiation of GPS pseudorange and carrier-phase observations as well as Doppler observations on L1 frequency for PRN 01 (Samsung S20).

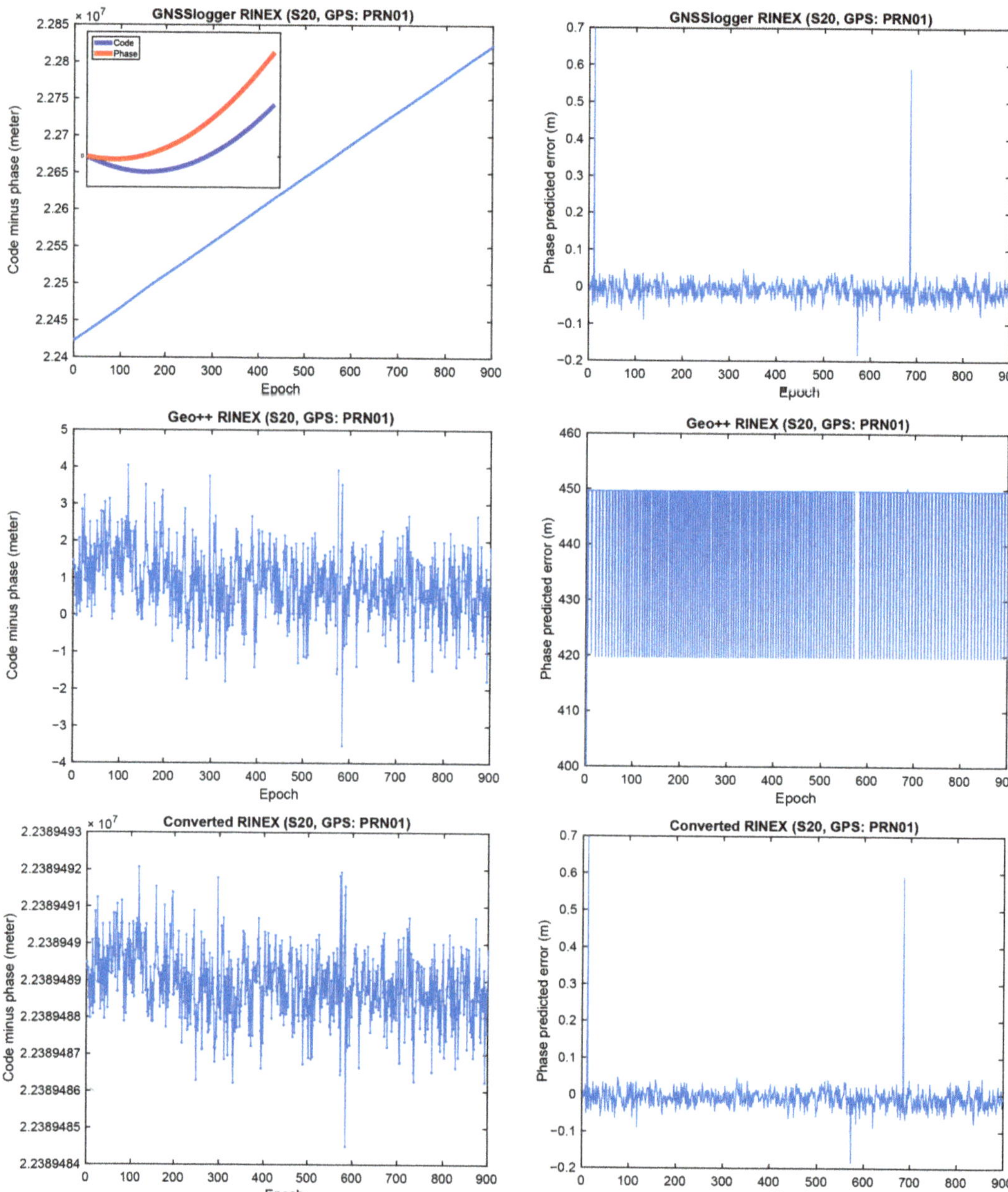

Figure 7. CMP carrier-phase predicted error for GPS PRNs 01 (Samsung S20).

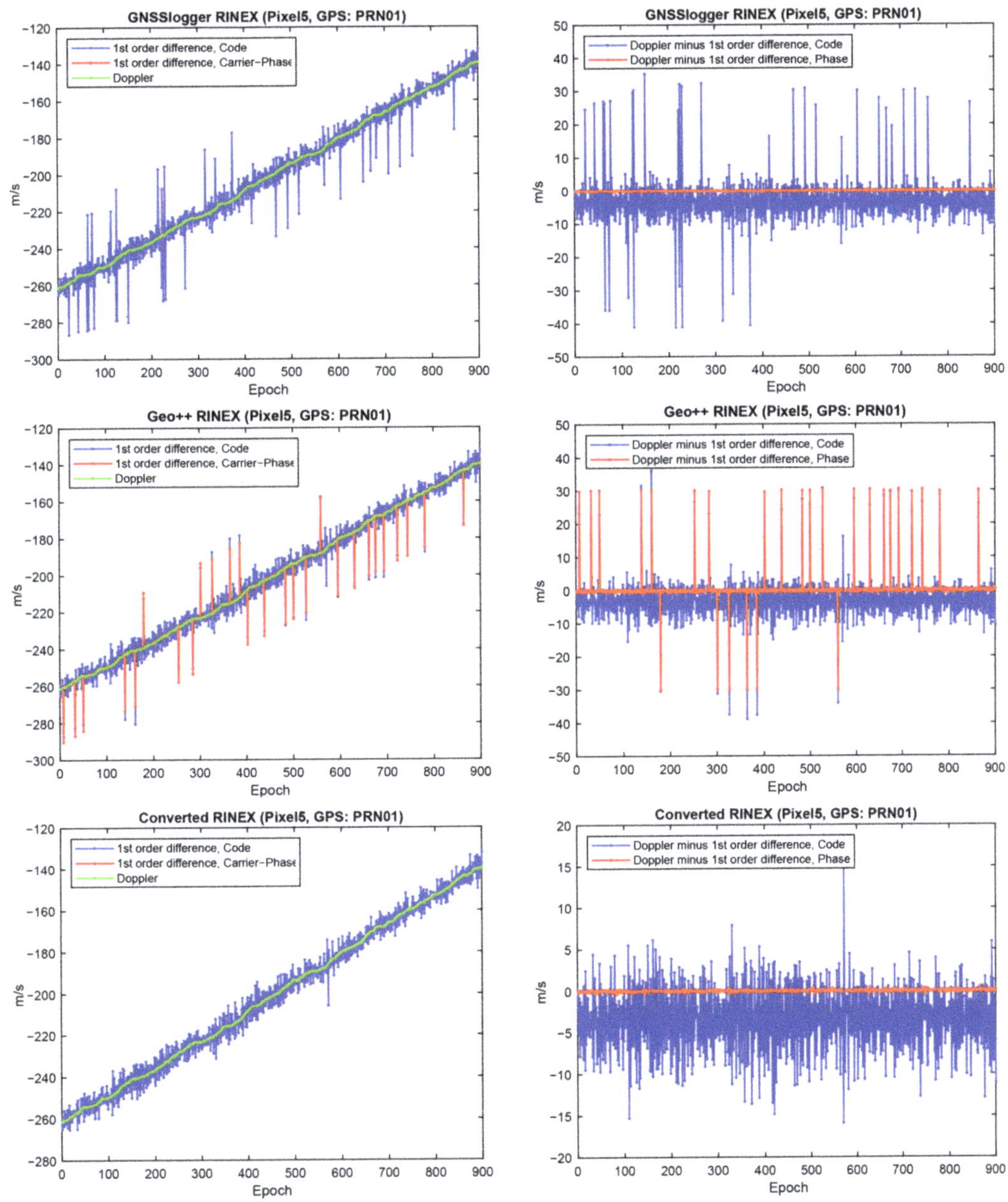

Figure 8. First-order differentiation of GPS pseudorange and carrier-phase observations as well as Doppler observations on L1 frequency for PRN 01 (Google Pixel 5).

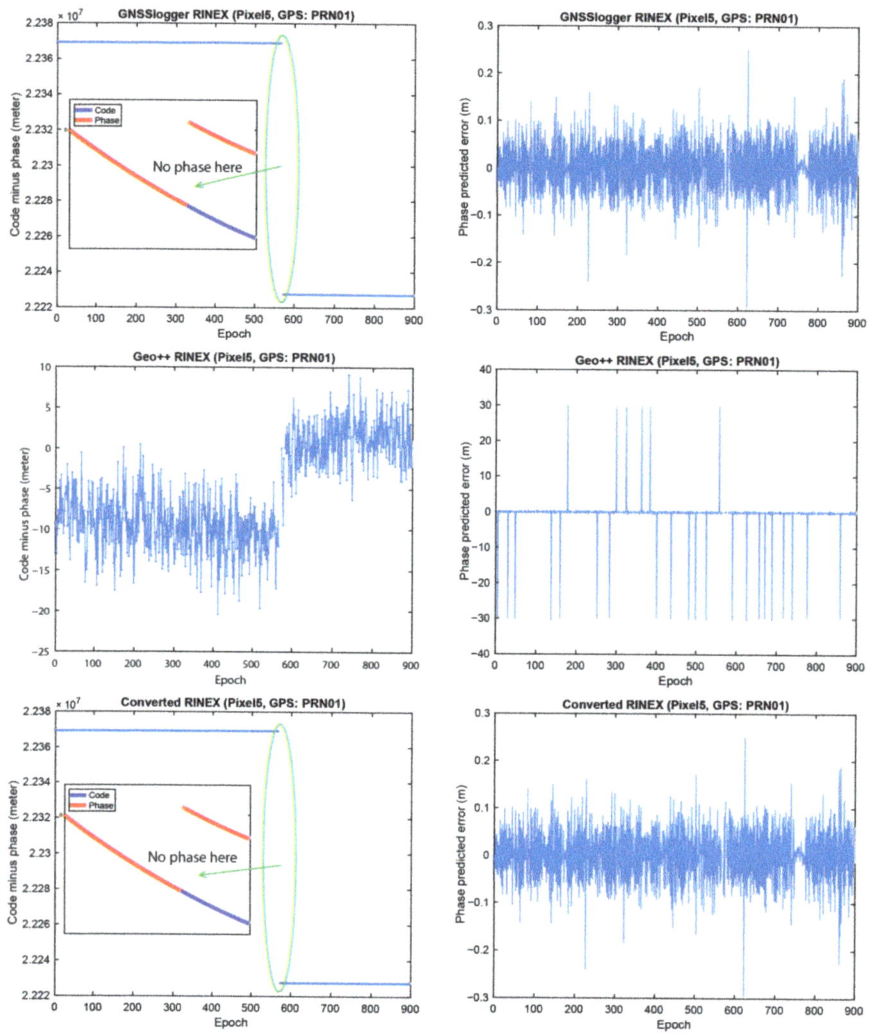

Figure 9. CMP and carrier-phase predicted error for GPS PRNs 01 (Google Pixel 5).

4.1.2. Carrier-Phase Observations with No Change over Time

Another problem was also observed in the RINEX files saved by the GnssLogger App; which is the existence of some carrier-phase observations with no changes over time. Such a phenomenon was observed for the Xiaomi Mi8 and Samsung S20, while this was not the case for the Google Pixel 5 dataset. Figure 10 provides the C/N0 records for the mentioned GPS PRNs along with the epochs in which the carrier-phase observations have no changes over time for the Xiaomi Mi8 and Samsung S20, depicted in the left and right panels, respectively. They are shown with the blue dots. The number of such satellites was more for the Xiaomi Mi8 compared to the Samsung S20. Those PRNs mostly belonged to the lower C/N0 values. A C/N0 mask is usually set to 15–25 dB-Hz, however, there are still some of those epochs with the C/N0 larger than the threshold. It therefore needs further attention than just masking the lower C/N0. As given in Equation (3), the carrier-phase observation can be obtained from *AccumulatedDeltaRangeMeters* variable (abbreviated as ADR) from the GNSSMeasurement class. As mentioned before, checking the validity of the carrier-phase measurements by using the AccumulatedDeltaRangeState variable is crucial. By looking at the ADR logs from the Xiaomi Mi8 and Samsung S20, it is revealed that these questionable epochs are mainly related to the invalid ADR. However, it is not observed in the Geo++ RINEX file, meaning that it properly handled the invalid ADR by excluding them. In the converted RINEX, such an issue cannot be seen.

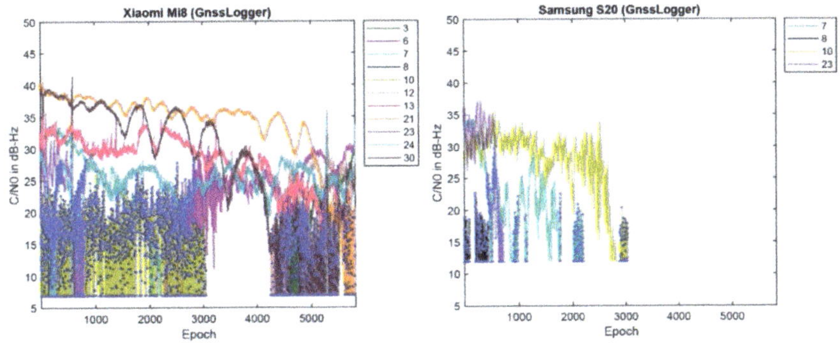

Figure 10. C/N0 records of GPS PRNs with problem in carrier-phase observations logs for Xiaomi Mi8 and Samsung S20 (Blue dots: epochs of carrier-phase observations with no change over time).

Concisely, the results confirmed the importance of evaluating the logging Apps before employing them since these logging Apps are the basics of any smartphone positioning algorithm development. In the next subsection, the UPPP positioning accuracy obtained from the three RINEX files are assessed in the post-processed mode within a kinematic experiment.

4.2. UPPP Positioning Accuracy Analysis

In this subsection, we provide the results of a kinematic test carried out by the same dual-frequency Xiaomi Mi8 device as the static experiment. A kinematic test was carried out on 22 April 2022 with a duration of almost one hour in mostly open-sky environment with overpasses, Calgary, Alberta, Canada. Figure 11 shows the kinematic test configuration and the reference vehicle's path in this experiment.

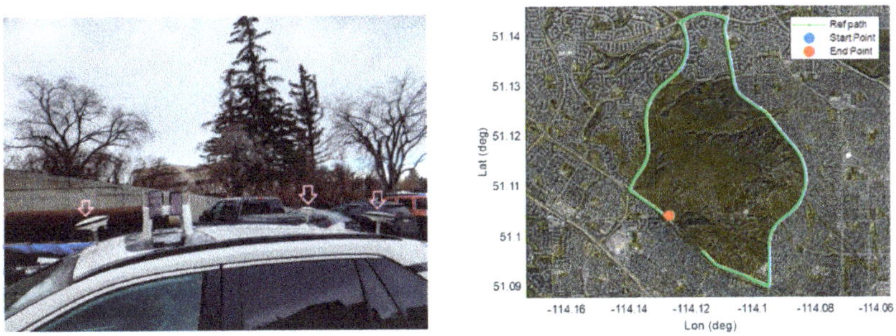

(**a**) (**b**)

Figure 11. Kinematic experiment done on 22 April 2022 (**a**): test configuration and (**b**): Reference trajectory).

The kinematic experiment involved three geodetic receivers (two U-blox F9P and one Septentrio AsteRx-m2), as shown by the three pick arrows in Figure 11, and six smartphones for our future research (here we only used the Xiaomi Mi8 Black dataset). The phones were placed on the vehicle roof. The offsets between all units were measured and applied prior to comparison. The reference trajectory of the vehicle during the kinematic experiment was obtained by the RTK fixed solutions from the three geodetic receivers as the rover receivers. A geodetic receiver on a geodetic pillar (with true position) on the rooftop of the Civil Engineering building, University of Calgary, was also selected as the base receiver. Table 5 provides GNSS data information and processing setting.

Table 5. GNSS data information and processing setting.

Device	Xiaomi Mi8
Measurements used	GPS (L1/L5), GLONASS (L1), Galileo (E1/E5a)
Mode	Kinematic
Date	22 November 2022
Duration	1 h
Sampling interval	1 s
Troposphere model	Saastamoinen model
Ionosphere model	Global ionospheric maps (GIM)
Functional model	UPPP model
Stochastic model	C/N0 and elevation weighting function
Elevation mask angle	10 deg
C/N0 mask	20 dB-Hz
Satellite orbit	CODE MGEX precise ephemerides (5 min interval)
Clock error	CODE MGEX precise clock (1 sec interval)
Satellite DCB correction	CAS DCBs in Bias SINEX (BSX) format

Figure 12 provides the positioning errors for Xiaomi Mi8 using the two RINEX files (RINEX by GEO++ RINEX logger and our converted RINEX) in the post-processed mode. The results of the RINEX file from the GnssLogger App were not provided since the obtained accuracy was at the single point positioning (SPP) accuracy level due to the frequent cycle slip detected. It should be noted that the root mean square (RMS) values provided in this figure have been computed using all epochs. In this figure, the cumulative distribution error (CDE) plots for the horizontal positioning error are also provided. The results confirmed the better performance of the converted RINEX in terms of East, North and Up RMS values and the 50th percentile error.

There are many studies devoted to the PPP smartphone positioning, among them we may refer to at least two, [7,43]. In Ref. [7], the Geo++ RINEX logger was used; while in the second study, the authors employed their own developed conversion tool in order to generate the RINEX file [43]. Wu et al. [7] employed the dual-frequency GPS (L1/L5) and Galileo (E1/E5a) observations from a Xiaomi Mi8 smartphone obtained from the Geo++ RINEX logger. Their numerical results showed that the positioning performance of the PPP algorithm employing the ionosphere-free combination was at the meter-level, in kinematic mode. Our positioning accuracy was better than their work, which might be due to differences in measurement environment and employed mathematical model, as well as considering GLONASS observations in our contribution. Chen et al. [43] also proposed a modified single-frequency PPP algorithm in which separate clock biases for pseudorange and carrier-phase observations are estimated. Using a Xiaomi Mi8 smartphone, the average horizontal and vertical RMS error were 0.81 m and 1.65 m, respectively. The difference in accuracy is acceptable since they used the single-frequency data and the predicted IGS products.

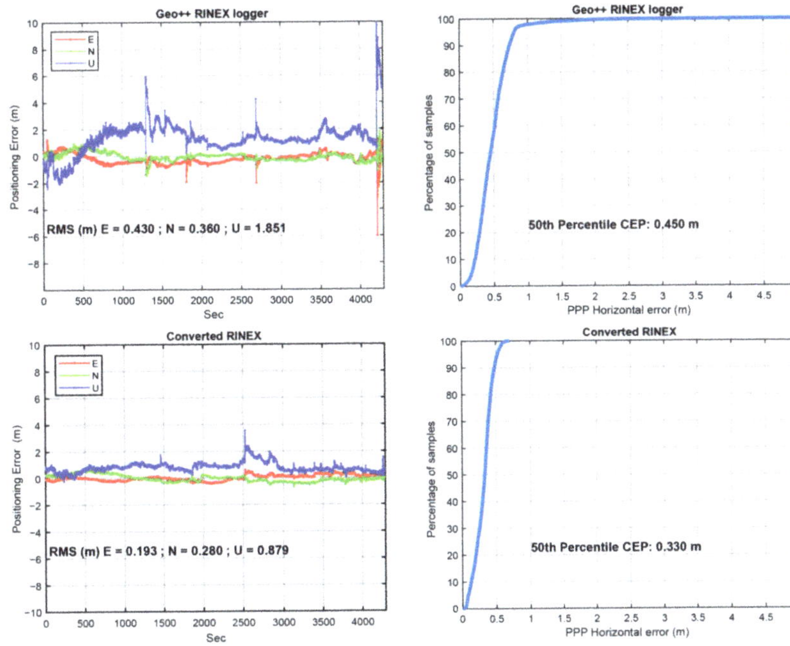

Figure 12. Horizontal positioning errors from Geo++ RINEX logger output and converted RINEX (**left**), cumulative distribution function plot for horizontal positioning error from Geo++ RINEX logger output and converted RINEX (**right**).

Finally, we should mention that, although there are several open-source Apps generating the typical GNSS observations from the Android location API and saving them into the RINEX format, we must still pay more attention to the generation of GNSS observations as we showed some possible issues in the generated observations.

5. Conclusions

In this study, we explored the performances of different open-source Apps in generating typical GNSS measurements. We also introduced our newly developed software (namely UofC CSV2RINEX) written in C++ for converting a CSV file into a RINEX file. The quality of raw GNSS observation logged by different smart devices and using different loggers was assessed from different aspects, including the inconsistency between the pseudorange, carrier-phase and Doppler measurements, presence of some carrier-phase observations without changes over time and its possible reasons, etc. Then, the positioning performance of our software was assessed using a kinematic experiment. The conclusions of our study are listed as follows:

1. Consistency between generated pseudorange, carrier-phase and Doppler observations from Android smartphone devices was not fully met in the RINEX outputs of the GnssLogger and Geo++ RINEX Logger Apps. As a highlight, in GnssLogger RINEX file, pseudorange and carrier-phase, observations were not consistent with each other while looking at the CMP combination. In Geo++ RINEX Logger output, the consistency between the carrier-phase and Doppler observations was not met. With our converter software, these three types of measurements were consistent;
2. GnssLogger App had an issue that some carrier-phase observations from the Xiaomi Mi8 and Samsung S20 devices (saved into the RINEX files) did not change over time. These epochs mainly belonged to the lower C/N0 values with invalid ADR states;

3. With our converter software, an improved positioning accuracy could be witnessed when compared with both Geo++ RINEX Logger and GnssLogger outputs. Using UofC CSV2RINEX output, the 50th percentile CEP was 0.330 m, which was 0.450 m for GEO++ RINEX Logger, and SPP-level accuracy for GnssLogger due to frequent cycle slip was detected.

Finally, it should be noted that, to obtain better understanding of the potential reasons of such misbehavior in the typical GNSS observations of the two Apps, a joint effort with their developers is recommended in the future to understand and assess the models and algorithms that have been used to generate the GNSS observations.

Author Contributions: All the authors have contributed to the presented work. The first author, F.Z., analyzed the static datasets, prepared the plots and wrote the draft of the manuscript. The second author, Y.J., collected kinematic dataset, generated the reference for it and wrote Introduction and Conclusion sections. F.Z. and Y.J. developed the UofC CSV2RINEX convertor. Y.G. supervised manuscript development and the direction of the research, revised the paper and provided further analysis. All authors participated in formulating the idea and in discussing the proposed approach and results. All authors have read and agreed to the published version of the manuscript.

Funding: The financial support from the Natural Sciences and Engineering Research Council of Canada (NSERC) is greatly acknowledged. The first author greatly acknowledges the funding support provided by the Izaak Walton Killam Memorial Scholarship.

Institutional Review Board Statement: Not applicable.

Informed Consent Statement: Not applicable.

Data Availability Statement: The data used in this study are available on request from the corresponding author.

Acknowledgments: We are grateful to Profound Positioning Inc. (PPI) for providing the Xiaomi Mi8 and Samsung S20 devices to collect data used in this paper.

Conflicts of Interest: The authors declare no conflict of interest.

References

1. Navarro-Gallardo, M.; Bernhardt, N.; Kirchner, M.; Musial, J.R.; Sunkevic, M. Assessing Galileo Readiness in Android Devices using Raw Measurements. In Proceedings of the ION GNSS 2017, Institute of Navigation, Portland, OR, USA, 25–29 September 2017; pp. 85–100.
2. Zhang, X.; Tao, X.; Zhu, F.; Shi, X.; Wang, F. Quality Assessment of GNSS Observations from an Android N Smartphone and Positioning Performance Analysis Using Time-differenced Filtering Approach. *GPS Solut.* **2018**, *22*, 70. [CrossRef]
3. Li, G.; Geng, J. Characteristics of raw multi-GNSS measurement error from Google Android smart devices. *GPS Solut.* **2019**, *23*, 90. [CrossRef]
4. Robustelli, U.; Baiocchi, V.; Pugliano, G. Assessment of Dual Frequency GNSS Observations from a Xiaomi Mi 8 Android Smartphone and Positioning Performance Analysis. *Electronics* **2019**, *8*, 91. [CrossRef]
5. Robustelli, U.; Paziewski, J.; Pugliano, G. Observation Quality Assessment and Performance of GNSS Standalone Positioning with Code Pseudoranges of Dual-frequency Android Smartphones. *Sensors* **2021**, *21*, 2125. [CrossRef] [PubMed]
6. Elmezayen, A.; El-Rabbany, A. Precise Point Positioning using World's First Dual-frequency GPS/GALILEO Smartphone. *Sensors* **2019**, *19*, 2593. [CrossRef]
7. Wu, Q.; Sun, M.; Zhou, C.; Zhang, P. Precise Point Positioning using Dual-frequency GNSS Observations on Smartphone. *Sensors* **2019**, *19*, 2189. [CrossRef]
8. Aggrey, J.; Bisnath, S.; Naciri, N.; Shinghal, G.; Yang, S. Multi-GNSS Precise Point Positioning with Next-Generation Smartphone Measurements. *J. Spat. Sci.* **2020**, *65*, 79–98. [CrossRef]
9. Shinghal, G.; Bisnath, S. Conditioning and PPP Processing of Smartphone GNSS Measurements in Realistic Environments. *Satell. Navig.* **2021**, *2*, 10. [CrossRef]
10. Zangenehnejad, F.; Gao, Y. Application of UofC Model Based Multi-GNSS PPP to Smartphones GNSS Positioning. In Proceedings of the 34th International Technical Meeting of the Satellite Division of the Institute of Navigation (ION GNSS+ 2021), St. Louis, MO, USA, 20–24 September 2021; pp. 2986–3003.
11. Dabove, P.; Di Pietra, V. Towards High Accuracy GNSS Real-time Positioning with Smartphones. *Adv. Space Res.* **2019**, *63*, 94–102. [CrossRef]

12. Dabove, P.; Di Pietra, V. Single-Baseline RTK Positioning using Dual frequency GNSS Receivers Inside Smartphones. *Sensors* **2019**, *19*, 4302. [CrossRef]
13. Zhang, K.; Jiao, W.; Wang, L.; Li, Z.; Li, J.; Zhou, K. Smart-RTK: Multi-GNSS Kinematic Positioning Approach on Android Smart Devices with Doppler-Smoothed-Code Filter and Constant Acceleration Model. *Adv. Space Res.* **2019**, *64*, 1662–1674. [CrossRef]
14. Heßelbarth, A.; Wanninger, L. Towards Centimeter Accurate Positioning with Smartphones. In Proceedings of the 2020 European Navigation Conference (ENC), Virtual, 23–24 November 2020; pp. 1–8.
15. Paziewski, J.; Fortunato, M.; Mazzoni, A.; Odolinski, R. An Analysis of Multi-GNSS Observations Tracked by Recent Android Smartphones and Smartphone-only Relative Positioning Results. *Measurement* **2021**, *175*, 109162. [CrossRef]
16. Yan, W.; Bastos, L.; Magalhães, A. Performance Assessment of the Android Smartphone's IMU in a GNSS/INS Coupled Navigation Model. *IEEE Access* **2019**, *7*, 171073–171083. [CrossRef]
17. Niu, Z.; Nie, P.; Tao, L.; Sun, J.; Zhu, B. RTK with the Assistance of an IMU-based Pedestrian Navigation Algorithm for Smartphones. *Sensors* **2019**, *19*, 3228. [CrossRef]
18. Bochkati, M.; Sharma, H.; Lichtenberger, C.A.; Pany, T. Demonstration of Fused RTK (Fixed)+ Inertial Positioning using Android Smartphone Sensors only. In Proceedings of the 2020 IEEE/ION Position, Location and Navigation Symposium (PLANS), Portland, OR, USA, 20–23 April 2020; pp. 1140–1154.
19. Paziewski, J. Recent Advances and Perspectives for Positioning and Applications with Smartphone GNSS Observations. *Meas. Sci. Technol.* **2020**, *31*, 091001. [CrossRef]
20. Zangenehnejad, F.; Gao, Y. GNSS Smartphones Positioning: Advances, Challenges, Opportunities, and Future Perspectives. *Satell. Navig.* **2021**, *2*, 24. [CrossRef]
21. Bahadur, B. A Study on the Real-time Code-Based GNSS Positioning with Android Smartphones. *Measurement* **2022**, *194*, 111078. [CrossRef]
22. Li, Y.; Cai, C.; Xu, Z. A Combined Elevation Angle and C/N0 Weighting Method for GNSS PPP on Xiaomi MI8 Smartphones. *Sensors* **2022**, *22*, 2804. [CrossRef]
23. Li, Y.; Cai, C. A Mixed Single-and Dual-Frequency Quad-Constellation GNSS Precise Point Positioning Approach on Xiaomi Mi8 Smartphones. *J. Navig.* **2022**, *75*, 849–863. [CrossRef]
24. Li, Z.; Wang, L.; Wang, N.; Li, R.; Liu, A. Real-time GNSS Precise Point Positioning with Smartphones For Vehicle Navigation. *Satell. Navig.* **2022**, *3*, 19. [CrossRef]
25. Li, X.; Wang, H.; Li, X.; Li, L.; Lv, H.; Shen, Z.; Xia, C.; Gou, H. PPP Rapid Ambiguity Resolution using Android GNSS Raw Measurements with a Low-Cost Helical Antenna. *J. Geod.* **2022**, *96*, 65. [CrossRef]
26. Xu, L.; Zha, J.; Li, M.; Yuan, Y.; Zhang, B. Estimation of Ionospheric Total Electron Content using GNSS Observations Derived From a Smartphone. *GPS Solut.* **2022**, *26*, 138. [CrossRef]
27. Zhu, H.; Xia, L.; Li, Q.; Xia, J.; Cai, Y. IMU-Aided Precise Point Positioning Performance Assessment with Smartphones in GNSS-Degraded Urban Environments. *Remote Sens.* **2022**, *14*, 4469. [CrossRef]
28. Yi, D.; Yang, S.; Bisnath, S. Native Smartphone Single-and Dual-Frequency GNSS-PPP/IMU Solution in Real-World Driving Scenarios. *Remote Sens.* **2022**, *14*, 3286. [CrossRef]
29. Bakuła, M.; Uradziński, M.; Krasuski, K. Performance of DGPS Smartphone Positioning with the Use of P (L1) vs. P (L5) Pseudorange Measurements. *Remote Sens.* **2022**, *14*, 929. [CrossRef]
30. Li, F.; Tu, R.; Han, J.; Zhang, S.; Liu, M.; Lu, X. Performance Research of Real-Time Kinematic/5G Combined Positioning Model. *Meas. Sci. Technol.* **2023**, *34*, 035115. [CrossRef]
31. Benvenuto, L.; Cosso, T.; Delzanno, G. An Adaptive Algorithm for Multipath Mitigation in GNSS Positioning with Android Smartphones. *Sensors* **2022**, *22*, 5790. [CrossRef]
32. Li, Y.; Mi, J.; Xu, Y.; Li, B.; Jiang, D.; Liu, W. A Robust Adaptive Filtering Algorithm for GNSS Single-Frequency RTK of Smartphone. *Remote Sens.* **2022**, *14*, 6388. [CrossRef]
33. Liu, Q.; Gao, C.; Shang, R.; Peng, Z.; Zhang, R.; Gan, L.; Gao, W. NLOS Signal Detection and Correction for Smartphone using Convolutional Neural Network and Variational Mode Decomposition in Urban Environment. *GPS Solut.* **2023**, *27*, 31. [CrossRef]
34. Miao, W.; Li, B.; Gao, Y. The Superiority of Multi-GNSS L5/E5a/B2a Frequency Signals in Smartphones: Stochastic Modelling, Ambiguity Resolution and RTK Positioning. *IEEE Internet Things J.* **2022**. [CrossRef]
35. Yong, C.Z.; Harima, K.; Rubinov, E.; McClusky, S.; Odolinski, R. Instantaneous Best Integer Equivariant Position Estimation Using Google Pixel 4 Smartphones for Single-and Dual-Frequency, Multi-GNSS Short-Baseline RTK. *Sensors* **2022**, *22*, 3772. [CrossRef]
36. Li, B.; Miao, W.; Chen, G.E.; Li, Z. Ambiguity Resolution for Smartphone GNSS Precise Positioning: Effect Factors and Performance. *J. Geod.* **2022**, *96*, 63. [CrossRef]
37. Fu, G.; Mohammed, K.; van Diggelen, F. Android Raw GNSS Measurement Datasets for Precise Positioning. In Proceedings of the 33rd International Technical Meeting of the Satellite Division of The Institute of Navigation (ION GNSS+ 2020), Virtual, 22–25 September 2020; pp. 1925–1937.
38. Geo++ GmbH. Geo++ GmbH. Logging of GNSS Raw Data on Android. Geo++. 2018. Available online: http://www.geopp.de/logging-of-gnss-raw-data-on-android/ (accessed on 12 December 2022).
39. Zangenehnejad, F.; Jiang, Y.; Gao, Y. Improving Smartphone PPP and RTK Performance Using Time-Differenced Carrier Phase Observations. In Proceedings of the 35th International Technical Meeting of the Satellite Division of The Institute of Navigation (ION GNSS+ 2022), Denver, CO, USA, 19–23 September 2022; pp. 2287–2300.

40. European Global Navigation Satellite Systems Agency, GSA. *Using GNSS Raw Measurements on Android Devices—Towards Better Location Performance in Mass Market Applications (White Paper)*; Publications Office of the European Union: Luxembourg, 2018. Available online: https://www.gsa.europa.eu/system/files/reports/gnss_raw_measurement_web_0.pdf (accessed on 12 December 2022).
41. Teunissen, P.J.G.; Kleusberg, A. *GPS for Geodesy*, 2nd ed.; Springer: Berlin/Heidelberg, Germany; New York, NY, USA, 1998.
42. Kouba, J.; Héroux, P. Precise Point Positioning using IGS Orbit and Clock Products. *GPS Solut.* **2001**, *5*, 12–28. [CrossRef]
43. Chen, B.; Gao, C.; Liu, Y.; Sun, P. Real-Time Precise Point Positioning with a Xiaomi MI 8 Android Smartphone. *Sensors* **2019**, *19*, 2835. [CrossRef]

sensors

MDPI

Article

A Comprehensive Analysis of Smartphone GNSS Range Errors in Realistic Environments

Jiahuan Hu, Ding Yi * and Sunil Bisnath

Department of Earth and Space Science and Engineering, York University, Toronto, ON M3J 1P3, Canada
* Correspondence: dingyi@yorku.ca

Abstract: Precise positioning using smartphones has been a topic of interest especially after Google decided to provide raw GNSS measurement through their Android platform. Currently, the greatest limitations in precise positioning with smartphone Global Navigation Satellite System (GNSS) sensors are the quality and availability of satellite-to-smartphone ranging measurements. Many papers have assessed the quality of GNSS pseudorange and carrier-phase measurements in various environments. In addition, there is growing research in the inclusion of a priori information to model signal blockage, multipath, etc. In this contribution, numerical estimation of actual range errors in smartphone GNSS precise positioning in realistic environments is performed using a geodetic receiver as a reference. The range errors are analyzed under various environments and by placing smartphones on car dashboards and roofs. The distribution of range errors and their correlation to prefit residuals is studied in detail. In addition, a comparison of range errors between different constellations is provided, aiming to provide insight into the quantitative understanding of measurement behavior. This information can be used to further improve measurement quality control, and optimize stochastic modeling and position estimation processes.

Keywords: smartphone range errors; realistic driving scenarios; environment classification; error distribution and correlation

Citation: Hu, J.; Yi, D.; Bisnath, S. A Comprehensive Analysis of Smartphone GNSS Range Errors in Realistic Environments. *Sensors* **2023**, 23, 1631. https://doi.org/10.3390/s23031631

Academic Editors: Jari Nurmi and Robert Odolinski

Received: 6 January 2023
Revised: 30 January 2023
Accepted: 31 January 2023
Published: 2 Febraury 2023

1. Introduction

Over the last decade, the proliferation of low-cost GNSS-enabled smartphones has boosted personalized Location-Based Services (LBS) businesses thanks to the evolution of GNSS systems and the maturity of microelectronic technologies. In 2016, Google released the Android 7.0 platform that allows smartphone users to access the code and phase measurements free of charge [1], enabling researchers to analyze data quality and refine positioning algorithms. This progress has, in turn, significantly expanded smartphone-based mass market applications such as cadastral surveying, asset management, mobile mapping, seismic monitoring, precise agriculture, lane-level navigation for unmanned vehicles, etc. [2–5].

In addressing demands from multiple industries, how to provide real-time continuous, accurate, and reliable smartphone navigation services has been seen as the major challenge faced by the GNSS community. Owing to hardware limitations, most early studies attempted different strategies to evaluate and improve positioning accuracy of single-frequency smartphones. Pesyna et al. [6] achieved centimetre-level accuracy with smartphone-quality antenna and single-frequency carrier-phase differential GNSS technology. In addition, after phase measurements are available, Sikiria et al. [7] evaluated the single point positioning (SPP) performance with Huawei P10 pseudorange measurements, and obtained horizontal and vertical rms of 10 m and 20 m, respectively. Similar smartphone single point positioning was also seen in [8,9]. However, the positioning accuracy of several metre-level cannot respond to growing market needs. In this context, a host of studies demonstrated that single-frequency smartphone RTK/NRTK solutions may achieve decimetre- to centimetre-level positioning accuracy under ideal environments [10–13]. Furthermore, Gill et al. [14] carried out single-frequency precise point positioning (PPP) static experiments, and the final solutions provide 37 cm and 51 cm rms in horizontal and vertical domains, respectively. In addition, such solutions can be further improved with the ionospheric-constrained PPP strategy [15].

In 2018, the launch of the first dual-frequency smartphone MI 8 equipped with a BCM47755 chip provided the opportunity to also utilize the L5/E5 frequency and better manage ionospheric delays [16–18]. Aggrey et al. [19] compared PPP performance for four smartphones under static and kinematic experiments, and the MI 8 achieved 40 cm rms in the horizontal direction, which was far better than other single-frequency smart devices. In addition, similar performance can be also seen from numerous recent contributions with real-time and final products [20–24]. Continuing this research, recent studies prove that dual-frequency smartphones can provide lane-level navigation processed with both RTK and PPP technologies in realistic driving environments [25,26], and the solutions can be further improved with the aid of smartphone native Inertial Measurement Unites (IMUs) [27,28].

Despite this considerable progress, the major limitations preventing smartphone-based precise applications are their noisy or non-existing measurements due mainly to the poor polarized antenna and multipath contamination [5,6,29]. To tackle these issues, several studies thus far have investigated smartphone signal strength, stochastic modeling, observation noise characterization, as well as measurement optimization strategies. Compared to geodetic GNSS receivers, it is observed that smartphone observations are prone to significantly lower carrier-to-noise density ratios (C/N_0) [30–32], especially for highly multipath-contaminated measurements, indicating that smartphone GNSS modules suffer from poor signal reception. Therefore, it is generally agreed in the literature that the C/N_0-based weighting scheme is more appropriate for smartphone positioning compared to the elevation-based weighting scheme [33–35]. To mitigate the negative impacts of native smartphone antennas, Riley et al. [36] revealed that smart devices with different grades of GNSS antennas are likely to perform significantly better in localization and provided insight into GNSS antenna replacement. Subsequent contributions confirm that smartphone centimeter-level solutions are achievable with an external geodetic antenna through relative positioning [37] and PPP-AR [38].

As unexpected smartphone measurement noise may reach a dozen meters [5,31,35,39], smartphone observation quality assessment has received considerable attention from researchers. Lachapelle et al. [40] evaluated the pseudorange quality from Huawei P10 through code and phase measurements difference, and results illustrate that smartphone code measurements are ten times noisier than geodetic GNSS receivers. Similar conclusions can be drawn from [30,32,41]. Moreover, Liu et al. [31] adopted a short baseline and the single-difference approach between the reference station and smartphone observations to assess the pseudorange residuals including multipath errors and observation noise, and this study showed that there are 10–30 m gross errors existing in smartphone code measurements compared to geodetic receivers, and vary between different constellations. Furthermore, by utilizing the multipath combination algorithm [42], it is confirmed that L5/E5a signals are significantly less influenced by the multipath effect compared to L1/E1 signals [43], owing to their higher transmission power level.

It is well recognized that receiver noise and multipath effects significantly limit the smartphone's precise positioning and, therefore, its applications. However, in view of all work that has been mentioned, few studies focus on smartphone data quality analysis in realistic, e.g., driving environments, and many experiments are stabilized on open rooftops or vehicle roofs, which do not correspond to consumer habits in daily life. To improve understanding of smartphone measurement behavior and further enhance the stochastic modeling and estimation processes, this study provides a detailed assessment of the actual smartphone range errors under different driving scenarios. Furthermore, no contributions have explored and the relationship between range errors with satellite geometry, signal strength, as well as pre-fit residuals under multiple multipath profiles, and the range errors and their distribution difference caused by the different smartphone mounting location are still unclear so far. Given that, the main contributions of this work are to answer the following questions: (1) How does the smartphone range error distribution behave based on different signal frequencies and constellations? (2) What is the relationship among range errors, satellite elevation angles, and SNR values? (3) How does range error vary under different multipath profiles in real driving environments? Do any differences exist with different mounting location? (4) What is the correlation between smartphone pre-fit residuals and range errors? The novelties of this paper are in analyzing the range errors under different environments, investigating the correlation between pre-fit residuals and range errors, and also comparing the range errors for smartphones mounted on a car roof and dashboard. It is also worth mentioning that, to determine actual range errors, the geodetic tightly-coupled post-processing kinematic (PPK)+inertial measurement unit (IMU) solutions are used as reference, which is different from the smartphone GNSS-based position estimates used in most literature.

This contribution is organized as follows: first, two different range error computation methodologies are comprehensively discussed in Section 2, followed by the measurement campaigns and experimental design in Section 3. Section 4 provided a comprehensive analysis of range errors to address the posed questions. Finally, this paper ends with conclusions and future work.

2. Methodology

To calculate the range errors for the smartphone, a collocated geodetic-grade receiver is needed as a reference, for which the range errors can be neglected compared with noisier smartphone measurements. Figure 1 illustrates the receiver sets (*A* and *B*) mounted on the car and with the tracked satellites *m* and *n*. The range errors can then be generated using measurement-differenced or state-differenced methods. In the measurement-differenced method, double-differencing is conducted with raw observations from two close receivers with consideration of the antenna lever arm correction, while in the state-differenced method, the state terms such as slant ionospheric effect, zenith troposphere delay are derived using the un-combined and un-differenced Precise Point Positioning (PPP) model from the reference receiver and applied to the smartphone as "true" values, and then single-differenced state residuals are computed to eliminate receiver-related terms, and hence the range errors are obtained. It should be noted that, due to the differencing procedure, the calculated range errors would be enlarged correspondingly.

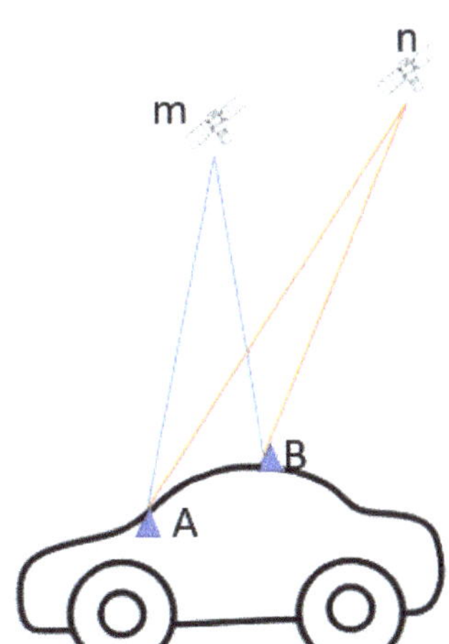

Figure 1. Illustration of receiver suits on a car and the tracked satellites.

The pseudorange measurement P on jth frequency for satellite m at receiver A can be given as Equation (1):

$$P^m_{A,j} = \rho^m_A + c(dt_A - dt^m) + I^m_{A,j} + T^m_A + b_{A,j} - b^m_j + \epsilon^m_{A.j} \tag{1}$$

where ρ denotes the geometry distance between satellite and receiver, c is the speed of light, dt_A and dt^m are the clock offset for receiver A and satellite m, respectively. $I^m_{A,j}$ is the slant ionospheric effect in meters on jth frequency, and T is the slant troposphere effect in meters. $b_{A,j}$ is the code hardware delay from the receiver antenna to the signal correlator, and b^m_j is the code hardware delay from the satellite signal generator to the satellite antenna. $\epsilon^m_{A.j}$ is the measurement noise, which contains the range errors that need to be derived.

With another receiver B, the double-differenced pseudorange measurement between receiver A and B and satellite m and n on jth frequency can be given as Equation (2):

$$\nabla\triangle P = (\rho^m_A - \rho^n_A) - (\rho^m_B - \rho^n_B) + \delta I + \delta T + (\epsilon^m_A - \epsilon^n_A) - (\epsilon^m_B - \epsilon^n_B) \tag{2}$$

where the ∇ denotes the between-receiver differencing and $\triangle$ denotes the between-satellite differencing, δI and δT are the ionosphere and troposphere residuals after double-differencing, respectively. For two close receivers whose lever arm is within several meters, the atmospheric delays can be eliminated by performing double-differencing, thus δI and δT can be ignored.

With two assumptions, that (a) the range errors for geodetic-grade receiver A can be ignored compared to those of smartphone B [44], and (b) the satellite with the highest elevation angle which is chosen as reference satellite when performing between-satellite differencing has significant lower range errors compared to lower-elevation satellites, the measurement-differenced range errors can therefore be derived as Equation (3):

$$rangeerror = \nabla\triangle P - (\rho^m_A - \rho^n_A) + (\rho^m_B - \rho^n_B) \tag{3}$$

where the geometry distance ρ^m can be given as a function of receiver position (x, y, z) for satellite $m(x^m, x^m, z^m)$ as Equation (4):

$$f(x,y,z) = \rho^m = ((x - x^m)^2 + (y - y^m)^2 + (z - z^m)^2)^{1/2} \tag{4}$$

In a realistic environment, the lever arm between A and B is constant in the vehicle body coordinate system (east, north, and up), and can be transformed to epoch-wised ECEF coordinate system using the approximate position of vehicles. Take one specific epoch for an example, with the coordinates of receiver $A(X_A, Y_A, Z_A)$, satellite $m(x^m, x^m, z^m)$ and the lever arm (X_0, Y_0, Z_0), Taylor's formula can be applied to ρ^m_B at the approximate position of A as Equation (5):

$$\rho^m_B|_{(x,y,z)=(X_A,Y_A,Z_A)} = \rho^m_A + f'(x,y,z) \cdot (X_0, Y_0, Z_0) + f''(x,y,z) \cdot \frac{(X_0, Y_0, Z_0)}{2} + \cdots \tag{5}$$

The second order of Equation (5) can then be derived as $\frac{1}{2}(\frac{X_0}{\rho^m_A} + \frac{Y_0}{\rho^m_A} + \frac{Z_0}{\rho^m_A})$ and can be ignored because of the meter-level lever-arm. In this way, only first order is taken into consideration, and hence Equation (3) can be further reparameterized as Equation (6):

$$\begin{aligned} rangeerror = \nabla\triangle P - ((&\frac{X_A - x^m}{\rho^m_A}X_0 + \frac{Y_A - y^m}{\rho^m_A}Y_0 + \frac{Z_A - z^m}{\rho^m_A}Z_0) \\ -(&\frac{X_A - x^n}{\rho^n_A}X_0 + \frac{Y_A - y^n}{\rho^n_A}Y_0 + \frac{Z_A - z^n}{\rho^n_A}Z_0)) \end{aligned} \tag{6}$$

It can be noted from Equation (6) that the coefficients of the lever-arm values are the line-of-sight values from receiver A to satellites, and corresponding coefficients can be determined reliably even with a ten-meter-accuracy approximate position. Therefore, the measurement-differenced range errors can be derived with only the pseudorange measurements and the fixed lever arm.

Aside from the measurement-differenced method based on Equation (1), there is another straightforward approach to estimate range errors through pseudorange residual differencing between satellites. This state-differenced method can better show the estimated state characteristics than the measurement-differenced method. Correspondingly, all receiver-dependent states are eliminated with the relative range errors remaining, and the state-differenced or PPP-based range errors can be expressed as Equation (7):

$$\triangle P = P^m_{A,j} - P^n_{A,j} = (\rho^m_A - \rho^n_A) + (I^m_{A,j} - I^n_{A,j}) + (T^m_A - T^n_A) + (\epsilon^m_A - \epsilon^n_A) \tag{7}$$

Compared to the aforementioned measurement-differenced range errors, the calculated range errors noise is only amplified once from the differencing. However, it requires accurate satellite-dependent states such as geometric ranges, ionospheric delays, as well as tropospheric delays during computation. Furthermore, other terms such as relativity corrections, Sagnac corrections, solid tide corrections, and satellite antenna corrections [45] need to be taken into consideration. A geodetic receiver, NovAtel OEM7, is utilized in the following processing, and satellite-related states are estimated through PPP processing.

3. Measurement Campaigns and Experimental Setup

The data for the analysis are collected in and around York University, Toronto, Canada. York University and adjoining areas cover all the aspects of multiple multipath profiles needed for the experiment. The profiles include highways, parking lots, sub-urban with vegetation, overpasses, and open sky environments, each having a unique multipath characteristic and contributing to better appreciating the range error behaviors. Figure 2 shows the aerial and detailed street view photos of these measurement campaigns, and blue arrows indicate the direction.

As shown in Figure 3, the experimental setup contains two sensor suites. The first contains two geodetic GNSS receivers to provide position trajectory references for tested smartphone. One NovAtel OEM7 receiver is mounted on a rooftop as a reference station within a 5 km baseline, whilst the other receiver is connected with a geodetic antenna and mounted inside the vehicle experimental box (see Figure 3a). The second sensor suite includes Xiaomi MI 8 phone, which is fixed with holders and mounted on the vehicle dashboard to mimic real-life applications (see Figure 3b). In addition, it is feasible to mount MI 8 on the top of the experimental box for in-depth comparison and further analysis. The lever arm between the smartphone and the referenced center has been carefully measured.

Table 1 highlights three datasets used in this study, including the test number, collections date, time duration, as well as phone mounted points. Road tests were carried out on two separate days with identical route and different traffic conditions, and the time durations for each test are 31, 28, and 29 min, respectively. These datasets are used and analyzed in the following section:

Table 1. Summary of collected smartphone datasets.

Test #	Collection Date	Duration	Mounted Points
1	8 September 2020	22:29–23:00	Roof
2	8 August 2021	1:14–1:42	Roof
3	8 August 2021	1:50–2.19	Dashboard

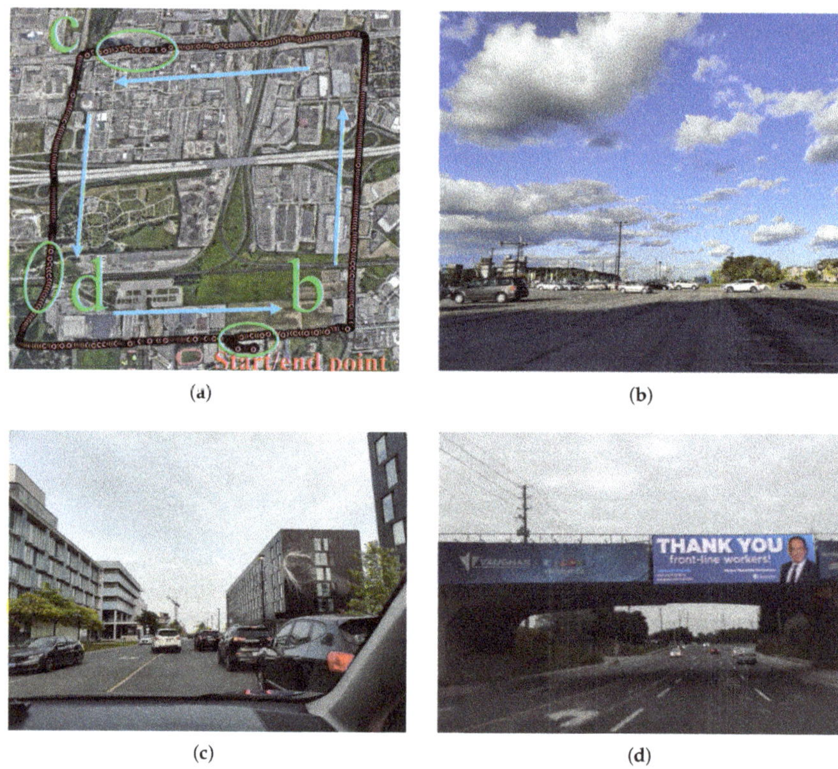

Figure 2. (**a**) aerial and street views of vehicle trajectory with (**b**) open-sky parking lots, (**c**) suburban road, and (**d**) short underpass.

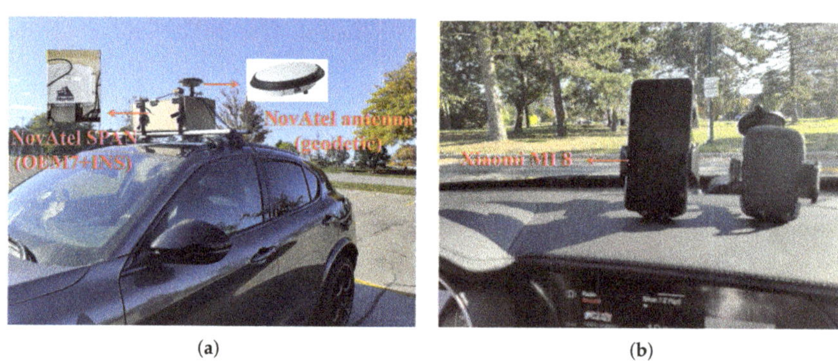

Figure 3. Photos of experimental setup. (**a**) experimental vehicle and setup, (**b**) experimental smartphones.

4. Results

The key to further improving the smartphone positioning performance is to understand the smartphone GNSS measurement characteristics. Therefore, analysis regarding range errors on different frequencies and the relationship between signal-to-noise ratio (SNR) values or elevation angles with range errors is first carried out, followed by the assessment of correlations between range errors and pre-fit residuals, range errors under different environments, and comparison of range errors when the smartphone is mounted on the car dashboard and roof. The smartphone data processing strategies have the assumption that the measurement errors should follow a similar distribution as geodetic receivers; therefore, the zero baseline ZIM2-ZIM3 MGEX stations are processed as a reference to provide a clear view of the characteristics of smartphone range errors compared with geodetic receivers. It is worth mentioning that the first (L1 for GPS and GLONASS, E1 for Galileo, and B1 for BDS) and third (L5 for GPS, and E5a for Galileo) frequencies are used in this analysis for both smartphone and geodetic receiver assessments. Hereafter, P1 and P5 are used to denote the first and third frequencies, respectively. In the result analysis, dataset 1 is used in Sections 4.1–4.4, and datasets 2 and 3 are used in Section 4.5.

4.1. Distribution of Range Errors

The distribution of range errors on the first and third frequencies is first analyzed. As shown in Figure 4, the upper subplots show the distribution of range errors for the smartphone, while the lower panel gives the results for MGEX stations. It can be observed that, for both smartphone and geodetic receivers, the standard derivation (STD) for the third frequency is smaller than the first frequency, with values of 7.3 m on P1 and 2.1 m on P5 for smartphones, and 0.5 m and 0.2 m for those of geodetic receivers, respectively, which indicates that the observation precision on the third frequency is higher. This difference may be due to the fact that the bandwidth of L5 is larger and the transmitted power of L5 is higher than the first frequency. Therefore, the anti-interference ability and anti-noise performance may also be significantly improved for L5 [31,46,47]. The range errors for the geodetic receivers are obeying a nearly zero-mean distribution, while for smartphones, the mean values are not zero, but rather 0.8 m and 0.6 m for the first and third frequencies, respectively. When further investigating the range errors, the STD of range errors for the geodetic receiver on P1 is 0.5 m, considering that the double-differenced measurements are formed when calculating the range errors, the original code precision should be 2 times smaller than the range errors by applying the error propagating law, which is 0.2–0.3 m. This value is consistent with the code precision used in MGEX data processing as in [48], which demonstrates the validity of the proposed range error derivation method.

To provide a clear view of the characteristics of range errors for different constellations, Figure 5 illustrates the mean values and STDs of each constellation for smartphone and geodetic receivers. For GLONASS and BDS, there are no observations for the test smartphone data, and there are only a few GLONASS observations for ZIM2 and ZIM3. It can be concluded from Figure 5 that, among the four constellations, the GLONASS pseudorange measurements have the largest STDs, with values of 11.8 m and 0.7 m for smartphone and geodetic receivers. Therefore, in the GNSS data processing procedure, GLONASS pseudorange measurements are usually de-weighted by two times, and these results on the first frequency are similar to what has been found in [31]. With the four constellations, except for Galileo, the STDs on P5 are nearly 2 times smaller than on P1, which infers that more weight can be put on the third frequency. For this smartphone dataset, the STDs are more than ten times larger compared to the corresponding constellations for geodetic receivers, and hence, for the analysis for range errors on different frequencies of different constellations, the stochastic models of the smartphone observation noise can be adjusted accordingly in the future research.

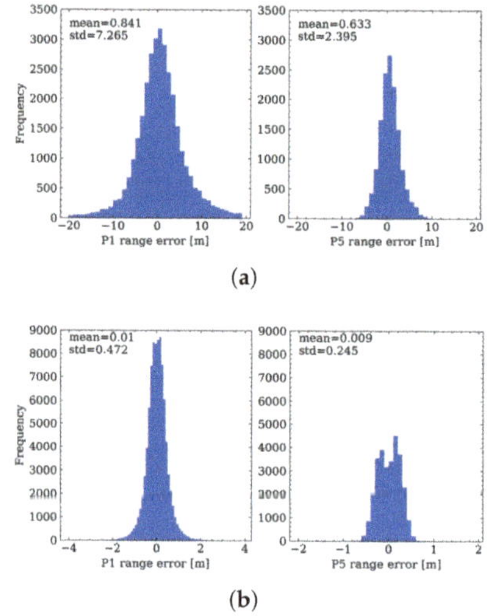

Figure 4. Distribution of range errors on P1 and P5 for different sensors. (**a**) distribution of range errors on smartphones; (**b**) distribution of range errors on MGEX stations.

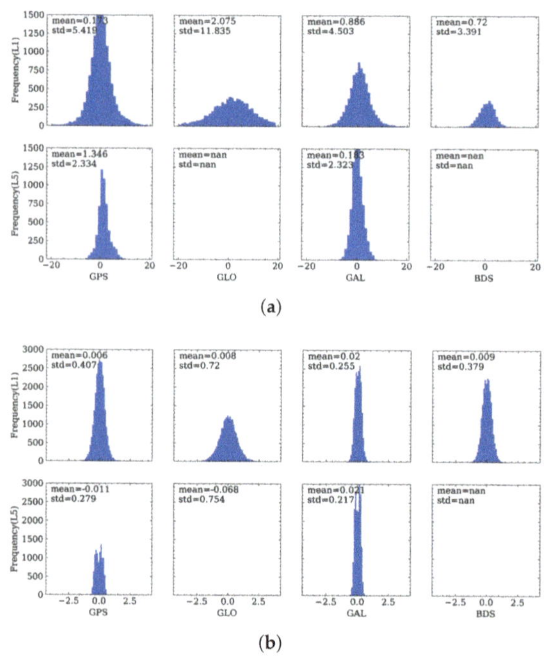

Figure 5. Distribution of range errors of different constellations for (**a**) smartphone and (**b**) ZIM2.

4.2. Range Errors Correlation with Elevation Angle and SNR

Two common weighting schemes are applied to smartphone and geodetic receiver observations, namely SNR-based and elevation angle-based weighting, respectively. These two weighting schemes indicate that measurement quality is highly related to the SNR and elevation angles. Therefore, in this subsection, the relationship between range errors, elevation angle, as well as SNR values, is assessed. Figure 6 depicts the temporal characteristics of range errors and the corresponding relationship with SNR and elevation angles, and different colors represent different satellites. It can be noted again that the range errors on the first frequency are larger than those on the third frequency from the left panel which shows the time series of the range errors. In addition, there is a significant decreasing trend when the SNR increases for smartphone data both for P1 and P5, and a similar situation can be observed when comparing the relationship between range errors and elevation angles for geodetic receivers. Due to the short observation time for the smartphone data (only 30 min), elevation angles are not fully distributed from 0–90 degrees. However, it can also be concluded that, when the satellite tracked by a smartphone is newly rising or falling, the range errors will increase slightly, and therefore the elevation cut-off angle should be set reasonably. For the test ZIM3 observation, no correlation between range errors and SNRs can be found; this may be the reason that the geodetic receiver is under an optimal observation environment, and the SNR values are at a high level, which less affects the range errors.

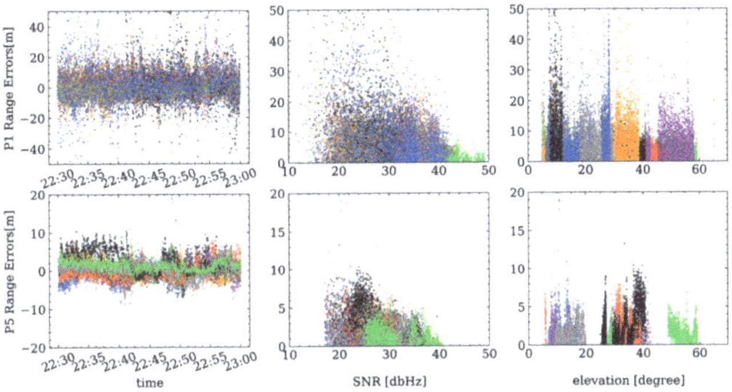

(a)

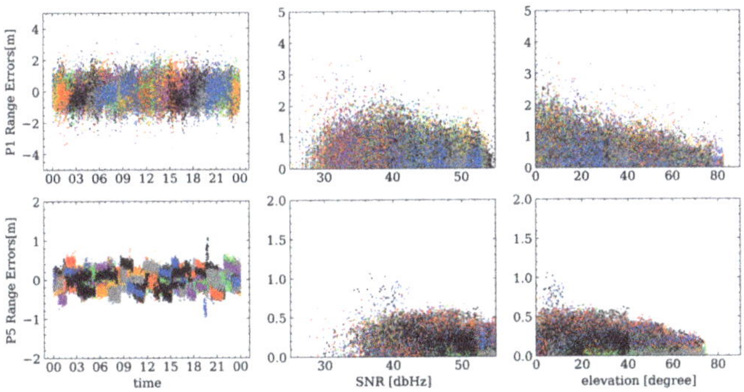

(b)

Figure 6. Temporal characteristics of range errors (left) and the relationship with SNR (middle) and elevation angles (right) for smartphone (**a**) and ZIM2 (**b**).

To further investigate range error correlations, error bars are plotted in Figure 7 to show the mean values and STDs of absolute range errors at different SNR values and elevation angles. The dots and lines denote the mean values and STDs of absolute range errors, respectively. As shown in Figure 7, it can be clearly observed that there is a significant correlation between smartphone range errors and SNR values, especially on the first frequency. Concurrently, a small correlation on the P1 range errors for geodetic receivers can be observed, which is similar to the results in [44]. The small STD for geodetic receivers at low SNR values (around 25) is due mainly to the limited observations with low SNRs. On the other hand, there is a decreasing STD trend on P1 for the geodetic receiver when the elevation angles are increasing, while the mean values and STDs seem to be stable for P5, which have not been mentioned in the previous research.

In view of the existing literature, the SNR cut-off strategy is widely adopted to screen out satellites that are supposed to contain noisy measurements [49]. However, in this work, it is interesting to identify that sometimes these measurements with lower SNR values are even less noisy and may be beneficial for positioning. Furthermore, range error trends vary with different signal frequencies, indicating that weighting schemes need to be tuned and adjusted when applied to different signals. Meanwhile, the elevation angle-based weighting scheme is usually used for geodetic data processing, which might be well suited to the first frequency. However, there is also a potential weighting scheme that can be proposed in future research, to consider both elevation angle and SNR values or to apply different weighting schemes for different frequencies.

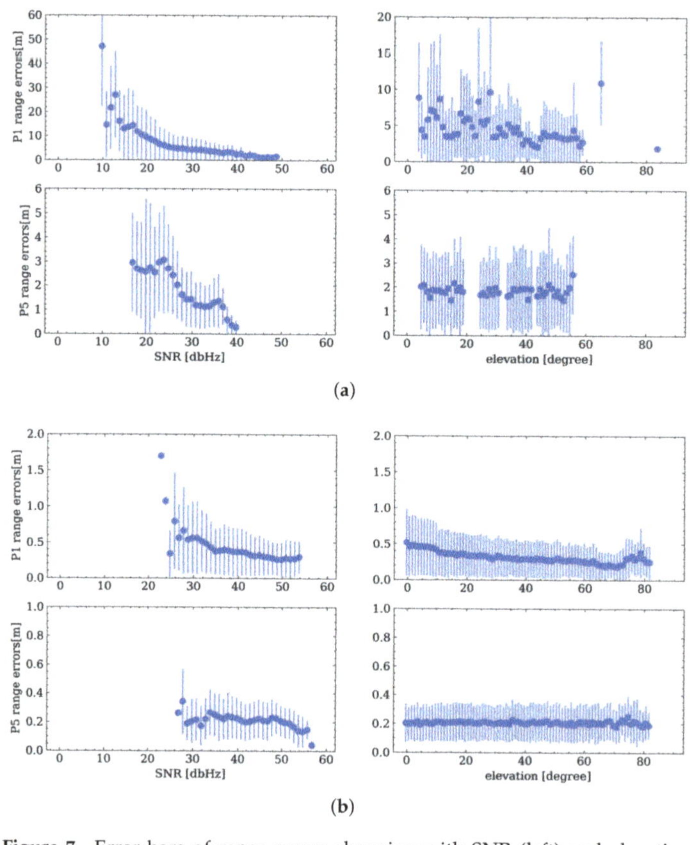

Figure 7. Error bars of range errors changing with SNR (left) and elevation angles (right) for smartphone (**a**) and ZIM2 (**b**).

4.3. Range Errors under Different Environments

GNSS measurements under environments such as urban canyons, overpasses and vegetation usually suffer from poor signal reception, low gain, large multipath errors, and noises, which result in variation and increases in range errors. To investigate this issue, the range errors under three scenarios are analyzed in this subsection to fathom the impact of the environment on it.

As shown in Figure 8, three typical epochs are selected for detailed analysis: (a) in a parking lot, and can be regarded as an open sky area; (b) in a suburban area, where the car was driving close to tall buildings; (c) in a vegetated area, where the car was close to trees on one side of the road. The sky plots on the right show the corresponding epoch, the second of the day (SoD), and range errors for different satellites are presented by different colors and sized dots (the larger the dots, the larger the range errors). It can be seen from Figure 8 that, owing to nearby buildings and trees limiting sky visibility, GNSS signals were blocked significantly, reducing satellite visibility. As expected, the range errors are at a low level when driving in the parking lot, except for satellite R17, whose elevation angle is under 5°. In contrast, range errors increase significantly when the vehicle was in the suburban area, even for signal azimuths without building blockage. While driving in the vegetation area, it seems that the signal blockage by the tree has limited impacts on the range errors of other tracked satellites.

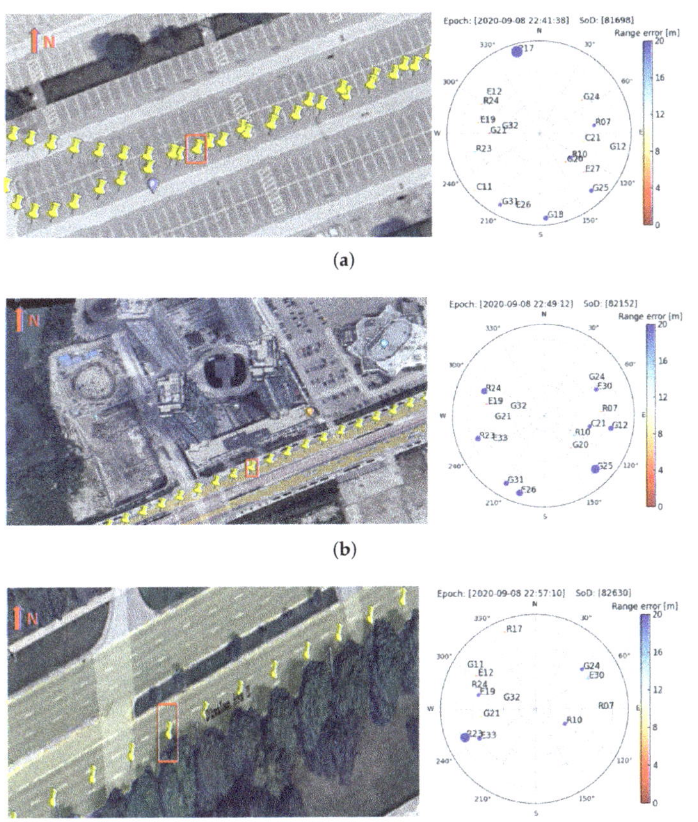

Figure 8. Sky plot of satellite range errors under (**a**) open sky, (**b**) sub-urban, and (**c**) vegetated environments.

To better understand the range errors under different environments, the whole trajectory of the smartphone data are classified into three scenarios, namely open sky, vegetation, and sub-urban, Table 2 summarizes the overall statistics of the range errors and the 95th per denotes the range errors of exact the 95th percentile. It can be noted that, in the sub-urban environment, the mean values of range errors and corresponding STDs for P1 and P5 are significantly larger than that in open sky and vegetation, which is expected because of the signal blockage and higher-level multipath in the sub-urban. Under the vegetation environment, the range errors are slightly larger than those in the open sky and are consistent with the analysis above, for which the vegetation does not have a significant impact on range errors in comparison to sub-urban.

Table 2. Statistics for satellite absolute range errors under different environments (unit: meter).

	P1				P5			
	Mean	**STD**	**95th per**	**Number**	**Mean**	**STD**	**95th per**	**Number**
Open-sky	0.8	7.3	14.5	32267	0.9	2.4	5.1	14198
Sub-urban	2.2	9.4	18.2	1443	1.2	3.8	5.9	658
Vegetation	1.5	7.5	17.1	354	1.4	1.9	3.9	176

4.4. Comparison between Range Errors and Pre-Fit Residuals

In smartphone GNSS data processing, quality control is vital and more challenging compared to geodetic data. Pre-fit outlier rejection is one effective quality control method. Therefore, in this subsection, the correlation between ranger errors and pre-fit residuals is assessed. As shown in Figure 9, different colors represent different satellites. It can be concluded that the range errors and pre-fit residuals for P1 have a higher correlation than P5, while the range errors and pre-fit residuals for P5 are at a lower level compared with P1. The reason for a higher correlation in P1 could be that the smartphone antenna is designed for the first frequency, and may not be optimized for other frequencies. To further evaluate the correlation satellite-by-satellite, Figure 10 illustrates the relationship among correlation coefficient, elevation angle, and SNR, where each dot represents one satellite and different colors denote the corresponding SNR values. With a lower elevation angle, the correlation coefficients tend to be smaller. Usually, GLONASS satellites have the highest correlation coefficients despite larger range errors. One possible reason is that, for other constellations, due to the high noise level itself for smartphones, the correlation between range errors and pre-fit residuals can be easily hidden within the noise; while there is a larger range error for GLONASS satellites, so the correlation can be more conspicuous. The results indicate that pre-fit residual rejection/de-weighting is worth further investigation and may provide potential solution improvement in future research, as it reflects the range errors with a high confidence level.

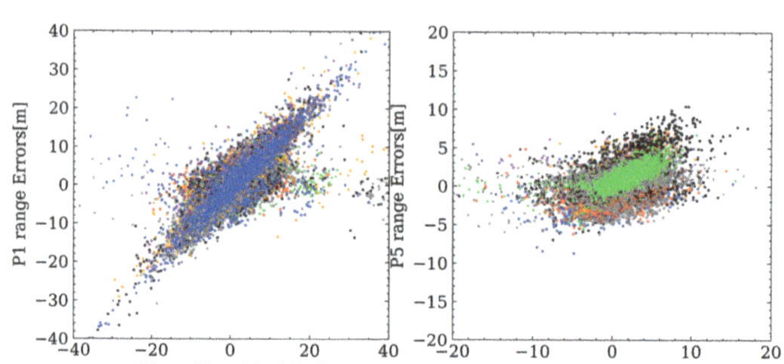

Figure 9. Correlation between range errors and pre-fit residuals.

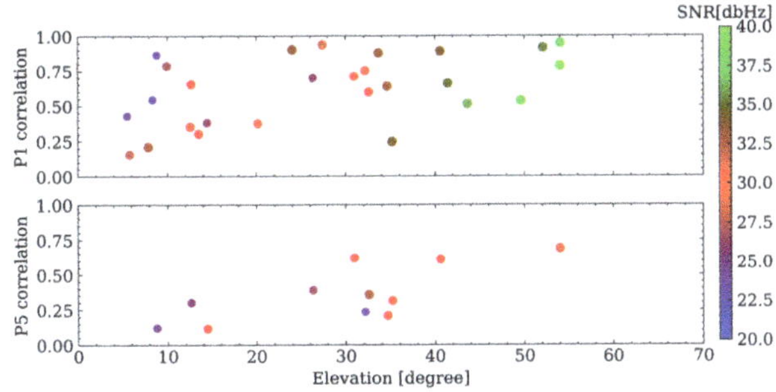

Figure 10. Relationship among correlation coefficients, elevation angle, and SNR values.

4.5. Range Errors Comparison on Car Dashboard and Roof

In a realistic application of smartphone positioning, drivers usually mount their devices on the dashboard for navigation or chatting purposes. However, much of the published research conducted experiments under open-sky environments or mounted smartphones on car roofs. Considering that the car roof may also influence the GNSS observation quality, this subsection assesses how the distribution of range errors will be affected when the smartphone is placed on the car roof compared to the car dashboard. Figure 11 depicts the probability distribution of range errors, in which the blue and green bins denote the range errors on the roof and dashboard, respectively. The STD values of range errors on the car roof are 5.1 m and 5.3 m on the dashboard for P1, while they are 2.2 m and 2.4 m for P5, respectively. A clearer and higher centralization for the range errors on the roof is observed compared to those on the dashboard from Figure 11, which infers that the GNSS observation quality is slightly better when the phone is placed on the car roof, and different code precisions may need to be applied when the smartphones are in different places of the car in the data processing.

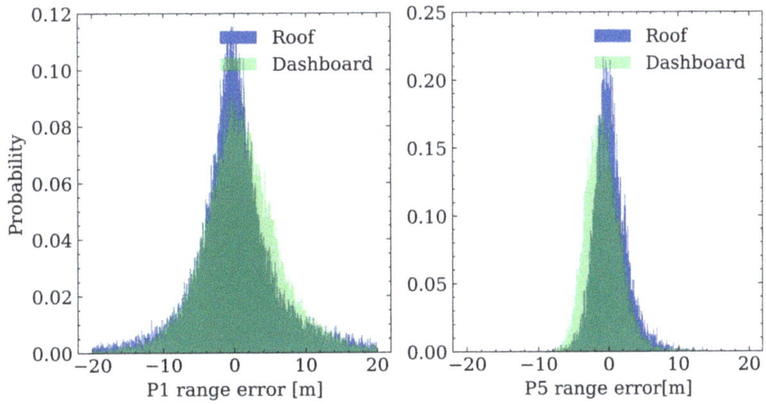

Figure 11. Range error distribution comparison when the smartphone is placed on the dashboard and roof.

5. Conclusions and Future Work

This study proposes a differencing method between a geodetic grade receiver and a smartphone with consideration of the lever arm, aiming to derive the actual range errors for smartphones under realistic usage. The methodology of the observation differencing is first introduced, and with several datasets collected, for which the smartphone is mounted on the car dashboard and roof, comprehensive investigation and analysis are carried out.

Firstly, question (1) proposed in Section 1 is addressed by assessing the distribution of the range errors and range error characteristics for the different constellations, and it is found that the range errors of smartphones on the first frequency are significantly larger than those of the third frequency, approximately ten times larger than a geodetic receiver/antenna. In addition, the range errors of GLONASS satellites have the largest STD, with values of 11.8 m and 0.7 m for smartphone and geodetic receivers, respectively. For GPS, Galileo, and BDS, the smartphone STD values of range errors are comparable.

Secondly, to answer question (2), the relationship among range errors, elevation angles, and SNR values is evaluated. Consistent with published research, the smartphone range errors are highly dependent on the SNR values, while the range errors of geodetic receivers are significantly correlated with satellite elevation angles. However, the trend on the third frequency is not as obvious as the first frequency, which is worth further investigation.

Thirdly, a vehicle trajectory is classified into different GNSS environments, namely open sky, sub-urban, and vegetation, to assess smartphone range error behaviors under different scenarios which is proposed in question (3). The results for the dataset analyzed indicate that range errors reach a peak in the suburban environment. While in the vegetation environment, the range errors are slightly larger compared to open-sky environments. Furthermore, for the data processed, the tree blockage seems to have little impact on affecting the measurement quality of other satellites.

Fourthly, a comparison between range errors and pre-fit residuals is carried out to address the question (4). A high correlation is found on the first frequency, while it is not significant for the third frequency. By computing correlation coefficient, it is observed that GLONASS measurements have larger correlation coefficients than others. The reason may be that, for other constellations, the correlation can be easily hidden in the observation noise, and it is similar when the third frequency appears to have a lower correlation. When evaluating the correlation satellite-by-satellite, lower elevation and lower SNR values can also cause a low correlation coefficient.

Finally, a range error comparison is conducted with the smartphone placed in different places. The results show that the STD values of range errors for the first frequency are 5.1 m and 5.3 m on the roof and dashboard, respectively, and are 2.2 m and 2.4 m for the third frequency, indicating that GNSS signals are further suppressed and interfered when the smartphone is mounted on the dashboard.

This paper provides a comprehensive assessment of the actual range errors for smartphones, and the results can further benefit the investigation on optimizing weighting schemes and quality control methods for future work, aiming for a higher level of smartphone positioning and application. The main purpose of this paper is to understand the actual range error distribution and characteristics; hence, some preliminary and commonly assessed performances are carried out first (e.g., relationship between range errors and SNR/elevation angle). However, the trends on different frequencies are slightly different, which can further be a reference of the optimization for quality control. In addition, there are new and interesting findings for the range error characteristics under different environments and the correlations between range error and pre-fit residuals, which can potentially benefit smartphone positioning algorithm development. e.g., different code precision and different weight may be applied to different constellations and different frequencies based on the analysis, and the SNR mask and elevation cut-off angle can be adjusted with different datasets. In addition, regarding different range error behavior under different environments, the weighting scheme considering different environments is worth investigating. Furthermore, the prefit residuals can partly reflect the range errors, different thresholds for prefit outlier rejection may be helpful according to this paper. Finally, different code precisions can also be applied when the smartphone is mounted in different places.

Author Contributions: Conceptualization, J.H. and S.B.; Data curation, D.Y.; Formal analysis, J.H.; Funding acquisition, S.B.; Investigation, J.H.; Methodology, J.H. and D.Y.; Software, J.H. and D.Y.; Supervision, S.B.; Validation, J.H. and D.Y.; Writing—original draft, J.H. and D.Y.; Writing—review and editing, J.H., D.Y. and S.B. All authors have read and agreed to the published version of the manuscript.

Funding: This research was funded by the Natural Sciences and Engineering Research Council of Canada (NSERC).

Institutional Review Board Statement: Not applicable.

Informed Consent Statement: Not applicable.

Data Availability Statement: The data showed in this study are available from the corresponding author on reasonable request.

Acknowledgments: The authors would like to thank their colleagues at the GNSS Laboratory for their support in collecting field data and discussing technical aspects of this work. In addition, the authors would like to acknowledge the data contribution from the International GNSS Service (IGS), German Research Centre for Geosciences (GFZ), and Centre National d'Etudes Spatiales (CNES).

Conflicts of Interest: The authors declare no conflict of interest.

References

1. Malkos, S. User location takes center stage in new Android OS: Google to provide raw GNSS measurements. *GPS World* **2016**, *27*, 36.
2. Zangenehnejad, F.; Gao, Y. GNSS smartphones positioning: Advances, challenges, opportunities, and future perspectives. *Satell. Navig.* **2021**, *2*, 24. [CrossRef] [PubMed]
3. Alsubaie, N.M.; Youssef, A.A.; El-Sheimy, N. Improving the accuracy of direct geo-referencing of smartphone-based mobile mapping systems using relative orientation and scene geometric constraints. *Sensors* **2017**, *17*, 2237. [CrossRef] [PubMed]
4. Kong, Q.; Allen, R.M.; Schreier, L.; Kwon, Y.W. MyShake: A smartphone seismic network for earthquake early warning and beyond. *Sci. Adv.* **2016**, *2*, e1501055. [CrossRef]
5. Paziewski, J. Recent advances and perspectives for positioning and applications with smartphone GNSS observations. *Meas. Sci. Technol.* **2020**, *31*, 091001. [CrossRef]
6. Pesyna, K.M.; Heath, R.W.; Humphreys, T.E. Centimeter positioning with a smartphone-quality GNSS antenna. In Proceedings of the 27th International Technical Meeting of the Satellite Division of The Institute of Navigation (ION GNSS+ 2014), Tampa, FL, USA, 8–12 September 2014; pp. 1568–1577.
7. Sikirica, N.; Malić, E.; Rumora, I.; Filjar, R. Exploitation of Google GNSS measurement API for risk assessment of GNSS applications. In Proceedings of the 2017 25th Telecommunication Forum (TELFOR), Belgrade, Serbia, 21–22 November 2017; pp. 1–3.
8. Szot, T.; Specht, C.; Specht, M.; Dabrowski, P.S. Comparative analysis of positioning accuracy of Samsung Galaxy smartphones in stationary measurements. *PLoS ONE* **2019**, *14*, e0215562. [CrossRef]
9. Paziewski, J.; Pugliano, G.; Robustelli, U. Performance assessment of GNSS single point positioning with recent smartphones. In Proceedings of the IMEKO TC-19 International Workshop on Metrology for the Sea, Naples, Italy, 5–7 October 2020.
10. Realini, E.; Caldera, S.; Pertusini, L.; Sampietro, D. Precise GNSS positioning using smart devices. *Sensors* **2017**, *17*, 2434. [CrossRef]
11. Dabove, P.; Di Pietra, V. Towards high accuracy GNSS real-time positioning with smartphones. *Adv. Space Res.* **2019**, *63*, 94–102. [CrossRef]
12. Odolinski, R.; Teunissen, P.J. An assessment of smartphone and low-cost multi-GNSS single-frequency RTK positioning for low, medium and high ionospheric disturbance periods. *J. Geod.* **2019**, *93*, 701–722. [CrossRef]
13. Sun, W.; Li, Y.; Duan, S. Xiaomi Mi 8 smartphone GNSS data quality analysis and single-frequency RTK positioning performance evaluation. *IET Radar Sonar Navig.* **2020**, *14*, 1410–1416. [CrossRef]
14. Gill, M.; Bisnath, S.; Aggrey, J.; Seepersad, G. Precise point positioning (PPP) using low-cost and ultra-low-cost GNSS receivers. In Proceedings of the 30th International Technical Meeting of The Satellite Division of the Institute of Navigation (ION GNSS+ 2017), Portland, OR, USA, 25–29 September 2017; pp. 226–236.
15. Wang, G.; Bo, Y.; Yu, Q.; Li, M.; Yin, Z.; Chen, Y. Ionosphere-constrained single-frequency PPP with an Android smartphone and assessment of GNSS observations. *Sensors* **2020**, *20*, 5917. [CrossRef]
16. Elmezayen, A.; El-Rabbany, A. Precise point positioning using world's first dual-frequency GPS/GALILEO smartphone. *Sensors* **2019**, *19*, 2593. [CrossRef]
17. Wu, Q.; Sun, M.; Zhou, C.; Zhang, P. Precise point positioning using dual-frequency GNSS observations on smartphone. *Sensors* **2019**, *19*, 2189. [CrossRef] [PubMed]

18. Dabove, P.; Di Pietra, V. Single-baseline RTK positioning using dual-frequency GNSS receivers inside smartphones. *Sensors* **2019**, *19*, 4302. [CrossRef] [PubMed]
19. Aggrey, J.; Bisnath, S.; Naciri, N.; Shinghal, G.; Yang, S. Multi-GNSS precise point positioning with next-generation smartphone measurements. *J. Spat. Sci.* **2020**, *65*, 79–98. [CrossRef]
20. Wang, L.; Li, Z.; Wang, N.; Wang, Z. Real-time GNSS precise point positioning for low-cost smart devices. *GPS Solut.* **2021**, *25*, 69. [CrossRef]
21. Robustelli, U.; Baiocchi, V.; Marconi, L.; Radicioni, F.; Pugliano, G. Precise Point Positioning with Single and Dual-Frequency Multi-GNSS Android Smartphones. In Proceedings of the ICL-GNSS (Work in Progress), Tampere, Finland, 2–4 June 2020.
22. Lachapelle, G.; Gratton, P. GNSS precise point positioning with Android smartphones and comparison with high performance receivers. In Proceedings of the 2019 IEEE International Conference on Signal, Information and Data Processing (ICSIDP), Chongqing, China, 11–13 December 2019; pp. 1–9.
23. Li, Z.; Wang, L.; Wang, N.; Li, R.; Liu, A. Real-time GNSS precise point positioning with smartphones for vehicle navigation. *Satell. Navig.* **2022**, *3*, 19. [CrossRef]
24. Chen, B.; Gao, C.; Liu, Y.; Sun, P. Real-time precise point positioning with a Xiaomi MI 8 android smartphone. *Sensors* **2019**, *19*, 2835. [CrossRef]
25. Seepersad, G.; Hu, J.; Yang, S.; Yi, D.; Bisnath, S. Changing Lanes with Smartphones Technology. In Proceedings of the 34th International Technical Meeting of the Satellite Division of The Institute of Navigation (ION GNSS+ 2021), St. Louis, MO, USA, 20–24 September 2021; pp. 3021–3036.
26. Seepersad, G.; Hu, J.; Yang, S.; Yi, D.; Bisnath, S. Performance Assessment of Tightly Coupled Smartphone Sensors with Legacy and State Space Corrections. In Proceedings of the 35th International Technical Meeting of the Satellite Division of The Institute of Navigation (ION GNSS+ 2022), Denver, CO, USA, 19–23 September 2022; pp. 2235–2255.
27. Yang, S.; Yi, D.; Vana, S.; Bisnath, S. Resilient Smartphone Positioning using Native Sensors and PPP Augmentation. In Proceedings of the 34th International Technical Meeting of the Satellite Division of The Institute of Navigation (ION GNSS+ 2021), St. Louis, MO, USA, 20–24 September 2021; pp. 4208–4222.
28. Yi, D.; Yang, S.; Bisnath, S. Native Smartphone Single-and Dual-Frequency GNSS-PPP/IMU Solution in Real-World Driving Scenarios. *Remote Sens.* **2022**, *14*, 3286. [CrossRef]
29. Humphreys, T.E.; Murrian, M.; Van Diggelen, F.; Podshivalov, S.; Pesyna, K.M. On the feasibility of cm-accurate positioning via a smartphone's antenna and GNSS chip. In Proceedings of the 2016 IEEE/ION Position, Location and Navigation Symposium (PLANS), Savannah, GA, USA, 11–14 April 2016; pp. 232–242.
30. Zhang, K.; Jiao, F.; Li, J. The assessment of GNSS measurements from android smartphones. In Proceedings of the China Satellite Navigation Conference, Harbin, China, 23–25 May 2018; Springer: Singapore, 2018; pp. 147–157.
31. Liu, W.; Shi, X.; Zhu, F.; Tao, X.; Wang, F. Quality analysis of multi-GNSS raw observations and a velocity-aided positioning approach based on smartphones. *Adv. Space Res.* **2019**, *63*, 2358–2377. [CrossRef]
32. Paziewski, J.; Sieradzki, R.; Baryla, R. Signal characterization and assessment of code GNSS positioning with low-power consumption smartphones. *GPS Solut.* **2019**, *23*, 98. [CrossRef]
33. Shinghal, G.; Bisnath, S. Conditioning and PPP processing of smartphone GNSS measurements in realistic environments. *Satell. Navig.* **2021**, *2*, 10. [CrossRef] [PubMed]
34. Banville, S.; Lachapelle, G.; Ghoddousi-Fard, R.; Gratton, P. Automated processing of low-cost GNSS receiver data. In Proceedings of the 32nd International Technical Meeting of the Satellite Division of The Institute of Navigation (ION GNSS+ 2019), Miami, FL, USA, 16–20 September 2019; pp. 3636–3652.
35. Zhang, X.; Tao, X.; Zhu, F.; Shi, X.; Wang, F. Quality assessment of GNSS observations from an Android N smartphone and positioning performance analysis using time-differenced filtering approach. *GPS Solut.* **2018**, *22*, 70. [CrossRef]
36. Riley, S.; Lentz, W.; Clare, A. On the path to precision-observations with android GNSS observables. In Proceedings of the 30th International Technical Meeting of The Satellite Division of The Institute of Navigation (ION GNSS+ 2017), Portland, OR, USA, 25–29 September 2017; pp. 116–129.
37. Geng, J.; Li, G.; Zeng, R.; Wen, Q.; Jiang, E. A comprehensive assessment of raw multi-GNSS measurements from mainstream portable smart devices. In Proceedings of the 31st International Technical Meeting of the Satellite Division of The Institute of Navigation (ION GNSS+ 2018), Miami, FL, USA, 24–28 September 2018; pp. 392–412.
38. Wen, Q.; Geng, J.; Li, G.; Guo, J. Precise point positioning with ambiguity resolution using an external survey-grade antenna enhanced dual-frequency android GNSS data. *Measurement* **2020**, *157*, 107634. [CrossRef]
39. Li, G.; Geng, J. Characteristics of raw multi-GNSS measurement error from Google Android smart devices. *GPS Solut.* **2019**, *23*, 90. [CrossRef]
40. Lachapelle, G.; Gratton, P.; Horrelt, J.; Lemieux, E.; Broumandan, A. Evaluation of a low cost hand held unit with GNSS raw data capability and comparison with an android smartphone. *Sensors* **2018**, *18*, 4185. [CrossRef]
41. Wanninger, L.; Heßelbarth, A. GNSS code and carrier phase observations of a Huawei P30 smartphone: Quality assessment and centimeter-accurate positioning. *GPS Solut.* **2020**, *24*, 64. [CrossRef]
42. Robustelli, U.; Pugliano, G. Code multipath analysis of Galileo FOC satellites by time-frequency representation. *Appl. Geomat.* **2019**, *11*, 69–80. [CrossRef]

43. Robustelli, U.; Baiocchi, V.; Pugliano, G. Assessment of dual frequency GNSS observations from a Xiaomi Mi 8 Android smartphone and positioning performance analysis. *Electronics* **2019**, *8*, 91. [CrossRef]
44. de Ponte Müller, F.; Steingass, A.; Strang, T. Zero-baseline measurements for relative positioning in vehicular environments. In Proceedings of the 6th European Workshop on GNSS Signals and Signal Processing, Neubiberg, Germany, 5–6 December 2013.
45. Kouba, J. A Guide to Using International GNSS Service (IGS) Products. 2009. Available online: http://acc.igs.org/UsingIGSProductsVer21.pdf (accessed on 1 March 2019).
46. Saleem, T.; Usman, M.; Elahi, A.; Gul, N. Simulation and performance evaluations of the new GPS L5 and L1 signals. *Wirel. Commun. Mob. Comput.* **2017**, *2017*, 7492703. [CrossRef]
47. Leclère, J.; Landry, R., Jr.; Botteron, C. Comparison of L1 and L5 bands GNSS signals acquisition. *Sensors* **2018**, *18*, 2779. [CrossRef] [PubMed]
48. Hu, J.; Li, P.; Zhang, X.; Bisnath, S.; Pan, L. Precise point positioning with BDS-2 and BDS-3 constellations: ambiguity resolution and positioning comparison. *Adv. Space Res.* **2022**, *70*, 1830–1846. [CrossRef]
49. Zangenehnejad, F.; Gao, Y. Application of UofC Model Based Multi-GNSS PPP to Smartphones GNSS Positioning. In Proceedings of the 34th International Technical Meeting of the Satellite Division of the Institute of Navigation (ION GNSS+ 2021), St. Louis, MO, USA, 20–24 September 2021; pp. 2986–3003.

sensors

MDPI

Article

Precise Position Estimation Using Smartphone Raw GNSS Data Based on Two-Step Optimization

Taro Suzuki

Future Robotics Technology Center, Chiba Institute of Technology, Chiba 2750016, Japan; taro@furo.org

Abstract: This paper presents a high-precision positioning method using raw global navigation satellite system (GNSS) observations from smartphones in the Google smartphone decimeter challenge (GSDC). Compared to commercial GNSS receivers, smartphone GNSS observations are noisy owing to antenna limitations, making it difficult to apply conventional high-precision positioning methods. In addition, it is important to exclude outliers in GSDC because GSDC includes data in environments where GNSS is shielded, such as tunnels and elevated structures. Therefore, this study proposes a smartphone positioning method based on a two-step optimization method, using factor graph optimization (FGO). Here, the velocity and position optimization process are separated and the velocity is first estimated from Doppler observations. Then, the outliers of the velocity estimated by FGO are excluded, while the missing velocity is interpolated. In the next position-optimization step, the velocity estimated in the previous step is adopted as a loose state-to-state constraint and the position is estimated using the time-differenced carrier phase (TDCP), which is more accurate than Doppler, but less available. The final horizontal positioning accuracy was 1.229 m, which was the first place, thus demonstrating the effectiveness of the proposed method.

Keywords: GPS; GNSS; smartphone; localization; time-differenced carrier phase

Citation: Suzuki, T. Precise Position Estimation Using Smartphone Raw GNSS Data Based on Two-Step Optimization. *Sensors* **2023**, *23*, 1205. https://doi.org/10.3390/s23031205

Academic Editor: Yang Gao

Received: 20 December 2022
Revised: 17 January 2023
Accepted: 19 January 2023
Published: 20 January 2023

1. Introduction

High-precision positioning with smartphones has become important for various applications, such as pedestrian navigation, vehicle lane-level navigation, and the increasing number of location-based games and virtual reality technologies. GNSS is conventionally used for location estimation in outdoor environments and as a means of location estimation in smartphones as it can estimate absolute positions on the Earth provided signals can be received from a satellite. Although smartphones have long been equipped with GNSS, users have only been able to access location information output using the GNSS receivers of smartphones. In 2016, the Android operating system released an application program interface to access raw GNSS measurement data from GNSS installed in smartphones [1]. Accordingly, raw GNSS data (pseudorange, pseudorange rate (Doppler), and accumulated delta range (carrier phase)), could be acquired. This facilitated the development of position-estimation algorithms for smartphones using raw GNSS measurements. Consequently, high-precision positioning at the decimeter and centimeter levels on smartphones has garnered significant attention [2–5].

However, there are several challenges associated with smartphone positioning compared with positioning using commercial GNSS receivers [4,5]. The GNSS antennas of smartphones exhibit lower performance than those of GNSS receivers for surveying and the noise in GNSS observations is very significant. Consequently, the application of existing high-precision positioning methods to smartphones, such as precise point positioning (PPP) [6,7] and real-time kinematic (RTK) GNSS [8,9], is challenging. The usual positioning accuracy of a smartphone is approximately 3–10 m, which results in significant challenges for advanced navigation and other applications, such as sidewalk- and lane-level navigation for pedestrians and vehicles.

To address this challenge, in 2020, Google released a dataset with GNSS raw measurements acquired by Android smartphones to develop positioning algorithms using smartphones [10]. Furthermore, in 2021, Google launched the Google smartphone decimeter challenge (GSDC) to further accelerate the development of high-precision positioning technology for smartphones [11]. In 2022, GSDC 2022 was held from 3 May to 30 July 2022, using a new dataset based on the raw GNSS measurements of smartphones [12]. This paper describes the winning solution at GSDC 2022.

2. Related Researches

Various high-precision positioning methods using GNSS measurements from smartphones have been studied. When applying PPP or RTK-GNSS to smartphones, noise from the built-in GNSS antenna of the smartphone becomes a problem [13]. The methods of installing a smartphone on a choke ring antenna [14,15], or using a re-radiating antenna or an external antenna [16,17], can achieve high-accuracy positioning with centimeter accuracy. However, PPP and RTK-GNSS are not effective when a smartphone is mounted on the dashboard of a vehicle and GNSS data is acquired by an internal antenna, as in the case of GSDC. Some research has been conducted to improve the accuracy of RTK-GNSS and PPP by combining GNSS observations from smartphones and IMUs [18,19]. However, time synchronization between Android-based IMU and smartphone's GNSS is an issue.

Graph-based optimization has been extensively studied in robotics; however, it has recently been actively studied in the GNSS field. The factor graph optimization (FGO) was proposed to model factorizations [20]. FGO can apply various complex non-linear constraints and simultaneously optimize all state variables (the entire driving trajectory). A factor graph is a graphical representation in which there is an unknown state variable (variable nodes) and a factor (factor nodes) that is a function of the state variable. The edges connecting the factor and variable nodes can be considered as constraints on the state variable by the factor. Hence, the state estimation problem is reduced to an optimization problem using the objective function constructed by the factor graph representation. One of the earliest studies proposed a robust optimization method using only GPS pseudorange observations based on factor graph optimization [21], which improved positioning accuracy better than least-squares-based positioning. Subsequently, extensions to real-time position estimation [22] and combinations with IMU and other sensors have been studied [23–25]. In [26], the authors discussed a comparative evaluation of GNSS positioning based on graph-based optimization with that based on least-squares-based positioning and extended Kalman filter. Several applications of time-differenced carrier phases (TDCP) to FGO have also been studied [27–29]. The study by [30] implemented precise point positioning using carrier phase measurements with graph optimization and compared it with Kalman-filter-based implementation. The results indicated that factor graph-based optimization exhibited better performance than conventional filtering methods, such as the extended Kalman filter. This is due to the global optimization with all the observations from past to present in the graph structure.

In this study, we propose a method to estimate location using FGO based on GNSS observations of smartphones, especially TDCP. The performance of positioning by graph-based optimization significantly depends on the graph structure. In most of the previous studies, the pseudorange of GNSS is adopted for graph construction. In addition, the FGO target is commercial GNSS receivers and GNSS observations of smartphones are yet to be used. This study is unique in that it adopts TDCP for graph optimization using GNSS observations from smartphones. Its optimization process is divided into two steps.

3. GSDC 2022 Overview

3.1. Dataset

At the GSDC, each dataset included raw GNSS measurements collected by several Android smartphone devices, together with the ground truth trajectories collected by a high-grade GNSS and an inertial navigation unit (IMU) integration system for reference [10,11].

The log data of multiple smartphones mounted on the dashboard of a vehicle were provided in two datasets: training data with reference position and test data for evaluation in the competition. The smartphone logs included observations such as the GNSS pseudorange, pseudorange rate, accumulated delta range, etc. Furthermore, the smartphone's IMU data, such as acceleration and angular velocity, were also included in the log data.

Driving data for five types of smartphones (Google Pixel4, Google Pixel5, Google Pixel6 Pro, Samsung Galaxy S20 Ultra, and Xiaomi Mi8) were provided at the GSDC 2022. Although the driving data for the San Francisco area was provided at GSDC 2021, that for the Los Angeles area was included at GSDC 2022. Figure 1 presents the GSDC 2022 data from the 36 runs included in the test dataset. At GSDC 2021, each run included data acquired simultaneously by multiple smartphone models; however, at GSDC 2022, only one type of smartphone was used for each run. This implies that the technique of improving accuracy by assembling the positions estimated independently for each smartphone was no longer available. All smartphones had a built-in dual-frequency GNSS receiver chip and included dual-frequency L1 and L5 observations relative to GPS and Galileo.

The driving data provided can be divided into two categories: highway driving in an open-sky environment and street driving in an area lined with trees and buildings. The Los Angeles driving data included several long tunnels and elevated sections, including several 10 s GNSS signal blockages. In approximately half of the runs included in the test data, the data were run on different courses not included in the training data. Figure 2 illustrates a section of the Los Angeles travel trajectory in the test data that stops under an elevated track or travels in a tunnel. There are certain sections where receiving GNSS signal is difficult, which results in large outliers and missing data.

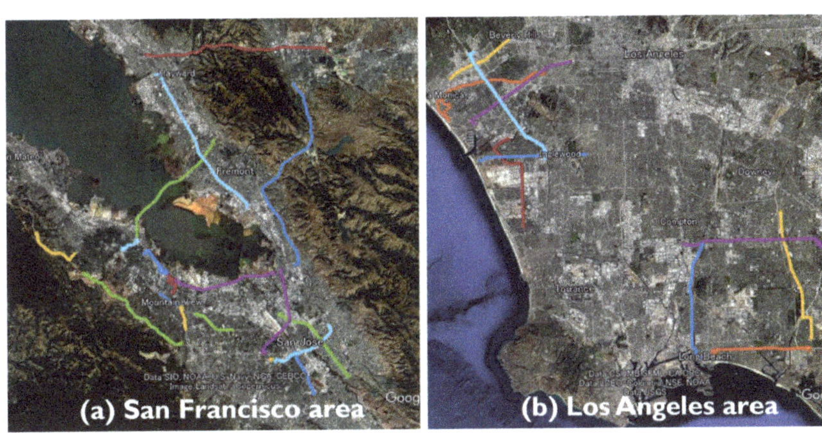

Figure 1. Driving trajectories included in test data provided at GSDC 2022. In total, 36 runs were provided and divided into two parts: one run in the San Francisco area and another in the Los Angeles area.

3.2. Score Metric

In the GSDC, the trajectories of all smartphones in the test dataset are estimated and evaluated using a score according to their accuracies. Here, the score is computed via the mean of the 50th and 95th percentile of the horizontal positioning errors. The estimated latitudes Φ_i and longitudes Λ_i of the i-th smartphone trajectory can be defined as follows.

$$\Phi_i = \{\phi_1 \cdots \phi_M\}, \quad \Lambda_i = \{\lambda_1 \cdots \lambda_M\} \tag{1}$$

where M denotes the total epochs of the smartphone's trajectory. $D_i = \{d_1 \cdots d_M\}$ represent the horizontal errors of the trajectory computed from the Haversine formula [31]. The horizontal error of j-th epoch d_j is represented as follows.

$$d_j = harversine((\phi_j, \lambda_j), (\phi_{GT,j}, \lambda_{GT,j})) \tag{2}$$

where *harversine* is a function that calculates the Haversine distance between two sets of latitudes and longitudes, while the subscript GT represents the ground truth. The score for N trajectories is calculated as follows.

$$score = \frac{1}{N}\sum_{i=1}^{N} \frac{percentile(D_i, 50) + percentile(D_i, 95)}{2} \tag{3}$$

where $percentile(D, p)$ computes a value that is greater than p percent of the values in D. Note that the altitude of the smartphone is ignored. In addition, the smartphone positioning can exhibit significantly large errors owing to multipath; however, in the afore-presented metric, the large top 5% positioning errors are ignored and do not affect the score.

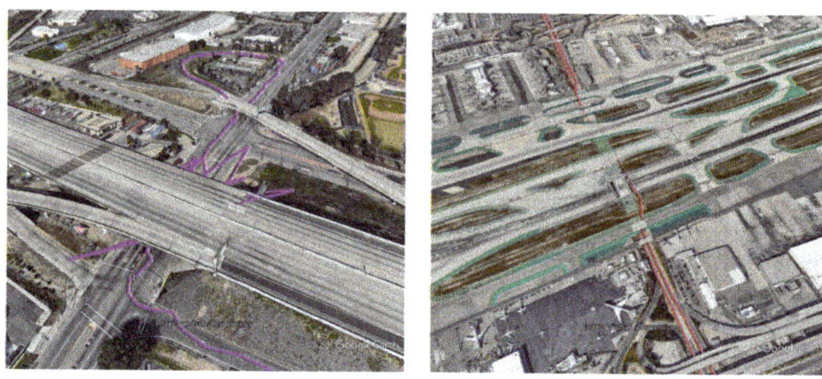

(a) Test/2022-04-01-US-LAX-3/]XiaomiMi8 (b) Test/2022-02-23-US-LAX-3/XiaomiMi8

Figure 2. Los Angeles vehicle driving trajectory included in GSDC 2022. (**a**) Stops under elevated tracks and (**b**) runs through long tunnels where GNSS signals are completely blocked.

3.3. Baseline Position

In addition to the raw GNSS data, the competition hosts provided a "baseline" of location estimation results for each smartphone. This baseline is the result of least-squares-based processing of the pseudorange observed by the smartphone in estimating the position. Table 1 presents the scores for the two categories of driving data in the GSDC 2022 training dataset. The highway represents an open-sky environment. The 3.6 m score is practical for positioning with pseudorange. However, in street areas, GNSS signals were blocked by roadside trees, which increased the positioning error to 4.6 m.

The satellite observation performance of smartphones has the following characteristics. First, smartphones cannot track satellite signals stably and only a small number of satellites can be observed continuously with dual-frequency signals. In addition, regarding satellite signal quality, satellites observed by smartphones generally exhibit a low carrier-to-noise ratio [5,32]. Furthermore, the positioning accuracy of smartphones is significantly impaired by multipath. The observed pseudorange of a smartphone is substantially noisier than that of a commercial GNSS receiver and GNSS code positioning using only the pseudorange has limited positioning accuracy.

Table 1. Score of baseline position for each course type in training dataset of GSDC 2022.

Course Type	Highway	Street	All
Score (50%) m	2.296	2.772	2.599
Score (95%) m	4.943	6.431	5.890
Mean m	3.620	4.602	4.245

4. Strategy

From the exploratory data analysis, GNSS observations, such as pseudorange and Doppler shift, from smartphones are more significantly noisy than commercial GNSS receivers. In addition, owing to the limitations of antennas on smartphones, the real-time kinematic (RTK)-GNSS technique, which is usually used for high-precision positioning, is difficult to utilize because it is difficult to solve the integer ambiguities in the carrier phase measurements using smartphone antenna. In addition, missing data, abnormal values, and unsynchronized observation time for each smartphone exist. Even in an open-sky environment, a 3-m accuracy cannot reach the decimeter level. Hence, we have to improve the absolute position accuracy beyond the limits of pseudorange accuracy. Because excessive noise and several outliers exist in the observations, a robust position-estimation method is required. Finally, because GSDC targets high-precision positioning by post-processing, methods specific to post-processing are beneficial.

At GSDC 2021, the author proposed a location estimation method based on smartphone observations using global optimization with FGO [33]. Here, the method developed in 2021 was modified to improve the location estimation accuracy, particularly in environments where the accuracy of pseudorange deteriorates under trees and elevated structures. The pseudorange of smartphones is noisy; hence, achieving highly accurate position estimation using the pseudorange alone is challenging. However, if the carrier phase can be tracked continuously, the relative position change (velocity) can be estimated with high accuracy from the TDCP [34]. Because of the limited availability of valid TDCPs, velocity can be estimated from more robust Doppler observations, although they are less accurate than TDCP. However, Doppler observations remain unavailable when satellites are shielded for long periods, such as in elevated structures or tunnels.

Therefore, we adopted a two-step optimization to estimate the position of the smartphone. Figure 3 illustrates the flow of the proposed method. First, we estimated the velocity, where the 3D velocity of the smartphone and receiver clock drift were estimated by employing FGO with Doppler observations, which are readily available but less accurate. From the 3D velocity estimated by the optimization, outliers in velocity were detected, excluded, and interpolated to obtain an estimate of the 3D velocity continuously. Subsequently, the states to be optimized were the 3D position and receiver clock bias of each satellite system. The 3D velocity and clock drift obtained in the previous step were adopted as loose constraints between the states in this step. In addition, if valid TDCPs were obtained, TDCP-based constraints between the states were added and, as an absolute position constraint, the pseudorange of the smartphone, whose error was corrected using the GNSS base-station, was utilized. Even when no TDCP was available, this method added a state-to-state constraint based on the velocity estimated using Doppler in the first step and the position could be obtained even in the tunnel.

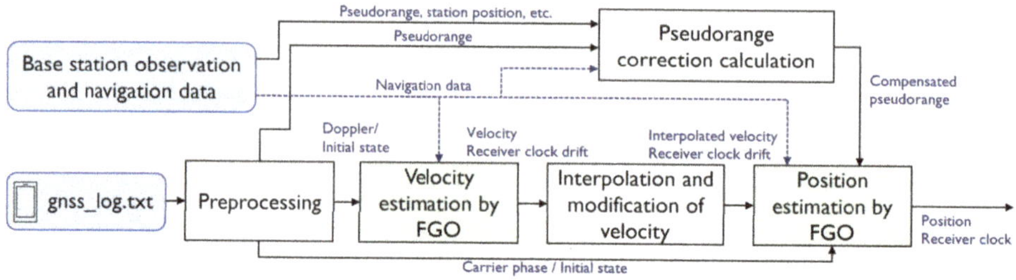

Figure 3. Flow of the proposed method comprising two optimization steps: velocity estimation and position estimation via factor graph optimization (FGO).

5. Preprocessing

In the preprocessing step, the initial values of the optimized state to be determined, including the GNSS pseudorange, carrier phase, and Doppler observations, are calculated from the smartphone logs. Here, various filtering processes are adopted to discard invalid GNSS observations and reject unreliable observations. Because smartphone observations contain larger amounts of unreliable data than commercial receivers, it is very important to screen GNSS observations in preprocessing.

5.1. Initial State Estimation

The estimated states in the optimization step are the 3D position and velocity in the Earth-centered Earth-fixed (ECEF) coordinate system and the receiver clock bias and drift. These initial values need to be pre-determined before optimization.

Regarding the initial 3D position for the optimization, the baseline position is utilized. However, the baseline position contains some large jumps and missing parts; hence, the following process is applied.

1. Convert to the east-north-up (ENU) coordinate system and detect large jumps in the altitude.
2. Delete the epoch detected in Step 1 as an outlier.
3. Interpolate the 3D position of missing epochs from the previous and next data.

The above process determines the initial values of 3D positions for all epochs. The initial value of 3D velocity is determined by simply calculating the difference in the initial 3D positions.

For the initial clock bias and drift, the log data from the smartphone contains estimates of the built-in GNSS receiver clock bias and drift, which are adopted as the initial values. Note that, among the smartphones, for the XiaomiMi8, the difference between the estimated clock bias is adopted as the initial value of the clock drift, because the clock drift is an invalid value. The missing data are also complemented with the previous and next data to calculate the initial values of the clock and clock bias for all epochs.

5.2. Selection of GNSS Observations

Similar to normal GNSS positioning, the satellite elevation angle and carrier-noise ratio (CNR) masks are used to reject all observations (pseudorange, Doppler, and carrier phase) for satellites with a low elevation angle and low signal strength. Here, the values of the satellite elevation angle and CNR masks are tuned and determined for each run. If the multipath indicator (`MultipathIndicator`) in the smartphone log is turned on, the observations for that satellite are rejected. The following is a description of the process for each observation.

5.2.1. Pseudorange Selection

Similar to [35], the observation values of valid GNSS signals are selected based on the pseudorange tracking status contained in the smartphone logs. In addition, the pseudorange residuals are calculated using the initial 3D position and clock bias described above. Thresholding in the pseudorange residual rejects pseudorange observations with large multipath errors.

5.2.2. Doppler Selection

Doppler observations (or pseudorange rates) $\dot{\rho}$ are the most robust observations in smartphones, even in adverse environments. To circumvent large errors in Doppler observations, the residuals of Doppler observations are calculated using the initial 3D velocity and clock drift and the aforementioned pseudorange; in addition, the Doppler observation values with large errors are rejected by thresholding.

5.2.3. Carrier Phase Selection

From the carrier-phase tracking status, the carrier phases continuously tracked with no cycle slip are reserved [35]. However, even in the carrier phases that have passed these processes, some carrier phases that contain cycle slips or errors still remain. Therefore, by comparing the Doppler observation and the aforementioned carrier phase, the outliers of the carrier phase are rejected by checking the consistency. The consistency of the carrier phase of the satellite k at the i-th ϕ_i^k and $i+1$-th epochs ϕ_{i+1}^k can be calculated from Doppler observation (in m/s) $\dot{\rho}_i^k$ and $\dot{\rho}_{i+1}^k$ as follows.

$$\left| \lambda(\phi_{i+1}^k - \phi_i^k) - \frac{(\dot{\rho}_i^k + \dot{\rho}_{i+1}^k)}{2}\Delta t \right| < D_{th} \tag{4}$$

where, λ, Δt, and D_{th} denote the wavelength, time step (1 s in the GSDC), and threshold value, respectively. The TDCP residuals of L1 signals in Equation (4), illustrated using the actual smartphone observations in the training dataset, are presented in Figure 4. The colors of the plots in Figure 4 represent different satellites and the increase in the residuals due to cycle slips in the carrier phase can be observed. Here, we used 1 m as the threshold value of D_{th} to exclude cycle slips.

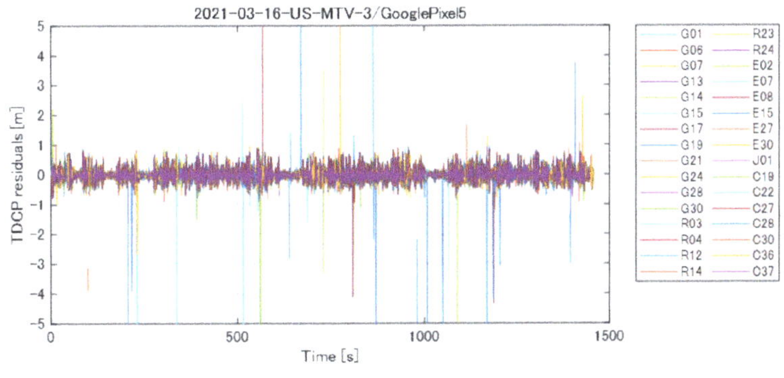

Figure 4. TDCP residual of L1 signals computed by Equation (4). The Doppler observation is used to detect and exclude cycle slips in the carrier-phase observation.

5.2.4. Example of Observation Selection

Using the aforementioned GNSS observation selection method, we preprocessed GNSS observations for the runs included in the GSDC 2022 training dataset. Figure 5 presents the change in the number of observations before and after selection of the GNSS pseudorange, Doppler, and carrier phase. The number of observations decreases significantly after the selection of the GNSS observations, thus verifying that the raw GNSS observations from smartphones contain a large amount of unreliable data with large errors. In addition, Doppler observations are more available than carrier-phase observations, and can be used even during times when the carrier phase contains cycle slips and is unavailable. The proposed method first estimates the 3D velocity using Doppler observations that are robust to adverse GNSS reception environments and then estimates accurate positions using less available, but more accurate, carrier phases in the next step.

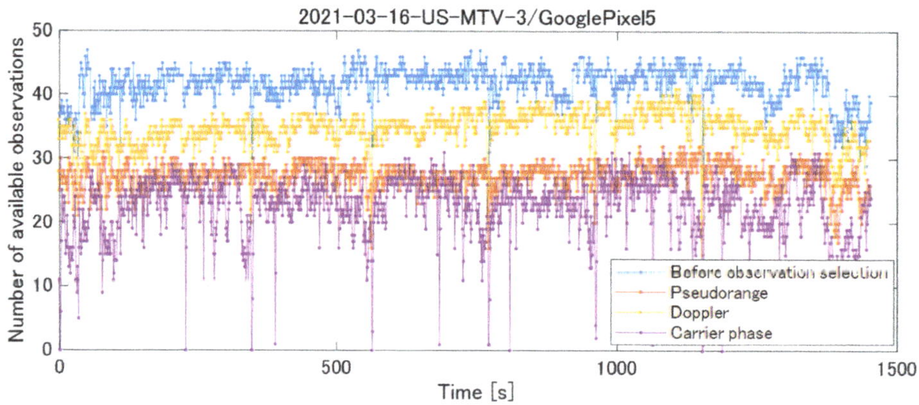

Figure 5. Selection of GNSS observations by preprocessing. Compared to the carrier phase, Doppler is highly available.

6. Velocity-Estimation Step

In the velocity-estimation step, the state **V** estimated in epoch i was the 3D velocity and receiver clock drift. The state **V** can be expressed as:

$$\mathbf{V}_i = \begin{bmatrix} \mathbf{v}_i \\ \dot{\mathbf{t}}_i \end{bmatrix} \tag{5}$$

$$\mathbf{v}_i = \begin{bmatrix} v_{x,i} & v_{y,i} & v_{z,i} \end{bmatrix}^T \quad \dot{\mathbf{t}}_i = \dot{t}_{\mathrm{gpsL1},i} \tag{6}$$

where $\dot{t}_{\mathrm{gpsL1},i}$ represents the receiver clock drift relative to the GPS time in m/s. The 3D velocity $\mathbf{v}_i$ is represented in the ECEF coordinate system. Here, only the clock drift computed from the GPS L1 signal was included in the state because the inter-system bias, including the receiver clock bias, between GPS L1 and other satellite systems can be considered constant.

The graph structure of the proposed FGO is presented in Figure 6. The circled markers in the figure represent the factor nodes, where the variable nodes are connected to the Doppler factors from each satellite and the motion factor is connected between variable nodes as an acceleration constraint.

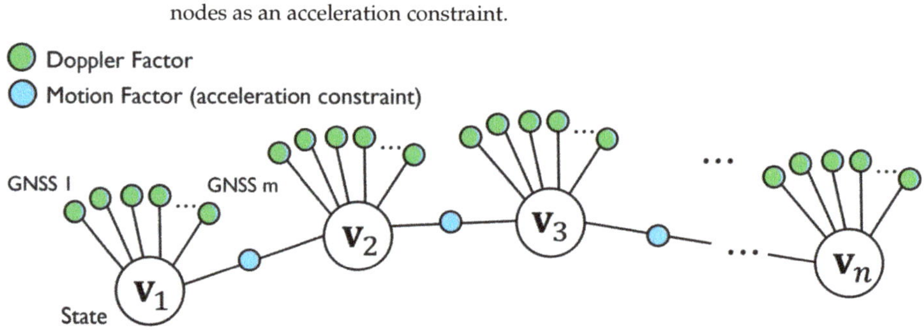

Figure 6. Graph structure of proposed method in velocity-estimation step. The Doppler and motion factors are adopted to estimate the velocity.

6.1. Doppler Factor

The observed Doppler frequency can be converted to a pseudorange rate. GNSS pseudorange rate from the Doppler of the satellite k in the i-th epoch ρ_i^k can be modeled as follows.

$$\dot{\rho}_i^k = \mathbf{u}_i^k \left(\mathbf{v}_{s,i}^k - \mathbf{v}_i \right) + \dot{\mathbf{t}}_i \tag{7}$$

where $\mathbf{v}_{s,i}^k$ represents the satellite velocity in the ECEF coordinate system. $\mathbf{u}_i^k = \left[u_{x,i}^k, u_{y,i}^k, u_{z,i}^k \right]$ denotes the unit line-of-sight vector from the receiver to the satellite k in ECEF.

The error function of the Doppler factor is represented as follows.

$$e_{d,i}^k = \mathbf{H}_{v,i}^k \mathbf{V}_i - \left(\dot{\rho}_i^k - \mathbf{u}_i^k \mathbf{v}_{s,i}^k \right) \tag{8}$$

Here, the measurement matrix $\mathbf{H}_{v,i}^k$ can be formulated as:

$$\mathbf{H}_{v,i}^k = \begin{bmatrix} \mathbf{u}_i^k & 1 \end{bmatrix} \tag{9}$$

Together with the error function, the optimization utilizes the information matrix $\Omega_{d,i}^k$, which represents the accuracy of the Doppler observations. Here, the information matrix is determined from a simplified model, where the error depends on the satellite elevation angle.

6.2. Motion Factor

The variable nodes in the graph are the velocity and clock drift and the constraints between the nodes function as acceleration constraints. Although acceleration observations can be obtained from the smartphone's IMU, they were not utilized in this study because they require a coordinate transformation based on the smartphone's orientation, including synchronization with GPS time. The motion factor uses the assumption that the velocity does not change significantly (i.e., acceleration is small) between successive variable nodes. The error function was simply defined as acceleration being close to zero, such that the velocity changed smoothly, as follows:

$$\mathbf{e}_{m,i} = \mathbf{V}_{i+1} - \mathbf{V}_i \tag{10}$$

Here, by appropriately providing $\mathbf{\Omega}_{m,i}$, the information matrix during optimization based on the maximum value of the vehicle's actual acceleration, large jumps in the velocity estimated by optimization can be suppressed and a continuous velocity can be estimated.

6.3. Optimization

The objective function to be optimized is presented as follows.

$$\hat{\mathbf{V}} = \underset{\mathbf{V}}{\operatorname{argmin}} \sum_i \|\mathbf{e}_{m,i}\|_{\mathbf{\Omega}_{m,i}}^2 + \sum_i \sum_k \left\| e_{d,i}^k \right\|_{\Omega_{d,i}^k}^2 \tag{11}$$

The M-estimator was used to reject the multipath error of Doppler measurements during the optimization [27]. The Huber function was adopted as the influence function of the M-estimator. The Huber function works remarkably for GNSS observations with several outliers, as adopted in [28,36]. The hyperparameters of the Huber function are determined by trial and error and the same values are used for all runs. Furthermore, the GTSAM [37] was utilized for the graph optimization backend, while RTKLIB [38] was used for GNSS general computation.

6.4. Outlier Removal and Interpolation

Interpolation was performed to discard outliers in the estimated velocity via FGO and also to obtain velocity estimates at all epochs. The estimated velocities in the ECEF coordinate system were transformed to the ENU coordinate system. In addition, the threshold for the estimated velocity in the up direction excluded outliers as the vehicle was bound on the ground.

Furthermore, to estimate the velocity at epochs where Doppler cannot be obtained, such as in tunnels, the velocity at all epochs was interpolated from the previous and next velocity information. Here, Makima interpolation [39] was adopted rather than simple linear or spline interpolation. Makima interpolation does not produce the overshoot produced by spline interpolation and it is suitable for velocity interpolation. Finally, this estimated velocity was adopted in the next position-optimization step.

7. Position-Estimation Step

In the position-estimation step, the state **X** estimated in epoch i was the 3D position relative to the initial position in ECEF coordinates and multi-GNSS clock biases. The state **X** can be expressed as:

$$\mathbf{X}_i = \begin{bmatrix} \mathbf{r}_i \\ \mathbf{t}_i \end{bmatrix} \tag{12}$$

$$\mathbf{r}_i = \begin{bmatrix} x_i & y_i & z_i \end{bmatrix}^T \quad \mathbf{t}_i = \begin{bmatrix} t_{\text{gpsL1},i} & t_{\text{glo},i} & t_{\text{galL1},i} & t_{\text{bds},i} & t_{\text{gpsL5},i} & t_{\text{galL5},i} \end{bmatrix}^T \tag{13}$$

where $t_{\text{gpsL1},i}$ denotes the receiver clock bias computed from GPS L1 signals, and $t_{\text{glo},i}$, $t_{\text{galL1},i}$, $t_{\text{bds},i}$, $t_{\text{gpsL5},i}$, and $t_{\text{galL5},i}$ represent the system biases including the time bias of the GLONASS, Galileo, and BeiDou relative to the GPS L1 signal, respectively. The state was defined by the difference in the initial values of the 3D position and clock bias for linearization. Figure 7 presents the proposed graph structure. Three types of factors, pseudorange, TDCP, and velocity/clock drift, were used.

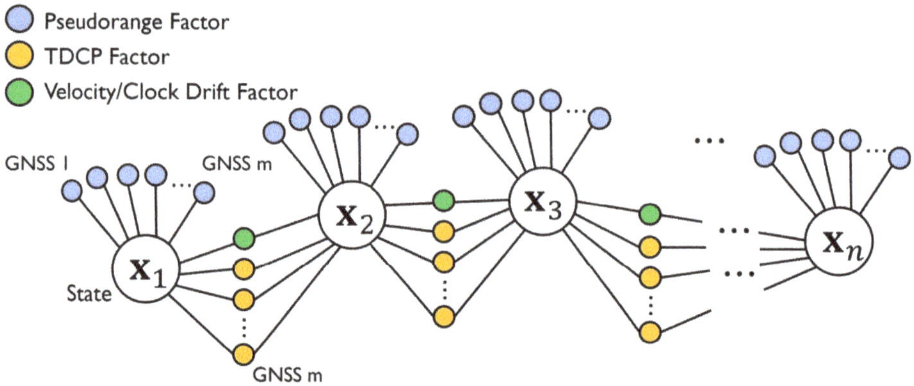

Figure 7. Graph structure of proposed method in position-estimation step. Pseudorange, TDCP, and velocity/clock drift factors were adopted.

7.1. Velocity/Clock Drift Factor

The velocity/clock drift factor is the relative constraint between time-series variable nodes. In the previous step, the 3D velocity and clock drift at all epochs were estimated based on Doppler observations; hence, the estimated velocity and clock drift were adopted as loose constraints between sequential states in the position-estimation step. If TDCP observations are not available, this velocity/clock drift factor will be the only constraint between sequential states. The error function of the velocity/clock drift factor is denoted as:

$$\mathbf{e}_{\mathrm{v},i} = (\mathbf{X}_{i+1} - \mathbf{X}_i) - \left(\frac{\tilde{\mathbf{V}}_i + \tilde{\mathbf{V}}_{i+1}}{2}\right)\Delta t \tag{14}$$

where $\tilde{\mathbf{V}}$ denotes the velocity and clock drift estimated in the previous step, with "0" added to match the dimension of $\mathbf{X}$. The velocity/clock drift factor was added between all nodes, such that, even in epochs where the pseudorange or TDCP was not observed at all (such as in a tunnel), the position was estimated based on the velocity estimated in the previous step.

7.2. Pseudorange Factor

In the pseudorange factor, the pseudorange compensated using the pseudorange error computed at a base-station was adopted to estimate the absolute position. The GNSS pseudorange of satellite k in the i-th epoch ρ_i^k can be modeled as follows.

$$\rho_i^k = r_i^k + t_i - \delta T_i^k + I_i^k + T_i^k \tag{15}$$

where r_i^k denotes the geometric satellite-to-receiver distance, which is calculated using the initial node. Further, δT_i^k, I_i^k, and T_i^k represent the clock bias of the satellite, ionospheric delay, and tropospheric delay, respectively.

To completely eliminate the satellite orbit, clock, tropospheric, and ionospheric delays in the smartphone pseudorange observations, the GNSS pseudorange error at the GNSS base-station was calculated to correct the pseudorange. The error function of the pseudorange factor is represented as follows.

$$e_{\mathrm{pr},i}^k = \mathbf{H}_{\mathrm{p},i}^k \mathbf{X}_i - \left(\rho_i^k - r_i^k - \varepsilon_i^k\right) \tag{16}$$

where ε_i^k denotes the pseudorange correction value, including the tropospheric and ionospheric delay, and the clock and orbit error of the satellite, calculated at a base-station whose position is known. Here, the measurement matrix $\mathbf{H}_{\mathrm{p},i}^k$ can be formulated as:

$$\mathbf{H}_{\mathrm{p},i}^k = \begin{bmatrix} \mathbf{u}_i^k & 1 & \delta_{\mathrm{glo},i}^k & \delta_{\mathrm{gal},i}^k & \delta_{\mathrm{bds},i}^k & \delta_{\mathrm{gpsL5},i}^k & \delta_{\mathrm{galL5},i}^k \end{bmatrix} \tag{17}$$

where $\delta_{\mathrm{glo},i}^k$, $\delta_{\mathrm{gal},i}^k$, $\delta_{\mathrm{bds},i}^k$, $\delta_{\mathrm{gpsL5},i}^k$, and $\delta_{\mathrm{galL5},i}^k$ are equal to "1" when the k-th GNSS measurement is GLONASS, Galileo, or BeiDou, respectively. For GSDC 2022, we selected the GNSS base-station closest to the trajectory from the NOAA CORS Network and utilized it as the base GNSS station. Regarding the information matrix of the pseudorange factor, we used an elevation angle-based error model and the Doppler factor.

7.3. TDCP Factor

Both Doppler and TDCP can be used to estimate the velocity. However, Doppler observations from smartphones are noisier than those from commercial GNSS receivers and velocity computed from TDCP measurement is significantly more accurate than those calculated from Doppler. The TDCP measurement $\Delta\Phi_i^k$ between sequential epochs i and $i+1$ is expressed as follows:

$$\lambda\Delta\Phi_i^k = \lambda\left[\Phi_{i+1}^k - \Phi_i^k\right] \simeq \Delta r_i^k + \Delta t_i \tag{18}$$

where λ, Φ_i^k, and r_i^k denote the signal wavelength, measured carrier phase in cycles, and satellite-receiver geometric distance, respectively. Δ represents the operator that computes the time difference. If the time difference is short, the ionospheric and tropospheric delays and satellite orbital clock errors in the carrier phase can be canceled. Here, $\lambda\Delta\Phi_i^k$ represents the exact receiver–satellite distance change between epochs i and $i+1$.

The error function of TDCP factor is as follows.

$$e_{\mathrm{td},i}^k = \mathbf{H}_{\mathrm{p},i}^k(\mathbf{X}_{i+1} - \mathbf{X}_i) - \left(\lambda\Delta\Phi_i^k - \Delta L_i^k\right) \tag{19}$$

Here, ΔL_i represents the change in distance owing to the satellite motion from the antenna position.

However, although TDCP is accurate, its availability is lower than the Doppler owing to cycle slip and half-cycle ambiguity challenges. If the carrier phase with cycle slip is not completely excluded, the TDCP factor will introduce relative position errors. Here, we address this cycle-slip problem with robust optimization using the M-estimator.

7.4. Optimization

The final objective function to be optimized, using all the factors, is presented as follows.

$$\hat{\mathbf{X}} = \underset{\mathbf{X}}{\operatorname{argmin}} \sum_i \|\mathbf{e}_{v,i}\|^2_{\Omega_{v,i}} + \sum_i \sum_k \left\| e^k_{\mathrm{pr},i} \right\|^2_{\Omega^k_{\mathrm{pr},i}} + \sum_i \sum_k \left\| e^k_{\mathrm{td},i} \right\|^2_{\Omega^k_{\mathrm{td},i}} \tag{20}$$

Similar to that in the velocity estimation step, robust estimation by the M-estimator with the Huber function was applied to the pseudorange and TDCP factors. Because the pseudorange observations contain multipath errors and the TDCP observations contain cycle slips, the M-estimator was adopted to exclude outliers. The hyperparameters of the Huber function were determined by trial and error. The smartphone position optimized in the position-estimation step was directly used as the final estimate.

8. Evaluation and Discussion

8.1. Evaluation Using Training Data

8.1.1. Evaluation of Positioning Accuracy

The training dataset, for which reference locations were provided, was used to evaluate the positioning accuracy of the proposed method. The GSDC 2022 training dataset contains 170 smartphone trajectories on 62 different routes. From these trajectories, we constructed a dataset for evaluation considering the following points:

1. Exclude runs for which the ground truth does not include altitude. Although the final score is determined by the horizontal positioning error, runs without ground truth for altitude were excluded. This is because we cannot compute position and velocity references in the ECEF coordinates.
2. Exclude runs for which the carrier phase has not been obtained. The `HardwareClock-Discontinue` flag in the log is only reported on Google Pixel 4, which includes some runs for which the carrier phase has not been obtained correctly. Because TDCP cannot be used in these cases, they were excluded from the evaluation.
3. During the competition, a participant pointed out that the ground truth for some runs was not sufficiently accurate. The competition host then made an announcement and published a list of the inaccurate ground truths. The runs containing these inaccurate ground truths were eliminated.

Using the above process, the proposed method is evaluated on the 55 runs extracted from the training dataset.

Three runs were selected from the entire evaluation data; the trajectories of each baseline are illustrated in Figure 8. Figure 8a–c present the data from the highway runs, street driving, and the area around Los Angeles, including the GNSS signal blockage for a long period of time, respectively.

Figure 9 presents the time-series horizontal positioning error. The blue line indicates the baseline position error and the red line indicates the proposed method. The colored areas in Figure 9 are the sections that traveled under the elevated tracks. In highway driving (Figure 9a), the TDCP factor estimated the relative position with high accuracy and the pseudorange factor corrected the absolute position. Furthermore, the baseline position sometimes exhibited an abrupt error of approximately 10 m, which occurred when the vehicle went under an elevated road while driving on the highway. However, the proposed method suppressed the increase in error by interpolating the velocity estimated from Doppler. Figure 9b presents the positioning error of street driving. The pseudorange and carrier-phase noise were large and the position-estimation accuracy was worse than that of highway driving. Furthermore, the baseline exhibited frequent position errors exceeding

10 m; however, the proposed method could estimate the accurate position. Figure 9c presents the positioning error of driving in the Los Angeles area. Although the error increased at the point where the vehicle went under the overpass, the proposed method significantly suppressed the increase in error when compared to the baseline.

(a) Train/2021-03-16-US-MTV-3/ GooglePixel5 (b) Train/ 2021-03-16-US-MTV-1/ SamsungGalaxyS20Ultra (c) Train/ 2021-12-07-US-LAX-1/ GooglePixel6Pro

Figure 8. Driving trajectory of three vehicles extracted from the training dataset for evaluation (**a**) highway environment, (**b**) street driving, and (**c**) driving with GNSS blockage for a long period of time.

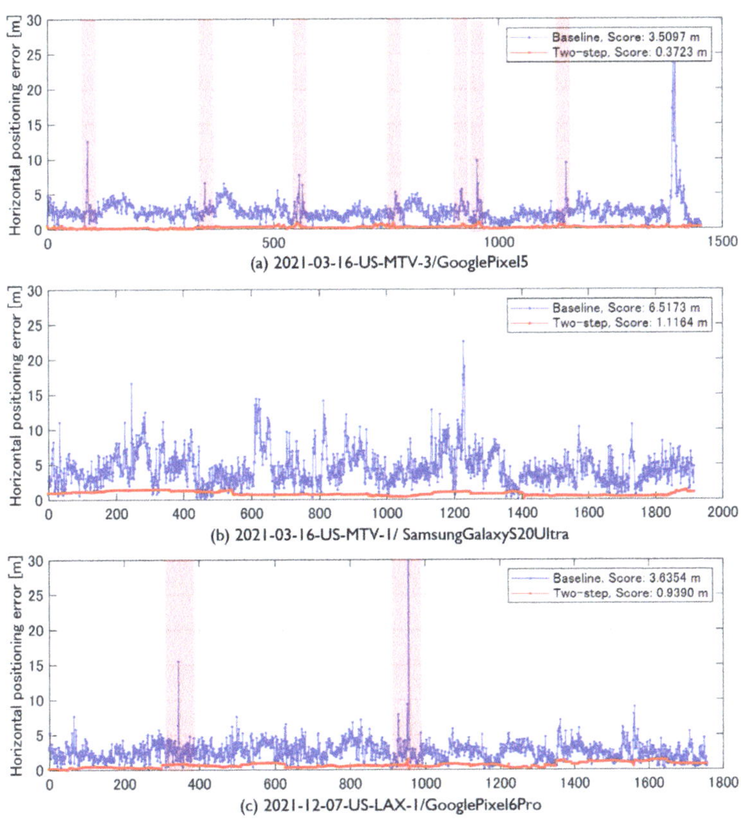

Figure 9. Comparison of horizontal positioning error between baseline (blue line) and proposed method (red line).

Figure 10 illustrates the horizontal cumulative distribution function (CDF) for each run and the entire evaluation dataset. Table 2 presents the 50 percentile error, the 95 percentile error, and the score calculated using Equation (3). The score was 0.372 m in the highway driving case, thus implying the achievement of decimeter accuracy. The score was 1.116 m in the street driving data. The final score for all runs included in the evaluation dataset was 1.023 m. The proposed method significantly improved the baseline scores provided by the host, thereby achieving decimeter accuracy when evaluated only on the highway data, but only slightly more than 1 m when evaluated on all runs.

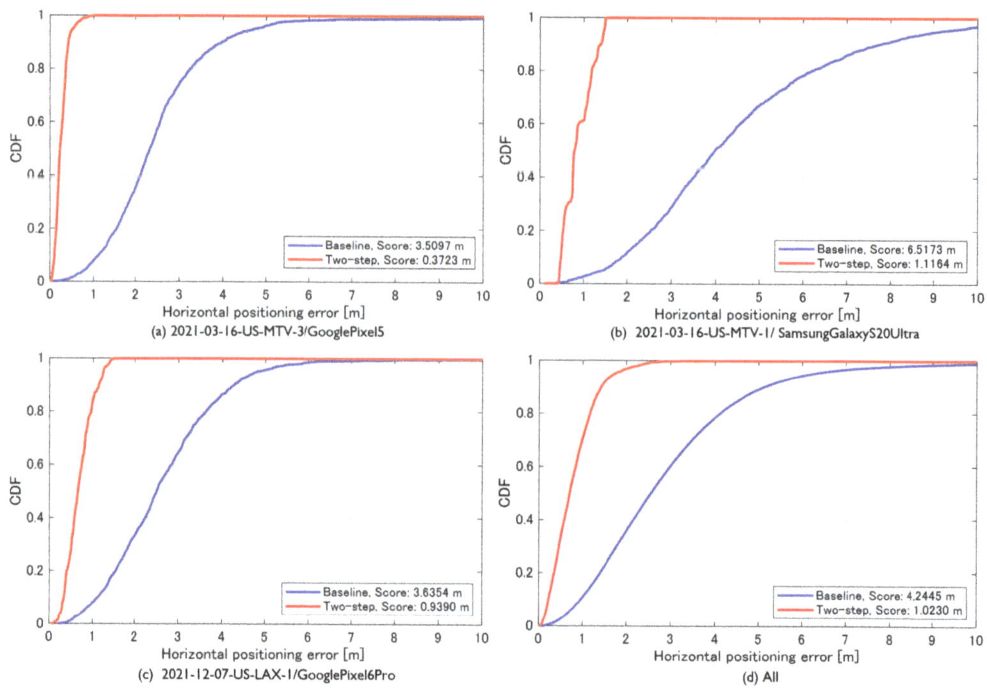

Figure 10. Comparison of horizontal cumulative distribution function (CDF) for each run and the entire evaluation dataset. The blue and red lines represent the baseline and proposed method, respectively.

Table 2. Comparison of positioning error between baseline and proposed method.

Course Phone	2021-03-16-US-MTV-3 GooglePixel5		2021-03-16-US-MTV-1 SamsungGalaxyS20Ultra		2021-12-07-US-LAX-1 GooglePixel6Pro		All	
	Baseline	Two-Step	Baseline	Two-Step	Baseline	Two-Step	Baseline	Two-Step
Score (50%) m	2.322	0.225	3.963	0.782	2.494	0.648	2.599	0.749
Score (95%) m	4.697	0.520	9.072	1.450	4.777	1.230	5.890	1.297
Mean m	3.510	0.372	6.517	1.116	3.635	0.939	4.245	1.023

8.1.2. Evaluation of Two-Step Optimization

The key contribution of the proposed method is that the velocity and position-estimation steps are separated and the position is estimated in a two-step optimization. To evaluate the effectiveness of the two-step optimization, we compared it with a one-step optimization method that simultaneously estimates position and velocity. The one-step optimization method is based on an implementation based on GSDC 2021, which adds 3D position, velocity, and receiver clock bias drift to the estimated state [36]. The GNSS observations used are completely equivalent to those of the two-step optimization method and equivalent parameters, such as the information matrix of the observations, were adopted.

Figure 11 compares the scores of the one-step and two-step optimizations for each run in the evaluation data. Figure 11 demonstrates that the proposed two-step optimization method outperforms the former in almost all runs. Table 3 presents the scores of the one-step and two-step optimization methods for all runs. The score of the one-step optimization is 1.122 m, thus indicating that the proposed two-step optimization method exhibits better performance.

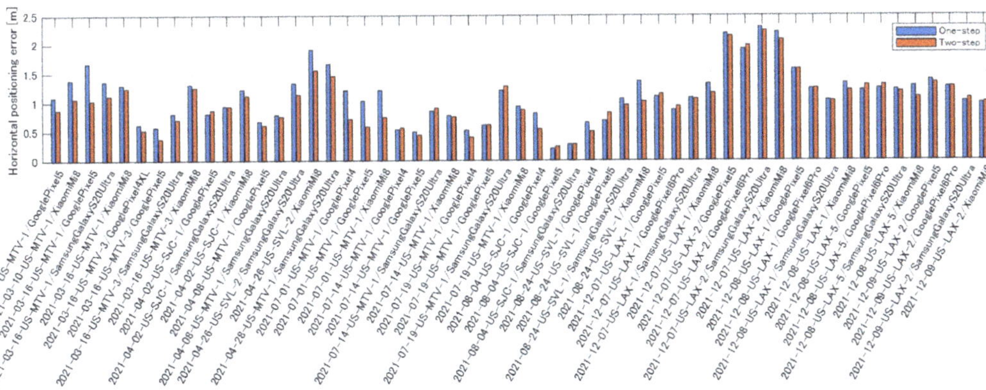

Figure 11. Comparison of positioning error between one-step and two-step optimization.

Table 3. Summary of positioning error comparison between one-step and two-step optimization.

Course	All	
	One-Step	**Two-Step**
Score (50%) m	0.801	0.749
Score (95%) m	1.442	1.297
Mean m	1.122	1.023

8.2. Evaluation Using Test Data

Using the above method, the position of the smartphone was estimated and the competition was addressed. Finally, the public score for the proposed method was 1.382 m, which ranked first. The final private score was 1.229 m, which was also in the first place. Hence, the proposed method can be used to estimate the position of smartphones with high accuracy. After the competition, the scores were calculated again with the final adjusted parameters (equivalent to those used in the evaluation of the training dataset above), resulting in public and private scores of 1.364 m and 1.177 m, respectively.

8.3. Discussion

The proposed two-step optimization method can estimate the position of the smartphones with an accuracy of 1.229 m in the GSDC 2022 test dataset. Here, the second place score was 1.499 m. Therefore, the proposed method is substantially the most accurate method for smartphone position estimation. The RTK-GNSS, which estimates the ambiguity of the carrier phase using the double-differenced GNSS observation from the base-station, is difficult to adopt with the noisy carrier phase of a smartphone; however, the TDCP observation, which is the time difference of its own carrier phase, can estimate the relative position with high accuracy.

The proposed two-step optimization was more accurate than the one-step optimization because the velocity was estimated first using Doppler observations, which are more robust, and the outliers of the estimated velocity were rejected beforehand and interpolated for missing values, which is considered to have an effect on accuracy.

The training data with position references for machine learning were provided; however, the proposed method did not use machine learning. Therefore, the incorporation of machine learning into the graph optimization framework represents a future challenge. In addition, IMU data from smartphones were also provided; however, the method described in this paper adopts only GNSS. Although there is a time synchronization problem between the smartphone IMU and GNSS [40], the combined IMU data are expected to improve accuracy in environments where GNSS signals are more shielded.

9. Conclusions

This paper described a method for estimating the position of a smartphone used in GSDC 2022. The proposed method adopted factor graph optimization to estimate the entire trajectory of a smartphone by creating various factors from the smartphone's GNSS observations. A two-step optimization method that estimates velocity and position in separate steps was proposed. The proposed method first estimated the velocity from the GNSS Doppler observation via graph optimization. The outliers of the estimated velocity were then excluded, interpolated, and subsequently used as loose constraints between states in the position-optimization step to make the position estimation more robust and accurate. Absolute constraints were added to the graph using corrected GNSS pseudoranges. In addition, the graph optimization process enabled highly accurate estimation of the 3D position of a smartphone.

Using the GSDC2022 training dataset, we compared the proposed two-step optimization with a one-step optimization that simultaneously estimates position and velocity and found that the two-step optimization improved the score from 1.12 m to 1.02 m. The results confirmed that estimating the velocity first improves the position estimation accuracy. Accordingly, from the proposed method presented at GSDC2022, the final private score was 1.229 m, which won the first place.

Funding: This research received no external funding.

Institutional Review Board Statement: Not applicable.

Conflicts of Interest: The authors declare no conflict of interest.

References

1. Humphreys, T.E.; Murrian, M.; Van Diggelen, F.; Podshivalov, S.; Pesyna, K.M. On the feasibility of cm-accurate positioning via a smartphone's antenna and GNSS chip. In Proceedings of the 2016 IEEE/ION Position, Location and Navigation Symposium (PLANS), Savannah, GA, USA, 11–14 April 2016; IEEE: Piscataway, NJ, USA, 2016; pp. 232–242.
2. Pesyna, K.M.; Heath, R.W.; Humphreys, T.E. Centimeter positioning with a smartphone-quality GNSS antenna. In Proceedings of the 27th International Technical Meeting of the Satellite Division of The Institute of Navigation (ION GNSS+ 2014), Tampa, FL, USA, 8–12 September 2014; pp. 1568–1577.
3. Zhang, X.; Tao, X.; Zhu, F.; Shi, X.; Wang, F. Quality assessment of GNSS observations from an Android N smartphone and positioning performance analysis using time-differenced filtering approach. *GPS Solut.* **2018**, *22*, 70. [CrossRef]
4. Li, G.; Geng, J. Characteristics of raw multi-GNSS measurement error from Google Android smart devices. *GPS Solut.* **2019**, *23*, 90. [CrossRef]
5. Zhang, K.; Jiao, F.; Li, J. The assessment of GNSS measurements from android smartphones. In Proceedings of the China Satellite Navigation Conference, Harbin, China, 23–25 May 2018; Springer: Berlin/Heidelberg, Germany, 2018; pp. 147–157.
6. Wu, Q.; Sun, M.; Zhou, C.; Zhang, P. Precise Point Positioning Using Dual-Frequency GNSS Observations on Smartphone. *Sensors* **2019**, *19*, 2189. [CrossRef] [PubMed]
7. Chen, B.; Gao, C.; Liu, Y.; Sun, P. Real-time Precise Point Positioning with a Xiaomi MI 8 Android Smartphone. *Sensors* **2019**, *19*, 2835. [CrossRef] [PubMed]

8. Dabove, P.; Di Pietra, V. Towards high accuracy GNSS real-time positioning with smartphones. *Adv. Space Res.* **2019**, *63*, 94–102. [CrossRef]
9. Dabove, P.; Di Pietra, V. Single-Baseline RTK Positioning Using Dual-Frequency GNSS Receivers Inside Smartphones. *Sensors* **2019**, *19*, 4302. [CrossRef]
10. Fu, G.M.; Khider, M.; van Diggelen, F. Android Raw GNSS Measurement Datasets for Precise Positioning. In Proceedings of the 33rd International Technical Meeting of the Satellite Division of The Institute of Navigation (ION GNSS+ 2020), Virtual, 22–25 September 2020; pp. 1925–1937.
11. Fu, G.M.; Khider, M.; van Diggelen, F. Google Smartphone Decimeter Challenge. 2021. Available online: https://www.kaggle.com/c/google-smartphone-decimeter-challenge/ (accessed on 15 December 2022).
12. Fu, G.M.; Khider, M.; van Diggelen, F. Google Smartphone Decimeter Challenge 2022. Available online: https://www.kaggle.com/competitions/smartphone-decimeter-2022/ (accessed on 15 December 2022).
13. Li, B.; Miao, W.; Chen, G.; Li, Z. Ambiguity resolution for smartphone GNSS precise positioning: Effect factors and performance. *J. Geod.* **2022**, *96*, 63. [CrossRef]
14. Sharma, H.; Bochkati, M.; Lichtenberger, C.; Pany, T.; Darugna, F.; Wübbena, J.B. Smartphone-Based GNSS Positioning—Today and Tomorrow. Available online: https://insidegnss.com/smartphone-based-gnss-positioning-today-and-tomorrow/ (accessed on 15 December 2022).
15. Zeng, S.; Kuang, C.; Yu, W. Evaluation of Real-Time Kinematic Positioning and Deformation Monitoring Using Xiaomi Mi 8 Smartphone. *Appl. Sci.* **2022**, *12*, 435. [CrossRef]
16. Siddakatte, R.; Broumandan, A.; Lachapelle, G. Performance evaluation of smartphone GNSS measurements with different antenna configurations. In Proceedings of the International Navigation Conference, Lausanne, Switzerland, 9–12 May 2017.
17. Wen, Q.; Geng, J.; Li, G.; Guo, J. Precise point positioning with ambiguity resolution using an external survey-grade antenna enhanced dual-frequency android GNSS data. *Measurement* **2020**, *157*, 107634. . [CrossRef]
18. Wang, L.; Li, Z.; Wang, N.; Wang, Z. Real-time GNSS precise point positioning for low-cost smart devices. *GPS Solut.* **2021**, *25*, 69. [CrossRef]
19. Bochkati, M.; Sharma, H.; Lichtenberger, C.A.; Pany, T. Demonstration of Fused RTK (Fixed) + Inertial Positioning Using Android Smartphone Sensors Only. In Proceedings of the 2020 IEEE/ION Position, Location and Navigation Symposium (PLANS), Portland, OR, USA, 20–23 April 2020; pp. 1140–1154. [CrossRef]
20. Kschischang, F.R.; Frey, B.J.; Loeliger, H.A. Factor graphs and the sum-product algorithm. *IEEE Trans. Inf. Theory* **2001**, *47*, 498–519. [CrossRef]
21. Sunderhauf, N.; Protzel, P. Switchable constraints for robust pose graph SLAM. In Proceedings of the IEEE International Conference on Intelligent Robots and Systems, Algarve, Portugal, 7–12 October 2012; pp. 1879–1884. . [CrossRef]
22. Sünderhauf, N.; Obst, M.; Wanielik, G.; Protzel, P. Multipath mitigation in GNSS-based localization using robust optimization. In Proceedings of the IEEE Intelligent Vehicles Symposium, Madrid, Spain, 3–7 June 2012; pp. 784–789. [CrossRef]
23. Chen, D.; Gao, G.X. Probabilistic graphical fusion of LiDAR, GPS, and 3D building maps for urban UAV navigation. *Navig. J. Inst. Navig.* **2019**, *66*, 151–168. [CrossRef]
24. Li, W.; Cui, X.; Lu, M. A robust graph optimization realization of tightly coupled GNSS/INS integrated navigation system for urban vehicles. *Tsinghua Sci. Technol.* **2018**, *23*, 724–732. [CrossRef]
25. Watson, R.M.; Gross, J.N. Robust navigation in gnss degraded environment using graph optimization. In Proceedings of the 30th International Technical Meeting of the Satellite Division of the Institute of Navigation, ION GNSS 2017, Portland, OR, USA, 25–29 September 2017; Institute of Navigation: Manassas, VA, USA, 2017; Volume 5, pp. 2906–2918. [CrossRef]
26. Wen, W.; Pfeifer, T.; Bai, X.; Hsu, L.T. Factor graph optimization for GNSS/INS integration: A comparison with the extended Kalman filter. *Navigation* **2021**, *68*, 315–331. .: 10.1002/navi.421. [CrossRef]
27. Suzuki, T. GNSS Odometry: Precise Trajectory Estimation Based on Carrier Phase Cycle Slip Estimation. *IEEE Robot. Autom. Lett.* **2022**, *7*, 7319–7326. [CrossRef]
28. Bai, X.; Wen, W.; Hsu, L.T. Time-Correlated Window-Carrier-Phase-Aided GNSS Positioning Using Factor Graph Optimization for Urban Positioning. *IEEE Trans. Aerosp. Electron. Syst.* **2022**, *58*, 3370–3384. [CrossRef]
29. Jiang, Y.; Gao, Y.; Ding, W.; Gao, Y. GNSS precise positioning for smartphones based on the integration of factor graph optimization and solution separation. *Measurement* **2022**, *203*, 111924. . [CrossRef]
30. Watson, R.M.; Gross, J.N. Evaluation of kinematic precise point positioning convergence with an incremental graph optimizer. In Proceedings of the 2018 IEEE/ION Position, Location and Navigation Symposium (PLANS), Monterey, CA, USA, 23–26 April 2018; IEEE: Piscataway, NJ, USA, 2018; pp. 589–596.
31. Robusto, C.C. The cosine-haversine formula. *Am. Math. Mon.* **1957**, *64*, 38–40. [CrossRef]
32. Robustelli, U.; Paziewski, J.; Pugliano, G. Observation Quality Assessment and Performance of GNSS Standalone Positioning with Code Pseudoranges of Dual-Frequency Android Smartphones. *Sensors* **2021**, *21*, 2125. [CrossRef]
33. Suzuki, T. First place award winner of the smartphone decimeter challenge: Global optimization of position and velocity by factor graph optimization. In Proceedings of the 34th International Technical Meeting of the Satellite Division of The Institute of Navigation (ION GNSS+ 2021), Tampa, FL, USA, 20–24 September 2021; pp. 2974–2985.
34. Freda, P.; Angrisano, A.; Gaglione, S.; Troisi, S. Time-differenced carrier phases technique for precise GNSS velocity estimation. *GPS Solut.* **2015**, *19*, 335–341. [CrossRef]

35. Everett, T.; Taylor, T.; Lee, D.K.; Akos, D.M. Optimizing the Use of RTKLIB for Smartphone-Based GNSS Measurements. *Sensors* **2022**, *22*, 3825. [CrossRef] [PubMed]
36. Suzuki, T. Robust Vehicle Positioning in Multipath Environments Based on Graph Optimization. In Proceedings of the 34th International Technical Meeting of the Satellite Division of The Institute of Navigation (ION GNSS+ 2021), Tampa, FL, USA, 20–24 September 2021; pp. 4223–4233.
37. Dellaert, F. *Factor Graphs and GTSAM: A Hands-on Introduction*; Technical Reports; Georgia Institute of Technology: Atlanta, GA, USA, 2012.
38. Takasu, T.; Yasuda, A. Development of the low-cost RTK-GPS receiver with an open source program package RTKLIB. In Proceedings of the International Symposium on GPS/GNSS, Jeju, Korea, 4–6 November 2009; pp. 4–6.
39. Akima, H. A New Method of Interpolation and Smooth Curve Fitting Based on Local Procedures. *J. ACM* **1970**, *17*, 589–602. [CrossRef]
40. Sharma, H.; Bochkati, M.; Pany, T. Time-Synchronized GNSS/IMU Data Logging from Android Smartphone and its Influence on the Positioning Accuracy. In Proceedings of the 34th International Technical Meeting of the Satellite Division of The Institute of Navigation (ION GNSS+ 2021), Tampa, FL, USA, 20–24 September 2021. [CrossRef]

sensors MDPI

Article

An Adaptive Algorithm for Multipath Mitigation in GNSS Positioning with Android Smartphones

Lorenzo Benvenuto [1,2], Tiziano Cosso [2] and Giorgio Delzanno [1,*]

1 DIBRIS—Department of Informatics, Bioengineering, Robotics and Systems Engineering, University of Genoa, 16146 Genoa, Italy; lorenzo.benvenuto@edu.unige.it
2 Gter s.r.l, 16128 Genoa, Italy; tiziano.cosso@gter.it
* Correspondence: giorgio.delzanno@unige.it

Abstract: We present a solution for improving the robustness of GNSS positioning with Android devices. The proposed method combines an acquisition phase performed in a dedicated Android app (thus working on the edge) and a processing phase, based on a modified version of the open source library RTKLIB, performed on a dedicated server. The processing phase applies an improved version of the RTK library based on an adaptive algorithm for mitigating the multipath effect on satellite radio signals received by smartphone's antennas. The algorithm is built on top of an extended version of the sigma-epsilon model in which weights associated to observables potentially affected by multipath errors are computed using logged data. In the paper, we will focus our attention on the architecture of the proposed solution and discuss preliminary experimental results obtained with the resulting system.

Keywords: GNSS positioning; data analysis; internet of things; mobile computing

Citation: Benvenuto, L.; Cosso, T.; Delzanno, G. An Adaptive Algorithm for Multipath Mitigation in GNSS Positioning with Android Smartphones. *Sensors* **2022**, *22*, 5790. https://doi.org/10.3390/s22155790

Academic Editors: Yang Gao, Jacek Paziewski, Michael Fu and Augusto Mazzoni

Received: 15 June 2022
Accepted: 29 July 2022
Published: 3 August 2022

1. Introduction

1.1. Background and Motivations

In May 2016, during the "Google I/O" conference, Google released an API to give Android developers access to GNSS raw measurements such as carrier phase, code measurements, and navigation messages. As stated in the white paper [1] of the GNSS Raw Measurement Task Force, coordinated by the European GNSS Agency (GSA), this new feature of the Android API offered new research directions. In particular, GNSS raw measurements can be used to optimise multi-GNSS and multi-frequency solutions, to select satellites based on their performance, to transfer processing techniques from GNSS receivers to smartphones, to combine GNSS raw data with data of other sensors that are available in smartphones, and to enable testing and post-processing analysis [2–5].

From a technical point of view, the use of GNSS raw measurements posed several challenges for both GNSS experts and software developers. Indeed, on one hand, GNSS standard formats, such as RINEX or NMEA, are not natively available on the Android platform. On the other hand, mobile app developers are not generally familiar with the complex algorithms and libraries used in GNSS positioning.

The GSA white paper [1] addressed the gap between the two fields, providing useful information, for example, for deriving the pseudoranges from Android. This important work opened a new research field aimed at developing low-cost applications for satellite-based positioning systems. In particular, after its publication, many authors started to analyse the quality of the raw measurements retrieved from smartphones and compare them with other types of low-cost devices. The main detected issue turned out to be the high noise of the GNSS observables. Indeed, smartphones are equipped with cellphone-grade GNSS chipsets and antennas, which have on average very low gain, resulting in a low and irregular signal-to-noise ratio (SNR) [5]. For this reason, smartphone positioning is very challenging, especially in harsh environments, such as urban areas, that are more vulnerable to multipath and other interferences (see e.g., [6–13]). Other issues were related to the duty cycle mechanism and to low values of the C/N_0 [14,15].

In [16], the authors demonstrated that it was possible to reach decimetric accuracy in terms of positioning performances following the post-processing approach, via a double difference of raw smartphone observations. Meanwhile, the authors of [17] first focused their attention on single-base RTK positioning and then demonstrated the possibility of obtaining centimetre-level accuracy through the use of NRTK corrections [17]. These results are supported by the authors in [18], who, employing a variometric approach, show decimetre accuracy in static conditions and sub-metre when used in an urban vehicle scenario. Very interesting and promising results have also been obtained by applying PPP (precise point positioning) techniques to smartphones devices. In [19], the authors analysed single-frequency PPP in static mode and reported that decimetre- to meter-level positioning accuracy can be achieved with the smartphone-grade hardware, while in [20], a smartphone software application called PPP WizLite is proposed. This application enacts PPP processing on smartphones using Doppler-smoothed code pseudoranges. The authors in [20] also showed that positioning accuracies at the sub-metre and metre level can be achieved in static and kinematic mode, respectively. In [21], the authors developed a new methodology for the PPP processing in smart devices, achieving sub-meter-level positioning accuracy. In particular, they wrote a software application based on the Android platform, named Smart-PPP, that implements a modified stochastic model in PPP processing to weight code and carrier phase measurements on the basis of SNR parameter. To achieve this goal, an improved uncombined PPP observation model is proposed in Smart-PPP. A modified C/N0-dependent weighting strategy is employed. Some data-processing strategies used in classical PPP with the geodetic receiver are improved and made more suitable for the data characteristics of smart devices. By applying the Smart-PPP approach to smart devices, the final positioning results can be smoother and more accurate.

Another milestone has been set by Broadcom, which announced, on 21 September 2017, the world's first mass-market, dual-frequency GNSS receiver device, the BCM47755. In May 2018, the Xiaomi Mi8 (Mi8) became the first smartphone in the world, employing a dual-frequency GNSS receiver L1/E1–L5/E5a. This led to the next series of studies in the investigation of smartphone-based positioning. Thanks to the double frequency introduced in [22], the multipath performance of the Xiaomi Mi8 device was investigated for both E1/L1 E5a/L5 signals using a proper linear combination. The results obtained were quite promising, but they also seem to indicate multipath as one of the main problems for smartphone positioning. Multipath effects on Android devices were also studied in [23], where the author performed positioning with the Nexus 9 tablet using a particular Eccosorb for multipath mitigation. The results shows that precise positioning with uncertainties lower than one metre was possible. In [24], the authors show encouraging results in ZTD estimation by smartphone devices.

Nowadays, raw GNSS measurements support is mandatory on devices that run Android 10 (API level 29) or higher, but unfortunately, the support for some of the raw GNSS measurement fields (e.g., pseudorange rate, ADR, AGC) is optional and can vary based on the type of GNSS chipset installed on the device. Furthermore, not all the smartphones present on the market support double frequency or multi-constellation [25]. For this reason, finding a robust use of GNSS raw measurements is still a topic of interest for the research communities working on GNSS and mobile computing.

1.2. Research Question

As mentioned before, the multipath effect, i.e., radio signals reaching the receiving antenna by two or more paths, is probably the major source of error in urban scenarios affecting the positioning quality. A series of tests carried out with smartphones equipped with dual-frequency receivers (Broadcom and Snapdragon chipsets) confirmed this hypothesis experimentally. Multipath is also a serious problem for the application of GNSS positioning algorithms that use raw measurements.

Our research question is whether multipath mitigation techniques used for GNSS receivers can be applied to RTK positioning with Android devices. RTK positioning is a class of algorithms that employ correction codes received from base stations. They are particularly interesting since they can reach centimetric accuracy without the need of positioning information from cellular and Internet networks. In particular, in this setting, our goal is to investigate different types of heuristics to increase robustness, in terms of precision and accuracy, of the RTK positioning in Android devices.

1.3. Our Contribution

Our first contribution is the design and implementation of a prototype system for applying multipath mitigation heuristics in RTK positioning with GNSS raw measurements. The proposed system is based on a pre-processing phase performed in a dedicated Android app (thus working on the edge) and on a real-time processing phase, based on a modified version of the open source library RTKLIB [26], performed on a dedicated server. The performance of the resulting system, including client-server latencies, is comparable to RTK positioning procedures for GNSS receivers and assisted GPS computing procedure. It is important to note that both other solutions also require network communication steps. The data acquisition phase, performed via an Android app developed during Lorenzo Benvenuto's PhD work [27], is aimed at cleaning, filtering, organizing, and delivering the data acquired via the GNSS raw measurements library. The app is in charge of collecting the satellite data of a given epoch, namely raw observations of different satellites and frequencies, converting them into a special message format, and sending the resulting message to the processing server. The app has several other functionalities, including that of receiving and visualizing the position inferred by our algorithms and providing options to control the acquisition phase (e.g., flags to enable/disable duty cycle and real-time plot of the SNR parameter).

The server-side processing phase exploits an improved version of the RTKLIB library, enabling RTK positioning from smartphones and increasing the solution robustness by means of multipath mitigation heuristics. In order to make RTKLIB work in real time with smartphones, a new API for processing GNSS data in the format defined for data collection was added to the original library. Concerning the multipath mitigation, the so-called MDP (multipath detection parameter) algorithm, conceived and patented by Gter, was implemented in the RTKLIB version adopted in this work. This algorithm performs multipath detection and mitigation in real time for single-frequency GNSS receivers. The MDP algorithm was improved and adapted for working with GNSS observables from Android devices. The main idea here is to weigh observations of each piece of satellite data with parameters associated to multipath (MDP variable) and signal noise (SNR) errors. The resulting weights are used to tune the RTKLIB positioning algorithms in order to assign low weights to unreliable observations. The algorithm has several possible configuration parameters and operating modes. In particular, one of the biggest improvements of our algorithms is in the adoption of adaptive thresholds that are inferred via statistical analysis of collected data.

The procedure described above was tested and validated using two different data sets: a static acquisition with multipath effect induced and a kinematic acquisition. For both case studies, several combinations of the MDP algorithm configuration parameters were tested. The results obtained seem very promising. Indeed, with the proposed algorithm, improvements in positioning accuracy were noted, especially for the period of induced multipath, meaning that the MDP algorithm is suitable for the mitigation of this effect. Furthermore, in the solution obtained with the MDP algorithm application, some positioning outliers are eliminated, and consequently, the solution's robustness is increased.

1.4. Originality and Reproducibility

The paper includes original work extracted from Lorenzo Benvenuto's PhD thesis [27], supervised by Tiziano Cosso and Giorgio Delzanno. The source code for the GNNS Base station server and the modified version of RTKLIB presented in this paper are available on GitHub (https://github.com/gtergeomatica/RTKLIB_Android, accessed on 7 February 2022). See also [27].

1.5. Plan of the Paper

In Section 2, we present a preliminary comparison between smartphones and low-cost GNSS receivers. In Section 3, we present the IoT RTK system In Section 4, we present the MDP multipath mitigation algorithm. In Section 5, we discuss experimental results obtained on static and kinetic tests. Finally, in Section 6, we address some conclusions and discuss future directions for our work.

2. Positioning Analysis with Android Devices

In this section, we present an analysis of the signal quality of smartphone receivers and compare it with that of geodetic and GNSS receivers.

The device considered in all tests is the Xiaomi Mi 8. This smartphone is equipped with the Broadcom 47755 dual-frequency GNSS chip capable of tracking GPS L1 C/A, GLONASS L1, BeiDou B1, Galileo E1, GPS L5, and Galileo E5a. The device has been chosen as it can provide both pseudo-range and carrier phase measurements, and navigation messages. It is worth mentioning that not all the smartphones on the market have access to all the GNSS raw measurements.

We consider here two types of tests: kinematic acquisition and static acquisition with multipath induced. The setup of each test is described in detail in the rest of the section.

2.1. Test 1: Kinematic Acquisition

Navigation is one of the main areas in which the GNSS receiver embedded into a smartphone is employed. Test 1 aims at making an assessment of the performance of the GNSS positioning from a smartphone in a typical scenarios such as pedestrian navigation. More specifically, test 1 consists of a kinematic pedestrian acquisition involving two GNSS receivers, the Xiaomi Mi 8 and the Stonex S500, used for comparison. The Stonex S500 is a single-frequency (L1) and multi-constellation (GPS, GLONASS, Galileo, and BeiDou) device typically used for GIS (geographic information system) and RTK applications based on a u−blox chipset. The test path, which was located in Genoa and is shown in the Figure 1, was identified by determining the vertices of a square. In order to get the vertexes' coordinates with high precision, the four vertexes were surveyed in NRTK modality with the Stonex S70G, a geodetic-level GNSS receiver, and exploiting the Ligurian NRTK network for getting the differential corrections. Once surveyed, the vertexes were materialised using four targets.

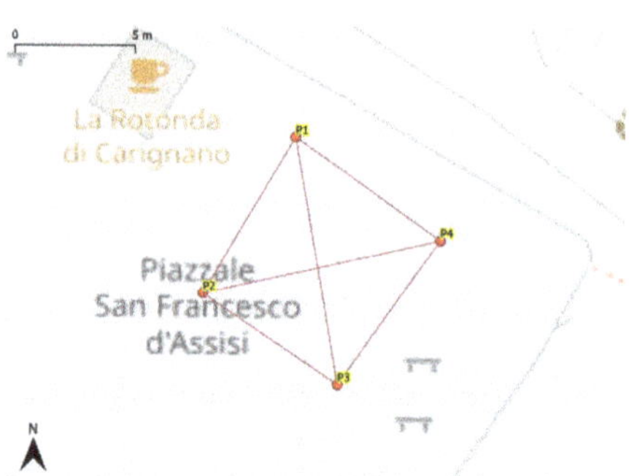

Figure 1. Trajectory used for test 1.

Test 1 started at vertex number 1 with a static acquisition of 2 min. After that, a walk between the four vertexes was performed along the squared sides, as depicted in Figure 1. Each side has been traversed in both directions (e.g., from vertex 1 to vertex 2 and from vertex 2 to vertex 1). The test ended at vertex 1 with a static acquisition of 2 min. Data from both the receivers were collected at 1 Hz rate. During the test, the two receivers were held at a height of about 1 m above the ground.

2.2. Test 2: Multipath Effect

Multipath and other interferences are one of the main sources of error in GNSS positioning in urban canyons, especially if mass market GNSS receivers are employed. Based on this consideration, test 4 aims at evaluating the impact of multipath in smartphone positioning.

Similarly to test 1, a second GNSS receiver is used for comparing positioning results. Test 2 consists of static acquisition, in which, for a specific interval, the multipath effect was reproduced by placing a metal plate behind the receivers (see Figure 2). The receivers used in this test are the Xiaomi Mi 8, and the u−blox ZED F9P coupled with the AN-MB-00 patch antenna. The u−blox ZED F9P is a mass market multi-frequency (L1/L2) and multi-constellation (GPS, GLONASS, Galileo, and BeiDou) GNSS receiver.

(a) (b)

Figure 2. Test 2 acquisition under standard condition (**a**), and with multipath induced (**b**).

The two receivers were placed at a point whose coordinates were determined with high precision in NRTK modality using the Stonex S70G receiver, and exploiting the Ligurian NRTK network for the differential corrections. The static acquisition started at 9:15 a.m. UTC time and lasted 1 h. From 9:45 to 10:00, the multipath effect was induced by placing the metal plate behind the receivers (see Figure 2). After 10:00, the plate was removed, and the acquisition ended at 10:15. Data from both the receivers were collected at 1 Hz rate.

2.3. Precision and Accuracy

Considerations regarding the accuracy and precision of the analysis of the acquired data sets can be made through the standard deviation (*STD*) and root mean square (*RMS*) values. Given a set of elements, *STD* and *RMS* can be defined as:

$$STD = \sqrt{\frac{1}{N} \cdot \sum_{k=1}^{N} (x_k - \mu_x)^2} \qquad RMS = \sqrt{\frac{1}{N} \cdot \sum_{k=1}^{N} (x_k - \tilde{x})^2} \tag{1}$$

where

- N is the total number of elements;
- x_k is a generic element belonging to the set;
- μ_x is the mean value;
- $\tilde{x}$ is a reference value for the elements (i.e., in this case, the precise coordinates of the point).

In general, *STD* is considered a precision indicator, while *RMS* is an accuracy indicator. In this paper, we will not consider sample *STD*/*RMS* since we analyse data sets consisting of a single data survey for each period. Given RMS for direction E and N, we will also consider a single measure $RMS\ 2D$, computed as follows:

$$RMS\ 2D = 2 * \sqrt{(RMS\ E)^2 + (RMS\ N)^2} \quad (2)$$

All the processing described in the rest of this section was performed using the RTKLIB version 2.4.3 b34 by Tomoji Tamasu, Sakura, Japan, distributed under a BSD 2-clause license. In order to evaluate both accuracy and precision, the static tests were processed as kinematic ones, that is, by obtaining a point cloud as output, with the coordinates computed epoch by epoch. Mainly two types of processing are performed: a stand-alone positioning, which requires only the observable of the receiver under examination, and a relative post-processing, which involves not only the observables of the receiver under examination but also the ones of a base station. This second processing is also called post-processed kinematics (PPK). Unless otherwise specified, the options used for processing are the following: The ionosphere and troposhere models used are the Klobuchar [28] (the parameters of which are transmitted with the navigation message) and Saastamoinen [29], respectively. For every test, only the broadcast ephemeris were considered, and an elevation mask of 15° is also set.

Concerning the PPK elaborations, the solution type combined was set. In this modality, the observation data is processed through a Kalman filter in the forward direction, that is, starting with the beginning of the data and continuing through to the end. Backward mode is the opposite: data is run through the filter starting with the end of the data and continuing to the beginning. In combined mode, the filter is run both ways, and the two results are combined into a single solution.

2.4. Test 1: Results

The performance of the smartphone GNSS receiver in a kinematic contest can be analysed at two different levels: the pre-processing level and the positioning level. In both cases, to better understand its performance, the data coming from the smartphone (Xiaomi Mi 8) are compared with those produced by a Stonex S500 receiver, which is considered as a benchmark.

Concerning the pre-processing level, the quality of the incoming signal for the two receivers is compared. A good indicator of the incoming signal quality is the signal-to-noise ratio (SNR), which is a measure of the strength of the desired signal relative to background noise. SNR expressed in DB-Hz presents high values for a good incoming signal, and low values for a bad signal. The SNR values for the two receivers used in test 1 are shown in Figure 3.

Figure 3 shows that the Stonex S500 receiver has a much better SNR than the Xiaomi Mi 8 receiver. In fact, an average difference of about 10 DB-Hz between the two receivers is observed. From the SNR trend, it is also possible to recognise the static and kinematic parts of the survey. The kinematic part, highlighted in the green dashed box, actually presents a noisier SNR trend than the static parts located at the beginning and at the end of the test.

The number of satellites observed by the Xiaomi Mi 8 is greater than that of the Stonex S500 (see Figure 4), but much more variable over time. In fact, the Xiaomi Mi 8 receiver also acquires very noisy satellites (SNR $<$ 25 DB-Hz) for short periods of time, which are ignored by the Stonex S500.

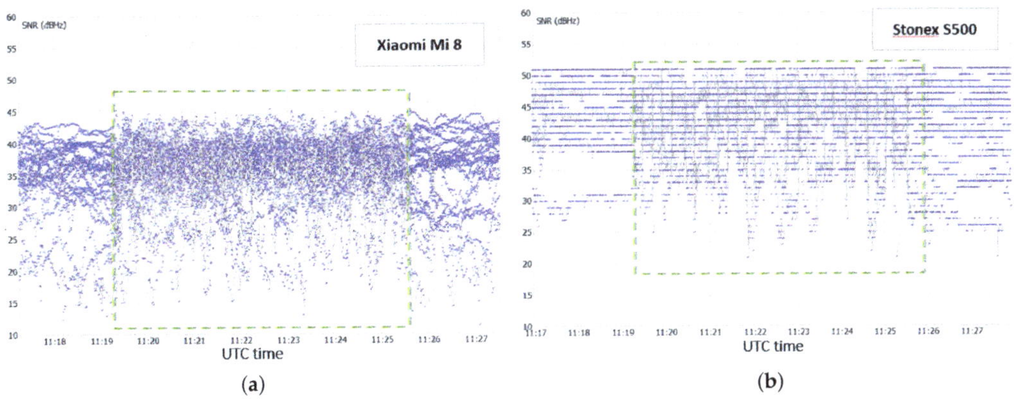

Figure 3. SNR for Xiaomi Mi 8 receiver (**a**) and for Stonex S500 receiver (**b**).

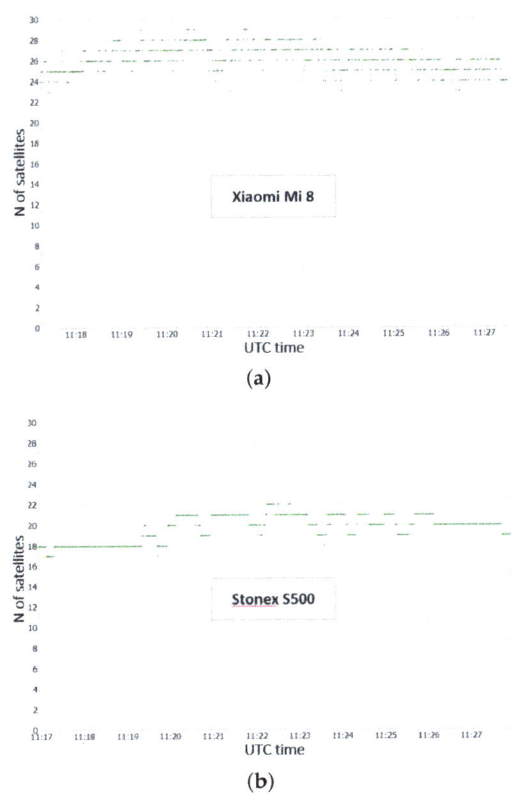

Figure 4. Number of satellite observed by Xiaomi Mi 8 receiver (**a**) and Stonex S500 receiver (**b**).

Concerning positioning, acquired data were post-processed with a common GNSS permanent station located about 200 m away from the test field, hereafter called LIGE [27]. The LIGE permanent station is equipped with the GNSS receiver u-blox ZED F9P coupled with the Hemisphere A45 antenna. For this case study, the solution type "forward" is set in order to simulate real-time conditions.

The positioning solutions obtained are shown in Figure 5.

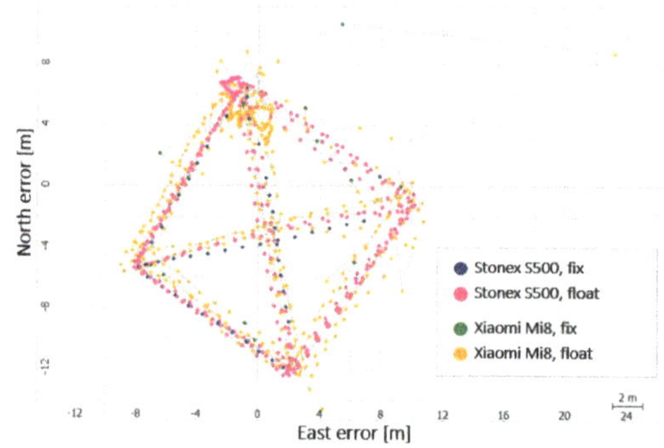

(**a**)

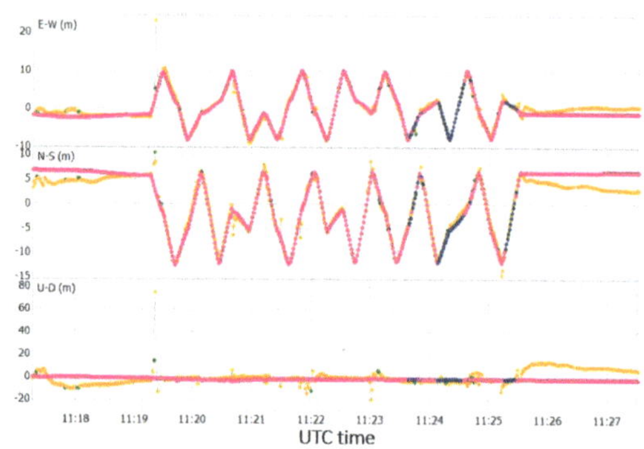

(**b**)

Figure 5. (**a**) Scatter plot of the positioning result for Xiaomi Mi8 and Stonex S500, and (**b**) time series of the positioning error for Xiaomi Mi8 and Stonex S500.

The Stonex S500 receiver solution is more accurate, especially when considering the North and Altitude components. The two solutions can be also compared in terms of number of computed positions and percentage of fix and float solutions, as shown in Table 1.

Table 1. Test 1: Solution quality.

Receiver	Number of Solutions	Fixed Solutions	Float Solutions
Xiaomi Mi 8	595	19 (3.2%)	576 (96.8%)
Stonex S500	617	47 (7.6%)	570 (92.4%)

The Xiaomi Mi 8 receiver presents fewer solutions than the Stonex S500. The poor quality of the GNSS measurement from the smartphone prevents the computation of the positioning for some epochs. Furthermore the Xiaomi Mi 8 solutions, coherently with other results obtained from the previous tests, present some outliers and false fixing solutions.

2.5. Test 2: Result

Similarly to test 1, for this case, the performance of the Xiaomi Mi8 receiver is compared to that of the other device involved, the u−blox ZED F9P, at two different levels: the pre-processing and the positioning level. Once again, in the pre-processing level, the two receivers are compared in terms of the quality of the incoming signal. For this purpose, the SNR trend for the two receivers is shown in Figure 6.

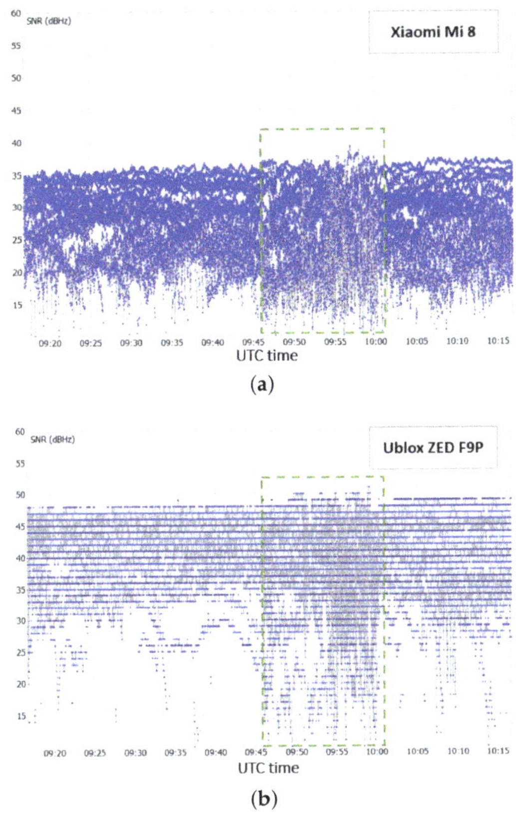

Figure 6. SNR for Xiaomi Mi 8 receiver and (**a**) for u−blox ZED F9P receiver (**b**).

The Xiaomi Mi 8 presents lower SNR values on average with respect to the u-blox ZED F9P. From the SNR trend of the two receivers, the period in which the multipath effect was induced with the metal plate is evident, that is, from 9:45 a.m. to 10:00 a.m. UTC (green dashed boxes in Figure 6). In this time interval, as expected, the signal acquired is degraded. This results in a general lowering of SNR values, and a significantly noisier trend.

Concerning the positioning part, the data were again post-processed with the LIGE permanent station. Similarly to test 1, for this case study the solution type chosen is also forward. Figure 7 shows the obtained positioning results in terms of errors with respect to the precise coordinates of the point.

The statistics values, in terms of RMS and standard deviation with respect to the precise coordinates of the receivers, are reported in Table 2.

In this case study, the solution for the Xiaomi Mi8 receiver has a convergence time of about 5 min. After this time interval, the standard deviations of the solution are 0.051 m, 0.08 m, and 0.093 m for the East, North, and Height components, respectively. The solutions therefore present a centimetre-level accuracy; nevertheless, the u-blox ZED F9P solution results were more accurate. After the convergence time, the Xiaomi Mi8 solution has average deviations from the precise point coordinates of −0.460 m, −0.549 m, and 0.390 m for the East, North, and Height components, respectively. It can be stated that the solution presents a decimetre precision level, but once again, the u-blox ZED F9P solution results are more precise.

Consistently with the results already obtained, for the other case studies, the Xiaomi Mi8 has a lower number of solutions (3473 against 3501) and a lower percentage of fixed solutions (2.4% against 81.3%).

For this case study, it is also interesting to analyse the solution in the time interval in which the multipath was induced (i.e., between the 9:45 and the 10:00 UTC). For the u-blox ZED F9P, the solution results are particularly degraded (see Figure 7b), and almost the totality of the float solutions obtained for this receiver are in this time interval. Concerning the Xiaomi receiver, the solution does not seem particularly degraded in this time interval with respect to the entire test period; nevertheless, in this interval, some false fixed solutions can be observed.

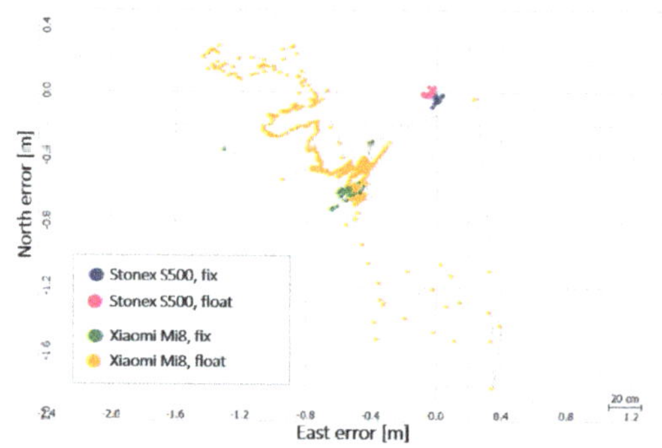

(a)

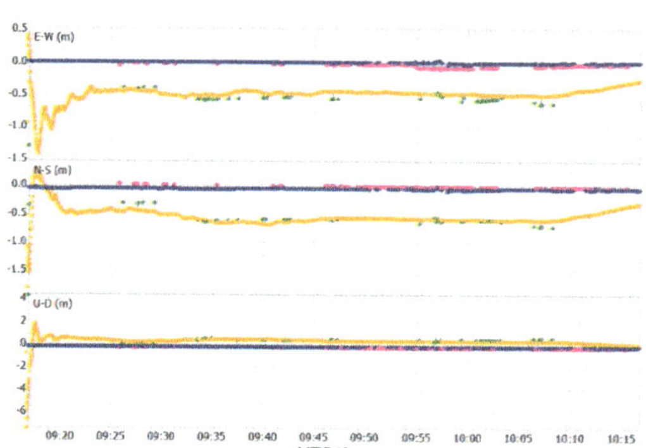

(b)

Figure 7. (**a**) Scatter plot of the positioning error for Xiaomi Mi8 and u−blox ZED F9P; (**b**) time series of the positioning error for Xiaomi Mi8 and u−blox ZED F9P.

Table 2. Test 2: Post-processing results.

Receiver	RMS E (m)	RMS N (m)	RMS H (m)	STD E (m)	STD N (m)	STD H (m)
Xiaomi Mi8	0.499	0.549	0.556	0.133	0.151	0.411
u-blox ZED F9P	0.021	0.043	0.020	0.179	0.017	0.018

3. The GNSS RTK IoT System

As discussed in Section 2, GNSS positioning with smartphones suffers from false fixed solutions, and outliers that compromise the robustness of the positioning itself. Those false fixed solutions can be caused by many factors, including the poor quality of the GNSS observables and the presence of external interferences such as the multipath effect. In this section, we present a solution to increase the robustness of GNSS RTK positioning. The architecture of the GNSS RTK positioning procedure that we used to process GNSS data coming from Android smartphones consists of the following components:

- A smartphone for data acquisition (the rover);
- A GNSS base station;
- A remote server for RTK processing.

The smartphone is supposed to be equipped with an Android app capable of reading GNSS raw observables, packing them in a proper way, and sending them to a server running RTKLIB through a TCP socket. The base station must be capable of sending, in real time, its observables together with its precise coordinates to the server. The server is supposed to be capable of reading the input streams from the smartphone and from the base station, processing them with RTKLIB, and then returning the coordinates to the smartphone. The resulting architecture is shown in Figure 8.

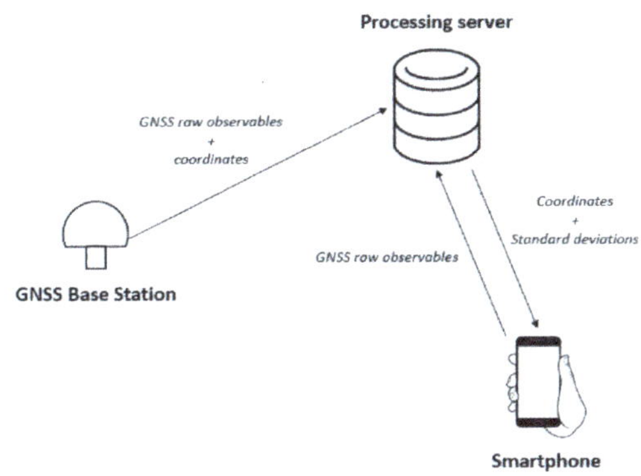

Figure 8. Proposed architecture for robust RTK positioning with smartphone.

The proposed system works with an ad-hoc GNSS base station assembled with a u-blox ZED F9 receiver, but it can be generalised by using an NRTK network instead.

The central component of the architecture, as shown in Figure 8, is a remote server for GNSS positioning. The server runs the RTKLIB library for data processing and communicates in real time with both the smartphone and the base station though TCP/IP sockets. The RTKLIB version installed in the server was properly modified in order to make it work with GNSS data coming from a smartphone in real time. This new version of RTKLIB not only can process data coming from a smartphone in real time but also implements an algorithm for detecting and mitigating the multipath effect. This algorithm, which is the main contribution of this work, is explained in depth in the following section.

4. Multipath Mitigation Algorithm

One of the major external interferences that degrade GNSS positioning is multipath. The effect of this interference is well described by its name: a satellite-emitted signal arrives at the receiver from different directions (i.e., following different paths). The multipath effect is mainly caused by reflecting surfaces near the receiver. For this reason, it frequently happens in urban canyons, a typical scenario in which smartphones are used. The multipath effect may also occur when using low-cost GNSS receivers equipped with low-quality antennas, such as the smartphone's. For this reason, a good strategy to improve the robustness in GNSS positioning from Android devices should aim at mitigating the multipath error.

Multipath mitigation can be performed at the antenna, receiver (signal processing), and navigation solution level [30].

In this work, multipath mitigation is approached at the software level, using an algorithm, called MDP (multipath detection parameter) conceived by Gter. In the following section, the algorithm is explained in detail, together with its implementation in RTKLIB.

4.1. The MDP Algorithm

The MDP algorithm consists of two main parts: the first one concerning multipath detection, and the second one concerning multipath mitigation. The multipath detection part is based on the SNR (signal-to-noise ratio) value and a new variable called MDP (multipath detection parameter). The aim of the MDP value is to create a variable representative for the multipath effect for single-frequency GNSS receivers, starting from the observation equation. The observation equations for GNSS are as follows:

$$\begin{aligned} P(t_1) &= \rho(t_1) + \Delta T_r^s(t_1) + Ion(t_1) + Trop(t_1) + Mult(t_1) + \epsilon_1 \\ L(t_1) &= \rho(t_1) + \Delta T_r^s(t_1) - Ion(t_1) + Trop(t_1) + mult(t_1) + \lambda A + \epsilon_2 \end{aligned} \quad (3)$$

where

- $P(t_1)$ is the pseudorange observable at instant t_1;
- $L(t_1)$ is the carrier phase observable at instant t_1;
- $\rho(t_1)$ is the geometric range between the satellite and the receiver at instant t_1;
- $\Delta T_r^s(t_1)$ is the global time unknown at instant t_1;
- $Ion(t_1)$ is the ionospheric effect on the signal at instant t_1;
- $Trop(t_1)$ is the tropospheric effect on the signal at instant t_1;
- λA is the phase ambiguity;
- $Mult(t_1)$ is the code multipath;
- $mult(t_1)$ is the phase multipath;
- ϵ_1 and ϵ_2 represent residual errors due to noise.

One common approach to isolate the multipath effect, from Equation (3), is to compute the difference [31]. The result is the so-called sentinel variable:

$$S(t_1) = P(t_1) - L(t_1) = 2Ion(t_1) - \lambda N + (Mult(t_1) - mult(t_1)) + \epsilon \quad (4)$$

where ϵ represents residual error. The definition of MDP variable requires two assumptions. Considering a sufficiently high acquisition rate (i.e., 1 Hz), we can assume that:

- The ionosphere effect is equal between two consecutive epochs;
- The phase ambiguity is equal between two consecutive epochs.

Based on these assumptions, the MDP variable is defined as follows:

$$MDP = S(t_2) - S(t_1) = \{[Mult(t_2) - mult(t_2)] - [Mult(t_1) - mult(t_1)]\} + \epsilon' \quad (5)$$

where ϵ' represents the difference of residual errors. Hence, as shown in Equation (5), the MDP variable is representative of the effect of the multipath effect and residual errors due to noise.

4.2. MDP: Detection Algorithm

The detection algorithm compares the SNR and MDP parameters to static and dynamic threshold values as explained below.

4.2.1. SNR Threshold

The multipath detection part proposed in the MDP algorithm also exploits the SNR value, which is not a specific indicator for multipath but for noise in general. The aim of checking the SNR mask as well is to identify outliers that can be caused by multipath or other external interference or can refer to corrupted data. More specifically, for the SNR parameter, a static threshold is set before running the positioning procedure. The value to be assigned to the SNR threshold depends on several factors, such as the quality of the

GNSS antenna and the environmental conditions of the survey. The optimal value to assign to the threshold is still an object of research. In this work, it is determined experimentally by trying different configurations, as explained in Section 5. Epoch by epoch, for each acquired observable, the SNR value is compared to the chosen threshold: if the SNR value is lower than the threshold, the observable is flagged with a so-called SNR flag.

4.2.2. MDP Threshold

The MDP is a parameter specifically designed to identify data affected by multipath. Under normal conditions, (i.e., absence of multipath), the variable MDP (measured in metres) is expected to have a white noise trend, and its values depend on the entity of residual errors due to noise. In case of incoming multipath, the MDP variable is expected to have some outliers. The aim of the MDP threshold is to identify those outliers, as they indicate multipath-affected data. Two different MDP thresholds were introduced in this work: a static one and a dynamic one. The static MDP threshold is used in a similar way to the SNR threshold. The optimal MDP threshold value is still an object of research, and in this work, it is determined experimentally by trying different configurations, as discussed in Section 5. Once the MDP static threshold is set, epoch by epoch, for each acquired observable, the MDP value is computed and compared to the threshold. The observable is then flagged with a so-called MDP flag if:

$$|MDP| \geq mdpthreshold \tag{6}$$

Concerning the MDP, an adaptive threshold was also proposed. This adaptive threshold is computed considering a statistical analysis on previous epochs. The heuristic is based on an additional parameter, N, that defines the length of the observation window (number of epochs to be analysed). As the aim of the MDP threshold is to detect outliers, Chebyshev's inequality is used for its definition. In fact, Chebyshev's inequality states that, considering a broad range of probability distributions, 88.89% of values lies within three standard deviations of the mean [32]. The MDP adaptive threshold is then defined as follows:

$$mdpthreshold = \mu \pm 3\sigma \tag{7}$$

where

- μ is the mean MDP value on N previous epochs;
- σ is the standard deviation of the N previous MDP values.

In this case, the observable is flagged if:

$$MDP \leq \mu - 3\sigma \vee MDP \geq \mu + 3\sigma \tag{8}$$

4.2.3. Initialisation Phase

The detection part of the algorithm, when the MDP dynamic threshold is chosen, requires an initialisation time equal to N times the acquisition rate. Therefore, for example, if the acquisition rate is 1 Hz and N is equal to 60, the detection algorithm needs 1 min to reach a complete operating status.

4.2.4. Detection Phase

Based on the current values of the MDP and SNR flags, the detection algorithm implements three different criteria to state if the observable is affected by multipath. The considered criteria are the following:

- Criterion 1: Only MDP flag;
- Criterion 2: MDP flag and SNR flag.

For each epoch, using this procedure, a set of observables potentially affected by multipath is identified. The detection phase is used then to trigger the mitigation procedure.

4.3. MDP: Mitigation Algorithm

The aim of the mitigation algorithm is to retrieve a more accurate and precise GNSS positioning, taking into account multipath-affected observables. One possible way to mitigate the multipath effect is to exclude the multipath-affected observables. This heuristic runs the risk of excluding too many GNSS observables, hence producing a poor-quality result due to missing redundancies in the observables. In order to avoid this scenario, a good way to proceed is to consider different associated weights to the multipath-affected GNSS observables.

Since the MDP is integrated in RTKLIB, the new proposed weight to be associated with the multipath-affected observables is based on the weight that the software associates with the observables. As previously described, RTKLIB uses a weight matrix defined as follows:

$$W = diag\left(\sigma_1^{-2}, \sigma_2^{-2}, \ldots, \sigma_n^{-2}\right) \tag{9}$$

In Equation (9), σ_n^2 is the variance associated to the n observable defined as:

$$\sigma_{meas}^2 = F^s R_r \left(a_\sigma^2 + \frac{b_\sigma^2}{sinEL_r^s}\right) \tag{10}$$

where

- F^s is the satellite system error factor, which is equal to 1 for GPS, Galileo, QZSS; equal to 1.5 for BeiDou; and equal to 3 for GLONASS;
- R_r is the code/carrier-phase error ratio;
- a_σ, b_σ are the carrier-phase error factors a and b in metres.

Furthermore, the software adds to this variance other contributions, so the final equation becomes:

$$\sigma_{obs}^2 = \sigma_{meas}^2 + \sigma_{eph}^2 + \sigma_{ion}^2 + \sigma_{trop}^2 + \sigma_{bias}^2 \tag{11}$$

where

- σ_{eph} is the standard deviation of ephemeris and clock error in metres;
- σ_{ion} is the standard deviation of ionosphere correction model error in metres;
- σ_{trop} is the standard deviation of troposphere correction model error in metres;
- σ_{bias} is the standard deviation of code bias error in metres.

In order to take into account the multipath effect, the variance of the observables identified as potentially affected by multipath is incremented, adding a new component. This component is based on the sigma-ϵ model [33,34] used to weight GNSS observables by means of their SNR value. A new term has been added to this model in order to also take into account the multipath effect by means of the MDP value. The resulting term, called MDP variance, is expressed by:

$$\sigma_{mdp}^2 = mdp^2 + C \cdot 10^{-\frac{SNR}{10}} \tag{12}$$

where C is a model parameter equal to 0.244 m^2 DB-Hz for the L1 frequency. Formula (11) is then revised as follows:

$$\sigma_{obs}^2 = \sigma_{meas}^2 + \sigma_{eph}^2 + \sigma_{ion}^2 + \sigma_{trop}^2 + \sigma_{bias}^2 + \sigma_{mdp}^2 \tag{13}$$

The proposed weight model is defined in such a way as to amplify the impact of the multipath effect (MDP variable) with respect to external noise (SNR). Since the weight associated to the observables is equal to the inverse of its variance (see Equation (9)), the weight of the observable decreases as its MDP value increase. Furthermore, using the proposed MDP variance, the multipath-affected observables are weighted in a different way depending on their MDP values: observables affected by a large multipath error (high MDP value) will have a lower weight with respect to observables with a lower multipath error (i.e., lower MDP value). Furthermore, the MDP variance takes into account the SNR value as well. As the SNR value decreases (i.e., the observable noise increases), the MDP variance associated to the variable increases, and consequently, its weight decreases. An experimental validation of the proposed algorithm will be presented in the next section.

5. Experimental Results

In this section, the effects on GNSS positioning by smartphones introduced by the application of the MDP algorithm are discussed. In particular, the performance of the algorithm, in terms of multipath detection and mitigation, under varying configuration parameters is detailed. For this purpose, both static and kinematic data sets are analysed. The selected case studies are used to test the main features of the proposed architecture. They are both processed using the LIGE base station, developed for this work. They require the use of our Android app installed on a device and the modified RTKLIB version for data processing. Although the proposed architecture has been designed for real-time applications, in order to evaluate the performance of the MDP algorithm, several post-processing elaborations were computed, all having the same input data and therefore being under the same contextual conditions.

The MDP algorithm present several input parameters to tune the computation phase. First of all, in the detection step, one out of three criteria must be selected. Moreover, the strategy for the MDP threshold computation must be chosen between the static and the adaptive. The value chosen as static threshold remains constant for the entire computation. For the adaptive threshold value, the number of previous epochs should be set. The MDP threshold will then depend on a statistical analysis computed on the previous MDP values following Equation (8). Finally, the SNR threshold must be set.

The values to be associated with the various parameters to achieve optimal results may depend on various factors (e.g., environmental conditions and hardware characteristics of the receiver) and are still an object of research. For the two case studies, some PPK processing was performed considering different combinations of the parameters, as explained later. Concerning the rest of the processing options used in RTKLIB for PPK positioning, the ionosphere and troposhere models used are the Klobuchar and Saastamoinen, respectively. Only the broadcast ephemeris were considered, and an elevation mask of 15° is also set. The solution type selected is "forward" for both case studies.

5.1. Static Test

The static data set refers to the acquisition described in Section 2. The MDP trend is presented for both the receivers used, with the purpose of understanding the MDP performance in detecting multipath errors for smartphone receivers with respect to a classic GNSS receiver. Figure 9 shows the MDP values detected by the Xiaomi Mi8 and the u-blox ZED F9P.

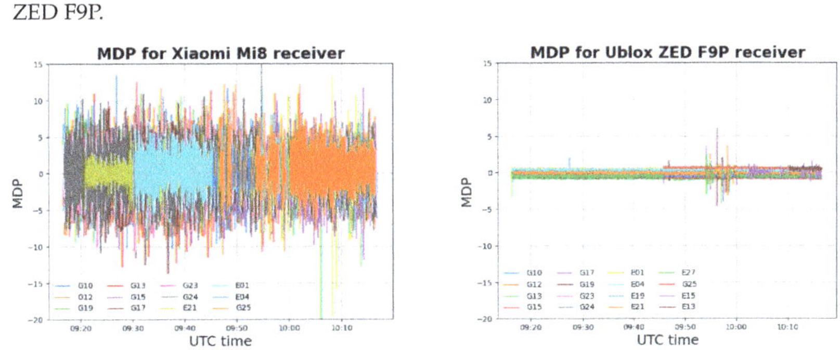

(a) (b)

Figure 9. MDP values for Xiaomi Mi8 receiver (**a**) and for Ublox ZED F9P receiver (**b**) for the static data set.

The MDP variable is representative of the multipath effect and residual errors due to external noise (see Equation (5)). Figure 9 shows that the MDP variable for the u-blox ZED F9P receiver has some peaks between the 9:45 and 10:00 UTC, that is, when the multipath effect was induced, as discussed in Section 2. The MDP variable can then be considered a reasonable multipath indicator for this kind of receiver. For the Xiaomi Mi8 receiver, the MDP variable presents a noisy trend with higher values in modulus. Differently from the u-blox ZED F9P, the Xiaomi Mi8 MDP trend has more diffuse peaks and is not particularly concentrated between 9:45 and 10:00 UTC. Nevertheless, a couple of peaks can be noted for the Xiaomi Mi8 in that interval. The noisier trend of the Xiaomi Mi8 MDP variable with respect to the u-blox ZED F9P can be explained by higher values of the external noise caused by the poor quality of the smartphone's antenna. This makes the MDP variable less effective in multipath identification for smartphones receivers.

For evaluating the performance of the MDP algorithm for the smartphone receiver, processing was computed under changing configuration parameters. Data were processed using all three criteria proposed for multipath detection. For every criterion, both static and adaptive MDP thresholds were used, as specified below.

- Concerning the static threshold, six values were selected on the basis of the MDP trend (Figure 9a). The chosen values of the MDP static threshold are 2.5, 3.0, 3.5, 4.0, 4.5, and 5 m. These values have been selected after several tests, as discussed in [27], and indications from the literature.
- For the adaptive threshold, six different windows (number of epochs) are selected: 10, 20, 30, 40, 50, and 60. The window size times the acquisition rate determines an initialisation time in which the algorithm cannot be applied. Considering that all smartphones have an acquisition rate of 1 Hz, an initialisation time no longer than 1 min (i.e., 60 epochs) was considered in the experiments.
- Finally, considering the SNR trend for the Xiaomi Mi8 receiver (Figure 6a), four values of the SNR threshold are considered: 20, 25, 30, and 35 DB-Hz. Again, these values have been selected after several tests, as discussed in [27], and indications from the literature.

Combining all these option values, we considered 108 different experimental tests. Hereafter, the most interesting results are reported. The interested reader may refer to Lorenzo Benvenuto's PhD thesis for a complete description of the experiments [27].

Considering criterion 1, the solutions obtained for the static MDP threshold values, compared with the solution without the MDP algorithm application, are shown in Figure 10.

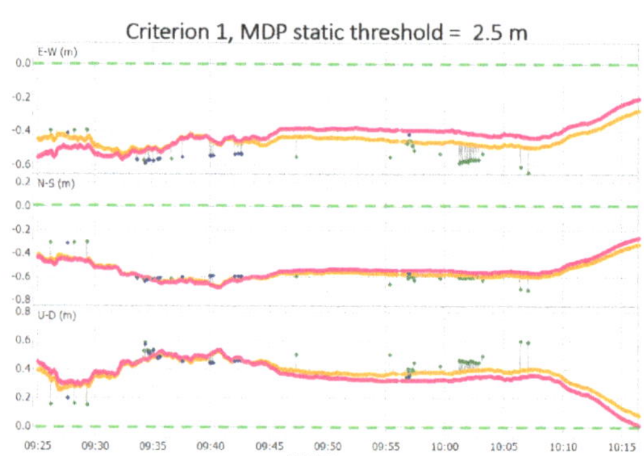

Figure 10. Results obtained after MDP algorithm application using criterion 1 and static MDP threshold equal to 2.5 m.

In Figure 10, the solutions obtained after the MDP algorithm is applied are depicted in pink and blue for float and fixed solutions, respectively, while the solution obtained without the application of MDP algorithm is depicted in yellow and green for float and fixed solutions, respectively. In the figure, the precise coordinates of the point are also highlighted with the dashed green line. The data set presents a convergence time of about 5 min, which is not considered for the results analysis.

The solution, after the application of the MDP algorithm, has slight improvements in terms of accuracy, especially in the case of MDP threshold equal to 2.5. Considering the planimetric accuracy, an improvement of 5 cm is obtained. Over all the threshold values considered for this criterion, this is the best result reached. The worst result obtained is the one with an MDP threshold value of 5.0 m (see [27]).

As additional experiment, the data set has been processed considering the adaptive MDP threshold with a time window containing 30 epochs. This window size has been selected after additional experiments (from 10 to 60 epochs) discussed in [27]. The positioning results concerning the processing executed with criterion 1 and the adaptive MDP threshold compared with the solution without the MDP algorithm application are reported in Figure 11.

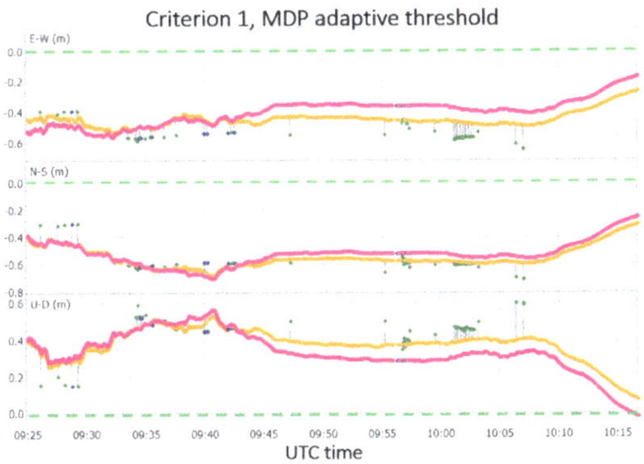

Figure 11. Results obtained after the MDP algorithm application using criterion 1 and adaptive MDP threshold.

The solution obtained with the application of the MDP algorithm is in pink and blue, and the solution obtained without its application is in yellow and green; the precise coordinates of the point are represented by the dashed green line. In this case, improvements are present in the solution accuracy not only with respect to the case without the MDP algorithm application, but also with respect to the case with the static MDP threshold equal to 2.5 m. The difference in terms of RMS for the three cases is reported in Table 3.

Table 3. Comparison between RMS obtained after MDP application, with criterion 1.

Processing	RMS E (m)	RMS N (m)	RMS H (m)	RMS 2D (m)
No MDP	0.450	0.554	0.387	1.428
MDP, crit 1 static thres.	0.427	0.538	0.370	1.374
MDP, crit 1 adaptive thres.	0.410	0.530	0.353	1.339

The results shown in Table 3 can be explained by the fact that the adaptive threshold is more selective than the static threshold in the identification of data potentially affected by multipath. When dealing with very noisy GNSS observables, such as those produced by smartphones, the usage of a static MDP threshold may lead to consideration of any data that are not affected by multipath as outliers in the MDP trend. Reducing the weights of those data still leads to some benefits in terms of positioning accuracy, but those benefits are lower with respect to the ones obtained if only multipath-affected observables (i.e., outliers in the MDP trend) will be assigned a reduced weight. Concerning the precision of the solutions, appreciable differences are noted in the STD of the solutions, so it can be stated that the MDP algorithm does not lead to improvement in positioning precision.

From Figures 10 and 11, it can be seen that the best improvements are obtained during the period of the induced multipath (i.e., from 9:45 to 10:00 UTC). The solution obtained for both adaptive and static threshold in that time interval are depicted in Figure 12. For the rest of the period, the solutions with and without the application of the MDP algorithm are quite similar.

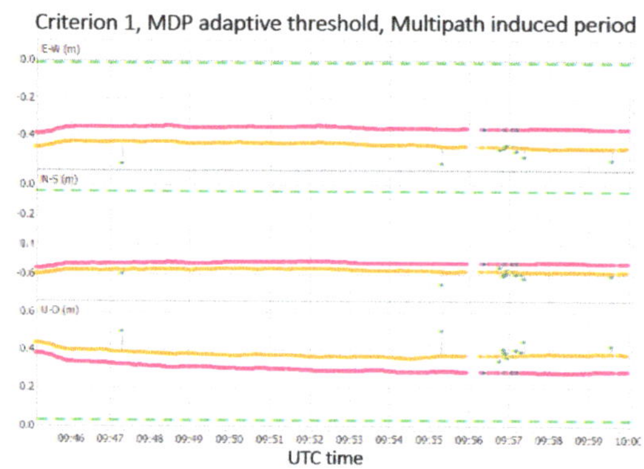

(**a**)

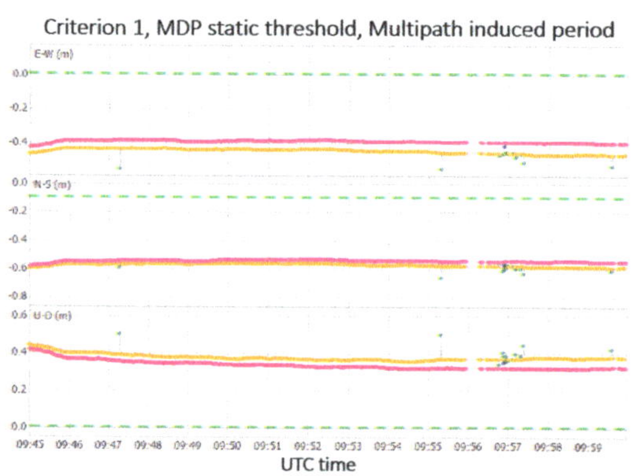

(**b**)

Figure 12. Results obtained after application of the MDP algorithm using criterion 1 with an adaptive MDP threshold (**a**) and static MDP threshold (**b**), referring to the multipath-induced interval.

In Figure 12, the solution obtained with the MDP algorithm presents an unstable set of fixed solutions. However, fixed solutions obtained without the MDP algorithm often turned out to be false solutions. Indeed, their RMS is in the order of tens of centimetres, while, for fixed solutions, the expected RMS should be in the order of few centimetres. In this sense, an unstable set of results that is still consistent with the rest of the obtained positions seems to be preferable to false fixed solutions. Therefore, the MDP heuristics seems to increase the robustness of the positioning process in the selected test battery.

The differences in terms of RMS for the adaptive threshold case and static one are reported in Table 4.

Similarly to the whole test period, in this time interval, the solution accuracy particularly increases when the adaptive threshold is considered. Considering that the multipath error was manually induced in this time interval, the MDP algorithm seems to be effective in mitigating this type of error.

In criterion 2, the observable is considered to be affected by multipath if its SNR value is lower than its threshold, and at the same time, the MDP value in modulus is higher than the current MDP threshold. Considering the static MDP threshold, the major improvements are obtained for MDP threshold equal to 2.5 m, so the impact of the SNR value is discussed only for this situation. The statistics in terms of the RMS referred to in our experiments are reported in Table 5.

Table 4. Comparison between RMS obtained after MDP application, criterion 1, during multipath-induced interval.

Processing	RMS E (m)	RMS N (m)	RMS H (m)	RMS 2D (m)
No MDP	0.444	0.567	0.384	1.441
MDP, crit 1 static thres.	0.387	0.5393	0.344	1.328
MDP, crit 1 adaptive thres.	0.356	0.519	0.309	1.259

Table 5. Comparison between RMS obtained after MDP application, criterion 2.

Processing	RMS E (m)	RMS N (m)	RMS H (m)	RMS 2D (m)
No MDP	0.450	0.554	0.387	1.441
MDP, crit 2, MDP thr = 2.5 m, SNR thr = 20 DB-Hz.	0.448	0.551	0.384	1.424
MDP, crit 2, MDP thr = 2.5 m, SNR thr = 35 DB-Hz.	0.397	0.509	0.333	1.250

In the case of the SNR threshold set to 20 DB-Hz, very low improvements can be noted in term of accuracy with respect to the solution without the MDP algorithm. This means that very few observations have the condition on MDP and SNR simultaneosly. Compared to the results in Table 3, the introduction of a very low SNR threshold worsens the performance of the algorithm in improving the accuracy of the solution. In addition, no significant improvements in terms of solution robustness can be observed in this case. Considering the case with SNR threshold equal to 35 DB-Hz instead, some improvements in terms of accuracy can be observed. Furthermore, if an SNR threshold equal to 35 DB-Hz is considered, the number of false fixed solutions is reduced with respect to the "no MDP" case. This means that the solution robustness is increased.

Some additional tests for criterion 2 considered the adaptive MDP threshold. In these cases, no significant differences were obtained with respect to the results for criterion 1. Indeed, as previously noted, the MDP adaptive threshold, differently from the static one, only selects a few outliers in the MDP trend. The SNR threshold instead selects many pieces of data, especially if its value is set to 35 DB-Hz. Considering that for criterion 2, the MDP and SNR condition must be verified at the same time, it is reasonable to deduce that the discarded data in this case are the same as those selected via criterion 1 where the SNR is not checked for the detection part of the algorithm.

Summarizing, it is possible to state that, considering criterion 2, the usage of the SNR values for the detection phase of the MDP algorithm has advantages for the mitigation performance of the algorithm itself, if a static MDP threshold is considered. Those benefits regard the position accuracy and robustness, and they are more evident for increasing values for the SNR threshold. No differences are introduced by the usage of SNR in the detection part if the MDP adaptive threshold is selected. No appreciable differences can be observed, for all the processing executed for this criterion, in terms of STD.

5.2. Kinematic Test

The kinematic data set refers to the pedestrian acquisition described in Section 2. As for the static case study, the MDP trend is presented for both receivers, with the purpose of understanding the MDP performance of detecting multipath for smartphone receivers compared to more traditional GNSS receivers.

Similarly to what observed for the static case study, the MDP variable for the smartphone receiver (Figure 13a) is noisier compared to of a classic GNSS receiver. The chosen values for the MDP static threshold are: 2.5, 3.0, 3.5, 4.0, 4.5, and 5 m. For the adaptive threshold, considering the results obtained for the static case study, only 30 previous epochs were considered for the threshold computation. Finally, considering the SNR trend for the Xiaomi Mi8 receiver (Figure 3a), four values of the SNR threshold were considered: 20, 25, 30, and 35 DB-Hz. Combining all these option values, 63 different processing results were obtained. Hereafter, the obtained results are reported.

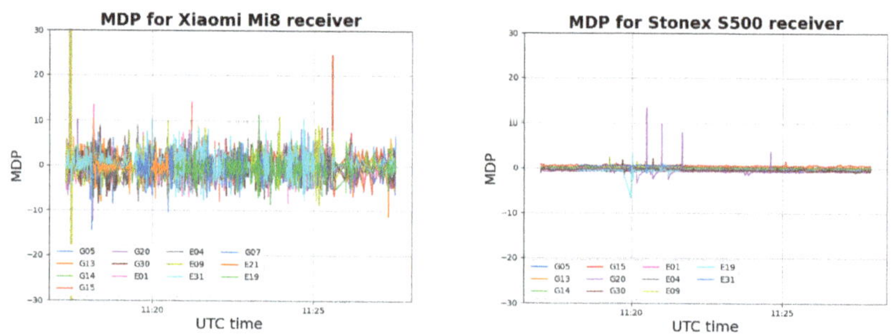

(a) **(b)**

Figure 13. MDP values for Xiaomi Mi8 receiver (**a**) and for Ublox ZED F9P receiver (**b**) for a kinematic data set.

The best results for criterion 1 were obtained considering the MDP threshold equal to 2.5 m. The RMS results for the static part are reported in Table 6. In order to justify the better improvements obtained with the MDP threshold equal to 2.5 m with respect to the other tested values, the worst result obtained (i.e., MDP threshold equal to 5.0 m) is also reported in the table.

Table 6. RMS obtained for criterion 1, varying static MDP threshold values.

Processing	RMS E (m)	RMS N (m)	RMS H (m)	RMS 2D (m)
No MDP	0.731	2.075	4.444	4.340
MDP, crit 1, MDP thr = 2.5 m	0.935	1.638	4.063	3.773
MDP, crit 1, MDP thr = 5.0 m	0.696	1.944	3.635	4.129

As shown in Table 6, some improvements in terms of accuracy are obtained after the MDP algorithm application. The MDP algorithm seems particularly effective in improving the accuracy of the solution when the MDP threshold equal to 2.5 m is considered.

Concerning the kinematic part, the application of the MDP algorithm produces some local benefits in the solution with the elimination of some outliers observed in the "NO MDP" solution. This means that in those intervals, the robustness of the solution is increased. Nevertheless, the MDP algorithm seems to introduce some outliers that were not observed in the "NO MDP" solution due to the fact that some data not affected by multipath were under-weighted by the algorithm. Using the static MDP threshold for multipath identification seems to work well if the receiver is static, but it seems to be not so effective if the receiver is moving.

Criterion 1 coupled with the adaptive MDP threshold was also tested for this case study. Unlike the results obtained for the static MDP threshold, the solution accuracy in this case is degraded if the static part of the data set is considered, as shown in Table 7.

Table 7. RMS obtained for criterion 1, adaptive MDP threshold.

Processing	RMS E (m)	RMS N (m)	RMS H (m)	RMS 2D (m)
No MDP	0.731	2.075	4.444	4.340
MDP, crit 1, adaptive MDP thr	0.867	2.845	3.521	5.949

This result highlights that for this case study, the adaptive SNR threshold seems to be less effective in multipath recognition. Nevertheless, a significant improvement in the position accuracy can be noted for the East and North components after 11:18 UTC. After this instant, the planimetric accuracy is improved by about 11 cm, as the 2D RMS passes from 3.642 m to 3.528 m. This means that the bad performance of the MDP multipath detection is limited to the first few epochs of the data set.

Concerning the kinematic part of the data set, a similar behaviour to the one observed for the static MDP threshold is obtained. Nevertheless, in this case, some improvements are shown. The adaptive MDP threshold solution does not in fact have some of the outliers that the static MDP solution presents, especially for the Height component. This means that if the criterion 1 is considered for kinematic data sets, the usage of the adaptive MDP threshold seems preferable to the usage of a static MDP threshold [27].

Hereafter, the results for criterion 2 are exposed with particular attention to the role played by the SNR threshold in the detection part of the MDP algorithm. Among all the MDP static threshold values considered, the best results were obtained for the threshold value equal to 2.5 m, which is consistent with the results observed for criterion 1. Varying the SNR threshold value, it was noted that the solution accuracy, at least for the static part, improves as the threshold value increases. Further processing was carried out by raising the SNR threshold value from 35 to 40 DB-Hz, but no significant differences were noted, as shown in Table 8.

Table 8. RMS obtained for criterion 2, MDP static threshold.

Processing	RMS E (m)	RMS N (m)	RMS H (m)	RMS 2D (m)
No MDP	0.731	2.075	4.444	4.340
MDP, crit 2, MDP thr = 2.5 m, SNR thr = 35 DB-Hz	0.542	1.722	3.929	3.611

From the SNR value reported, we observe that the accuracy of the solution in this interval is improved by about 19 cm, 35 cm, and 52 cm for the East, North, and Height components, respectively. The planimetric accuracy of the solution is improved by 73 cm. It is also interesting to note the difference in RMS between this solution and the one obtained for criterion 1 (for comparison, see Table 6), that is, without the usage of SNR for the multipath detection part. Concerning the kinematic part, no substantial differences are observed with respect to the criterion 1 case. It can be stated then that the solution obtained for criterion 2 is better in terms of accuracy compared to the one obtained for criterion 1, in the case of an MDP static threshold, especially for the static part of the test. Therefore, in static condition, considering both the SNR and MDP values for the detection part of the MDP algorithm increases the performance of the algorithm itself. The situation is different for the kinematic part, in which no differences are noted. In this case, it seems that using the SNR value does not provide any advantage to the detection part of the MDP algorithm.

The test with criterion 2 also involved the usage of the adaptive MDP threshold. In this cases, varying the SNR threshold value, no significant differences were reported with respect to the results already obtained for criterion 1, so the results are not shown. It can then be stated that for criterion 2, if the adaptive MDP threshold is considered, the usage of the SNR value does not improve the performance of the detection part of the MDP algorithm.

The kinematic part of this data set shows behaviour analogous to that obtained for the processing carried out for criteria 1 and 2. No further improvements are observed for this case study, and the same considerations, already discussed for the other cases, are valid also for this obtained result.

6. Conclusions and Perspectives

The obtained results are very encouraging and promising. In particular, they confirm that the proposed solution is capable of increasing both accuracy and robustness in RTK positioning from Android devices. Nevertheless, the validation process of this solution shall continue applying the MDP algorithm in several other tests, considering a wider range of boundary conditions.

The MDP algorithm detection capabilities need to be studied in deeper detail, especially for Android GNSS receivers that present very noisy observables. Regarding this topic, in this work, both a static and adaptive MDP threshold were tested. Concerning the static MDP threshold, it was observed that the performance of the algorithm improves as the threshold value decreases. The optimal threshold value found for both the data sets is 2.5 m. Regarding the multipath effect identified by outliers in the MDP trend (Figures 9a and 13a), and also considering the high noise of the smartphone's GNSS observables (which makes the MDP variable higher in modulus), the observables with MDP values lower than 2.5 m cannot be considered affected by multipath. The adaptive MDP threshold was also tested, showing interesting results. In this case, each observable has its own threshold value, and multipath identification seems to be more effective. Lastly, the effects of SNR were tested for the detection part of the MDP algorithm combining SNR and MDP thresholds by means of a different strategy, that is, criterion 2. With criterion 2, the usage of SNR provides benefits to the detection capabilities of the algorithm. The solution accuracy and robustness are increased, especially when high values of SNR threshold (i.e., 35 DB-Hz) are considered. A less interesting criterion, with MDP or SNR, has been considered in [27].

Among the future developments of this work, it is worth mentioning a deeper investigation of the adaptive version of the MDP algorithm and, in particular, dealing with epochs with several missing observations for some specific satellites.

Another future development of this work is related to GNSS and INS (inertial navigation system) integration using data from the inertial sensor embedded in the smartphone. Integrating the positioning derived from the application of the MDP algorithm with inertial data through a loosely coupled approach should further increase the robustness of the resulting solution.

Author Contributions: Conceptualization, data interpretation, and analysis, writing original draft preparation, all authors; tests conceptualization and realization, L.B.; IoT system implementation, L.B.; writing, review and editing, G.D., L.B. All authors have read and agreed to the published version of the manuscript.

Funding: This research was funded by Regione Liguria on the POR-FSE 2014-2020 programme (Programma Operativo Regione Liguria, Fondo Sociale Europeo).

Institutional Review Board Statement: Not applicable.

Informed Consent Statement: Not applicable.

Data Availability Statement: Data used in the experiments will be made available upon request made to the authors.

Conflicts of Interest: The authors declare no conflict of interest.

References

1. Agency, E.G. *Using GNSS Raw Measurements on Android Devices*; Publications Office of the European Union: Luxembourg, 2017.
2. Hofmann-Wellenhof, B.; Lichtenegger, H.; Wasle, E. *GNSS — Global Navigation Satellite Systems*; Springer: Vienna, Austria, 2008.
3. Teunissen, P.J.; Kleusberg, A. *GPS for Geodesy*; Springer: Berlin/Heidelberg, Germany, 1998.
4. Gunter, S. *Satellite Geodesy*, 2nd ed.; Walter de Gruyter: Berlin, Germany, 2003.

5. Zangenehnejad, F.; Gao, Y. GNSS smartphones positioning: Advances, challenges, opportunities, and future perspectives. *Satell. Navig.* **2021**, *2*, 24. [CrossRef] [PubMed]
6. Masiero, A.; Guarnieri, A.; Pirotti, F.; Vettore, A. A Particle Filter for Smartphone-Based Indoor Pedestrian Navigation. *Micromachines* **2014**, *5*, 1012–1033. [CrossRef]
7. Wang, L.; Groves, P.D.; Ziebart, M.K. Smartphone Shadow Matching for Better Cross-street GNSS Positioning in Urban Environments. *J. Navig.* **2015**, *68*, 411–433. [CrossRef]
8. Wang, L.; Li, Z.; Zhao, J.; Zhou, K.; Wang, Z.; Yuan, H. Smart Device-Supported BDS/GNSS Real-Time Kinematic Positioning for Sub-Meter-Level Accuracy in Urban Location-Based Services. *Sensors* **2016**, *16*, 2201. [CrossRef] [PubMed]
9. Wang, L.; Li, Z.; Yuan, H.; Zhao, J.; Zhou, K.; Yuan, C. Influence of the time-delay of correction for BDS and GPS combined real-time differential positioning. *Electron. Lett.* **2016**, *52*, 1063–1065. [CrossRef]
10. Adjrad, M.; Groves, P.D. Intelligent Urban Positioning: Integration of Shadow Matching with 3D-Mapping-Aided GNSS Ranging. *J. Navig.* **2018**, *71*, 1–20. [CrossRef]
11. Fissore, F.; Masiero, A.; Piragnolo, M.; Pirotti, F.; Guarnieri, A.; Vettore, A. Towards Surveying with a Smartphone. In *New Advanced GNSS and 3D Spatial Techniques*; Cefalo, R., Zieli, J.B., Barbarella, M., Eds.; Springer International Publishing: Cham, Switzerland, 2018; pp. 167–176.
12. Presti, L.; Al-Azizi, J.I.; Shafri, H.Z.M. Performance Evaluation of Pedestrian Locations Based on Contemporary Smartphones. *Int. J. Navig. Obs.* **2017**, *2017*, 6750346. [CrossRef]
13. Angrisano, A.; Gaglione, S. Smartphone GNSS Performance in an Urban Scenario with RAIM Application. *Sensors* **2022**, *22*, 786. [CrossRef]
14. Gogoi, N.; Minetto, A.; Linty, N.; Dovis, F. A Controlled-Environment Quality Assessment of Android GNSS Raw Measurements. *Electronics* **2019**, *8*, 5. [CrossRef]
15. Wanke, L.; Xiang, S.; Feng, Z.; Xianlu, T.; Fuhong, W. Quality analysis of multi-GNSS raw observations and a velocity-aided positioning approach based on smartphones. *Adv. Space Res.* **2019**, *63*, 2358–2377. [CrossRef]
16. Realini, E.; Caldera, S.; Pertusini, L.; Sampietro, D. Precise GNSS Positioning Using Smart Devices. *Sensors* **2017**, *17*, 2434. [CrossRef] [PubMed]
17. Dabove, P.; Di Pietra, V. Single-Baseline RTK Positioning Using Dual-Frequency GNSS Receivers Inside Smartphones. *Sensors* **2019**, *19*, 4302. [CrossRef] [PubMed]
18. Pirazzi, G.; Mazzoni, A.; Biagi, L.; Crespi, M. Preliminary Performance Analysis with a GPS + Galileo Enabled Chipset Embedded in a Smartphone. In Proceedings of the 30th International Technical Meeting of the Satellite Division of THE Institute of Navigation (ION GNSS+ 2017), Portland, OR, USA, 25–29 September 2017; pp. 101–115. [CrossRef]
19. Gill, M.; Bisnath, S.; Aggrey, J.; Seepersad, G. Precise Point Positioning (PPP) using Low-Cost and Ultra-Low-Cost GNSS Receivers. In Proceedings of the 30th International Technical Meeting of the Satellite Division of THE Institute of Navigation (ION GNSS+ 2017), Portland, OR, USA, 25–29 September 2017.
20. Laurichesse, D.; Rouch, C.; Marmet, F.X.; Pascaud, M. Smartphone Applications for Precise Point Positioning. In Proceedings of the 30th International Technical Meeting of the Satellite Division of THE Institute of Navigation (ION GNSS+ 2017), Portland, OR, USA, 25–29 September 2017; pp. 171–187.
21. Wang, L.; Li, Z.; Wang, N.; Wang, Z. Real-time GNSS precise point positioning for low-cost smart devices. *GPS Solut.* **2021**, *25*, 69. [CrossRef]
22. Robustelli, U.; Baiocchi, V.; Pugliano, G. Assessment of Dual Frequency GNSS Observations from a Xiaomi Mi 8 Android Smartphone and Positioning Performance Analysis. *Electronics* **2019**, *8*, 91. [CrossRef]
23. Hakansson, M. Characterization of GNSS observations from a Nexus 9 Android tablet. *GPS Solut.* **2019**, *23*, 21. [CrossRef]
24. Benvenuto, L.; Dabove, P.; Ferrando, I.; Sguerso, D. Preliminary Results on Tropospheric ZTD Estimation by Smartphone. *Remote. Sens.* **2021**, *13*, 4567. [CrossRef]
25. Darugna, F. Improving Smartphone-Based GNSS PositioningUsing State Space Augmentation Techniques. Ph.D. Thesis, Leibniz University of Hannover, Hannover, Germany, 2021.
26. Takasu, T.; Yasuda, A. Development of the low-cost RTK-GPS receiver with an open source program package RTKLIB. In *International Symposium on GPS/GNSS*; International Convention Center: Jeju, Korea, 2009.
27. Benvenuto, L. A Solution for Improving Robustness of GNSS Positioning from Android Devices. Ph.D. Thesis, University of Genoa, Genoa, Italy, 2022.
28. Klobuchar, J.A. Ionospheric Time-Delay Algorithm for Single-Frequency GPS Users. *IEEE Trans. Aerosp. Electron. Syst.* **1987**, *AES-23*, 325–331. [CrossRef]
29. Saastamoinen, J. Atmospheric Correction for the Troposphere and Stratosphere in Radio Ranging Satellites. In *The Use of Artificial Satellites for Geodesy*; American Geophysical Union: Washington, DC, USA, 1972; pp. 247–251. [CrossRef]
30. Bhuiyan, M.Z.H.; Lohan, E.S. Advanced Multipath Mitigation Techniques for Satellite-Based Positioning Applications. *Int. J. Navig. Obs.* **2010**, *2010*, 412393. [CrossRef]
31. Bisnath, S.; Langley, R. Pseudorange Multipath Mitigation By Means of Multipath Monitoring and De-Weighting. In Proceedings of the KIS 2001, Banff, AB, Canada, 5–8 June 2001; pp. 1–9.
32. Kvanli, A.H.; Pavur, R.J.; Keeling, K.B. *Concise Managerial Statistics*; South-Western/Thomson Learning: Mason, OH, USA, 2006.

33. Brunner, F.; Hartinger, H.; Troyer, L. GPS signal diffraction modelling: The stochastic SIGMA-Δ model. *J. Geod.* **1999**, *73*, 259–267. [CrossRef]
34. Wieser, A.; Brunner, F. An extended weight model for GPS phase observations. *Earth Planets Space* **2000**, *52*, 777–782. [CrossRef]

sensors

MDPI

Article

A Combined Elevation Angle and C/N0 Weighting Method for GNSS PPP on Xiaomi MI8 Smartphones

Yanjie Li, Changsheng Cai * and Zhenyu Xu

School of Geosciences and Info-Physics, Central South University, Changsha 410083, China; lee_yanjie@163.com (Y.L.); northernbaldy@163.com (Z.X.)
* Correspondence: csucai@csu.edu.cn

Abstract: Traditionally, an elevation-angle-dependent weighting method is usually used for Global Navigation Satellite System (GNSS) positioning with a geodetic receiver. As smartphones adopt linearly polarized antenna and low-cost GNSS chips, different GNSS observation properties are exhibited. As a result, a carrier-to-noise ratio (C/N0)-dependent weighting method is mostly used for smartphone-based GNSS positioning. However, the C/N0 is subject to the effects of the observation environment, resulting in an unstable observation weight. In this study, we propose a combined elevation angle and C/N0 weighting method for smartphone-based GNSS precise point positioning (PPP) by normalizing the C/N0-derived variances to the scale of the elevation-angle-derived variances. The proposed weighting method is validated in two kinematic PPP tests with different satellite visibility conditions. Compared with the elevation-angle-only and C/N0-only weighting methods, the combined weighting method can effectively enhance the smartphone-based PPP accuracy in a three-dimensional position by 22.7% and 24.2% in an open-sky area, and by 52.0% and 26.0% in a constrained visibility area, respectively.

Keywords: GNSS; smartphone; precise point positioning; elevation angle; C/N0

Citation: Li, Y.; Cai, C.; Xu, Z. A Combined Elevation Angle and C/N0 Weighting Method for GNSS PPP on Xiaomi MI8 Smartphones. *Sensors* **2022**, *22*, 2804. https://doi.org/10.3390/s22072804

Academic Editors: Sandra Verhagen and Robert Odolinski

Received: 19 February 2022
Accepted: 31 March 2022
Published: 6 April 2022

1. Introduction

The Global Navigation Satellite System (GNSS)-based navigation and positioning applications on smartphones have greatly aided our lives. The demand for precise positioning services on smartphones is increasing daily and attracting great attention in the GNSS community. For the GNSS positioning, its stochastic model is vital to determine the positioning accuracy when estimating the position parameters [1]. For geodetic-type GNSS receivers, an elevation-angle-dependent weighting method is usually used due to a strong correlation between the elevation angle and observation noise [2]. However, for smartphone-based GNSS receivers, due to hardware condition limitations and the complexity of the application environment, their observations exhibit different properties, such as large observation noise, drastic C/N0 variations and being prone to suffer from the multipath effect and even observation outages [3–5]. The observation noise and multipath effect on smartphones are more correlative to the carrier-to-noise ratio (C/N0) than the satellite elevation angle [6–8]. Consequently, the C/N0-dependent weighting method is mostly used in the smartphone-based GNSS positioning.

Several scholars evaluated the precise point positioning (PPP) performance on smartphones using different weighting methods. Chen et al. [9] used a Xiaomi MI8 smartphone to perform single-frequency PPP in a static mode based on an elevation-angle-dependent stochastic model, and the results showed that the average root mean square (RMS) of positioning errors can reach 0.81 m and 1.65 m in the horizontal and vertical directions, respectively. Furthermore, Wu et al. [10] used an elevation-angle-dependent stochastic model for static dual-frequency PPP on the Xiaomi MI8 smartphone and achieved a converged position accuracy of 0.22 m, 0.04 m and 0.11 m in the east, north and up directions, respectively. Nevertheless, the positioning accuracy decreased to 3–4 m in the kinematic mode due to insufficient dual-frequency observations. In addition, using the Xiaomi MI8 smartphone, Shinghal and Bisnath [11] utilized the C/N0-dependent stochastic model to improve the static PPP three-dimensional (3D) accuracy by about 27.0% over the elevation-angle-dependent stochastic model. Zhu et al. [12] used a Huawei Mate 30 smartphone to test the PPP in a kinematic mode based on the C/N0-dependent stochastic model to achieve a positioning accuracy of 0.93 m, 0.62 m and 2.17 m in the east, north and up directions, respectively; the accuracy was improved by about 26.2%, 20.5% and 20.4% when compared with the elevation-angle-dependent stochastic model. From the existing research, the advantage of the C/N0-dependent stochastic model is obvious for smartphone-based GNSS positioning. Although the C/N0-dependent weighting methods are mostly used in the smartphone-based GNSS positioning, the C/N0 is subject to the effects of the observation environment and severely fluctuates in the dynamic process. As a result, the C/N0-derived observation weights are prone to instability. Recently, a combined elevation angle and C/N0 weighting method was applied to smartphone based kinematic PPP by simply dividing the C/N0-derived variances by the elevation-angle-derived variances. A sub-meter positioning accuracy can be achieved based on the combined weighting method [13]. However, its performance was not compared with the other weighting methods and thus its advantage is unclear.

In this study, a new combined elevation angle and C/N0 weighting method is proposed for smartphone-based PPP by normalizing the C/N0-derived variances to the scale of the elevation-angle-derived variances. The performance of the proposed weighting method is compared with the elevation-angle-only and C/N0-only weighting methods in two kinematic smartphone-based PPP tests. The paper is outlined as follows: firstly, the correlations among the code multipath and noise (CMN), C/N0 and elevation angle are analyzed; secondly, a combined elevation angle and C/N0 weighting method for the smartphone-based PPP is presented; and finally, the proposed weighting method is evaluated in two kinematic PPP experiments.

2. Correlation of CMN, Elevation Angle and C/N0

To analyze the correlation of the CMN with the elevation angle and C/N0, a Xiaomi MI8 smartphone equipped with a BCM47755 GNSS chip and a linearly polarized antenna is used as the experimental device. The Xiaomi MI8 smartphone is the first-released dual-frequency Android smartphone. Figure 1 shows the GNSS observation on the roof of the Mining Building at Central South University, China on 15 November 2020. The smartphone application of GEO++ RINEX Logger 2.1.3 is used for the data collection [14]. The observation lasts about 5 h from GPS Time 9:00 to 14:00 with a sampling rate of 1 HZ. The Xiaomi MI8 smartphone can receive dual-frequency GPS L1/L5 signals and Galileo E1/E5a signals, as well as single-frequency GLONASS G1 and BDS B1 signals.

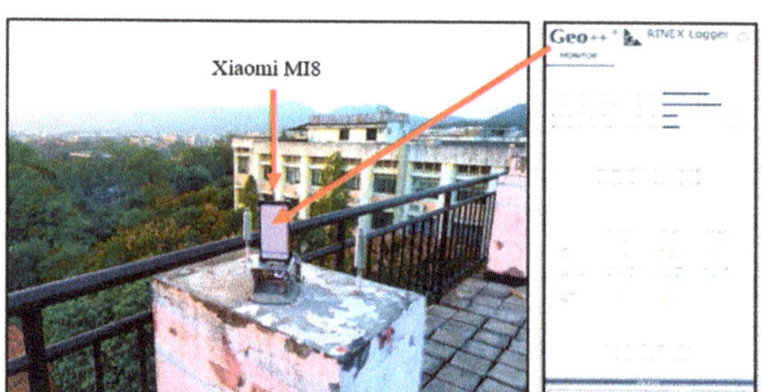

Figure 1. Smartphone data collection in an open sky area on 15 November 2020.

The CMN is a major error source on smartphones due to the built-in linear polarization antenna [3,4]. The multipath effect at an epoch (M_i) can be estimated using the multipath combination below [15,16]:

$$M_i = P_i - \frac{f_i^2 + f_j^2}{f_i^2 - f_j^2}\varphi_i\lambda_i + \frac{2f_j^2}{f_i^2 - f_j^2}\varphi_j\lambda_j \tag{1}$$

where i and j ($i \neq j$) denote two different frequencies. P is the pseudorange observation. φ is the carrier phase observation. λ is the wavelength at the corresponding frequency, f. M_i contains multipath effect, code noise, ambiguity term and hardware delay biases. The latter two items are stable and thus can be obtained by calculating the mean value of M_i at a certain number of epochs free of cycle slips [17], which is denoted as $\overline{M}_i$. Therefore, the CMN can be derived as:

$$CMN_i = M_i - \overline{M}_i \tag{2}$$

The correlation of the CMN to the satellite elevation angle and C/N0 was examined; Figure 2 shows the time series of the CMN, elevation angle and C/N0 for the GPS G26 and Galileo E27 satellites at both the L1/E1 and L5/E5a frequencies. In terms of the CMN amplitudes, the G26 satellite has a stronger CMN than the E27 satellite, and both satellites at the L1/E1 frequencies have a stronger CMN than the L5/E5a frequencies. These results are in line with the conclusions drawn by [3], which indicate that the Galileo satellites have a better CMN suppression ability than the GPS satellites and that the L5/E5a signals have a stronger CMN suppression ability than the L1/E1 signals. In addition, the CMN has a strong negative correlation with the C/N0 at both the L1/E1 and L5/E5a frequencies. When the CMN dramatically changes, the corresponding C/N0 vastly varies by over 10 dB-Hz. In contrast, the correlation with the elevation angle is relatively poor at the L1/E1 frequencies. In spite of this, their correlation is still obvious. When the elevation angle decreases, the CMN increases. However, at the L5/E5a frequencies, no obvious correlation is found between the CMN and the elevation angle, which might be caused by the irregular gain pattern of smartphones [18].

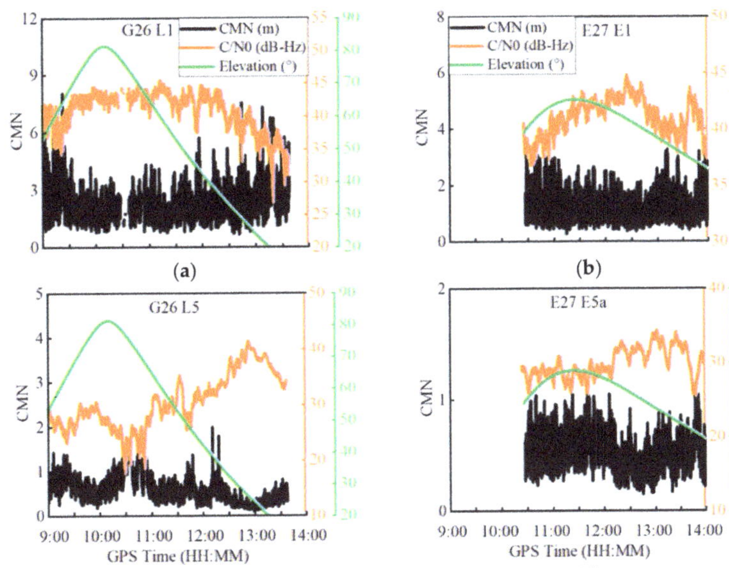

Figure 2. Time series of the code multipath and noise (CMN), carrier-to-noise ratio (C/N0) and elevation angle for G26 satellite at L1/L5 frequencies and E27 satellite at E1/E5a frequencies.

In order to further analyze the correlation of the CMN to the satellite elevation angle and C/N0 at the L1/E1 frequencies, the CMN for all dual-frequency GPS and Galileo satellites is plotted against the elevation angle and the C/N0, respectively, as shown in Figure 3. A second-order polynomial fitting curve is also plotted to reflect the variation trend in panels (a) and (b). The fitting curves illustrate that the CMNs have a negative correlation with the elevation angle and the C/N0. Their correlation coefficients are −0.20 and −0.37, respectively.

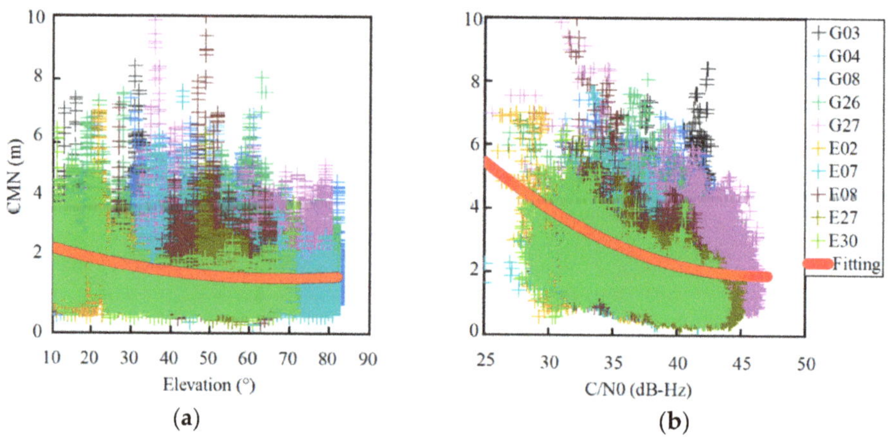

Figure 3. CMN of dual-frequency GPS and Galileo satellites against elevation angles (**a**) and C/N0 (**b**) at L1/E1 frequencies.

Figure 4 shows the CMN mean value statistics against the elevation angle per five degrees and against the C/N0 per 2 dB-Hz. In summary, it is obvious that the CMNs have a negative correlation with the elevation angle and C/N0. In contrast, their correlation to the C/N0 is stronger than the elevation angle.

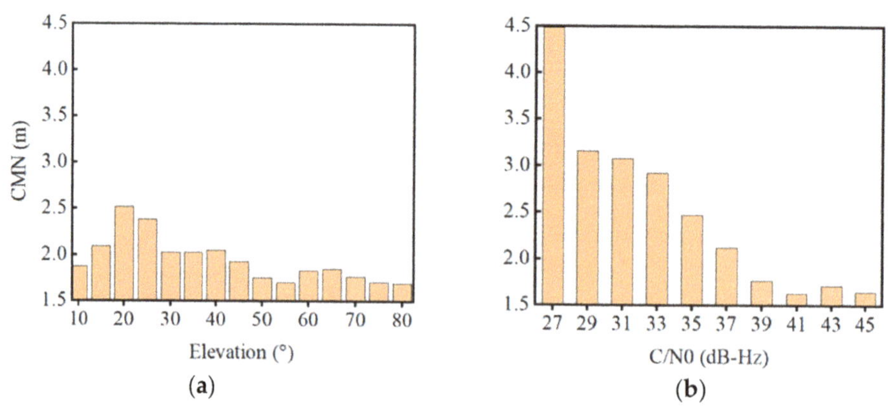

Figure 4. CMN mean value statistics against the elevation angle (**a**) and C/N0 (**b**).

Figure 5 shows the relationship between the C/N0 and elevation angle for quad-constellation GNSS at L1/G1/B1/E1 frequencies. A second-order polynomial fitting curve is employed to reflect the correlation between the C/N0 and elevation angle. As can be seen, their correlation is obvious since the C/N0 increases as the elevation angle increases.

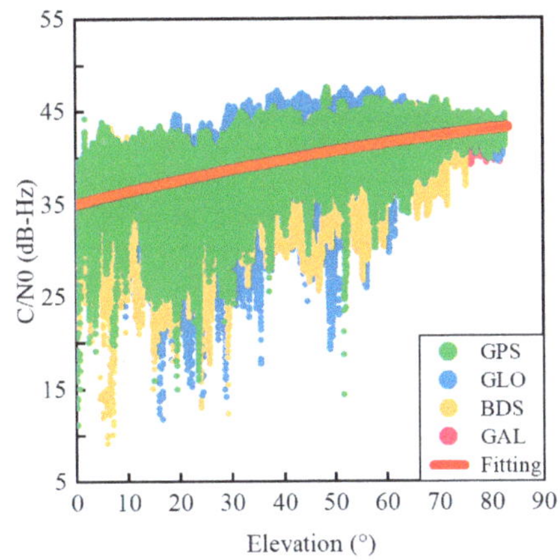

Figure 5. C/N0 of quad-constellation GNSS against elevation angles at L1/G1/B1/E1 frequencies.

3. Combined Elevation Angle and C/N0 Weighting Method

Generally, the observation weight matrix W can be depicted as [19]:

$$W = \text{diag}\left(\sigma_1^{-2}, \sigma_2^{-2}, \ldots, \sigma_m^{-2}\right) \tag{3}$$

where m is the number of observations; σ^2 is the observation variance, which can be expressed as:

$$\sigma^2 = \sigma_{obs}^2 + \sigma_{eph}^2 + \sigma_{ion}^2 + \sigma_{trop}^2 \tag{4}$$

where σ_{obs}^2 is the receiver-related variance term and σ_{eph}^2, σ_{ion}^2 and σ_{trop}^2 are the variances of the satellite orbit and clock error, ionosphere correction model error and troposphere correction model error, respectively. In this study, we mainly focus on the effect of the receiver-related variance term on the observation weight for smartphone-based GNSS positioning. Therefore, the following mentioned observation variance only refers to the receiver-related variance term σ_{obs}^2.

The traditional observation weighting methods for smartphone-based GNSS positioning mainly rely on the elevation angle or C/N0 [20]. The commonly used elevation-angle-dependent weighting model is depicted as [21,22]:

$$\sigma_{ele}^2 = \sigma_0^2 / \sin^2(ele) \tag{5}$$

where σ_{ele}^2 is the observation variance; σ_0^2 is a constant variance value; sin is the sine function and ele is the satellite elevation angle.

A commonly used C/N0 weighting model is expressed as [23,24]:

$$\sigma_{C/N0}^2 = \sigma_0^2 \times 10^{\frac{max(\mathrm{MAX} - C/N0,\, 0)}{10}} \tag{6}$$

where $\sigma_{C/N0}^2$ is the observation variance; max is the maximum function and MAX is a preset maximum C/N0.

To jointly use the satellite elevation angle and C/N0 information to determine the observation weight, we propose a combined elevation angle and C/N0 weighting method by normalizing the C/N0-derived variances to the scale of the elevation-angle-derived variances. The combined weighting method is established following the steps below.

Firstly, the correlation between the satellite elevation angle and the C/N0 at the L1/G1/B1/E1 frequencies is obvious, as analyzed in Section 2. Thus, a second-order polynomial function is employed to fit the C/N0 against the satellite elevation angle, as depicted below:

$$C/N0\_cal = a + b \times ele + c \times ele^2 \quad (7)$$

where $C/N0\_cal$ is the fitting C/N0 value and a, b and c are the fitting coefficients. According to the correlation between the satellite elevation angle and the C/N0 shown in Figure 5, the fitting coefficients of a, b and c can be taken as 35.0833 dB-Hz, 0.1365 dB-Hz/° and −0.0005 dB-Hz/(°)^2, respectively and may be used as empirical values.

Secondly, the C/N0-derived variances are converted to the scale of the elevation-angle-derived variances to unify the variance scale of the two indicators. For the L1/G1/B1/E1 signals, the elevation-angle-derived variances are used as the main part to determine the observation weight, while the C/N0-derived variances, after subtracting the fitting-C/N0-derived variances, are used as a supplement for observation weight determination, as depicted in Equation (8). Such a weighting method can respond to the instantaneous variation in observation accuracy and simultaneously avoid obtaining unstable observation weights. For the L5/E5a signals, since the correlation between the CMN and the elevation angle is not obvious, as analyzed in Section 2, only the C/N0-derived variances are used to determine the observation weight. To keep consistent with the observation variances at the L1/G1/B1/E1 frequencies, the C/N0-derived variances are also normalized to the scale of the elevation-angle-derived variances:

$$\begin{cases} \sigma_{L1}^2 = \left|\sigma_{C/N0,\ L1}^2 - \sigma_{C/N0\_cal}^2\right| \times \frac{\sigma_{\mathrm{ele\_MAX}}^2 - \sigma_{\mathrm{ele\_MIN}}^2}{\sigma_{\mathrm{C/N0\_MAX,\ L1}}^2 - \sigma_{\mathrm{C/N0\_MIN,\ L1}}^2} + \sigma_{ele}^2 \\ \sigma_{L5}^2 = \sigma_{C/N0,\ L5}^2 \times \frac{\sigma_{\mathrm{ele\_MAX}}^2 - \sigma_{\mathrm{ele\_MIN}}^2}{\sigma_{\mathrm{C/N0\_MAX,\ L5}}^2 - \sigma_{\mathrm{C/N0\_MIN,\ L5}}^2} \end{cases} \quad (8)$$

where σ_{L1}^2 and σ_{L5}^2 are the observation variances at the L1/G1/B1/E1 frequencies and the L5/E5a frequencies, respectively; $C/N0$ and $C/N0\_cal$ are the measured C/N0 and fitted C/N0, respectively; C/N0_MIN and C/N0_MAX are the minimum and maximum C/N0 values, which are set to 25 dB-Hz and 45 dB-Hz at the L1/G1/B1/E1 frequencies, and 20 dB-Hz and 40 dB-Hz at the L5/E5a frequencies, respectively and ele_MIN and ele_MAX are the minimum and maximum elevation angles, which are set to 10° and 90°, respectively.

The observation weight value is an inverse ratio of the observation variance. Once the observation variance is determined following Equation (8), the observation weight can be acquired.

4. Experimental Results and Discussion

In this section, the quad-constellation GPS/GLONASS/BDS/Galileo PPP processing strategy is provided in detail. Then, two kinematic smartphone-based GNSS PPP experiments are conducted to evaluate the combined elevation angle and C/N0 weighting method with comparisons to the conventional elevation-angle-only and C/N0-only weighting methods.

4.1. Quad-Constellation PPP Processing Strategy

An undifferenced and uncombined observation model is adopted for the smartphone-based quad-constellation PPP [25,26] to validate the proposed weighting method. The 2-Day predicted Global Ionospheric Map (GIM) products from the Chinese Academy of Sciences (CAS) are used as pseudo-observables to reduce the effect of ionospheric delay errors on the single-frequency observations [27,28]. The satellite PCO (Phase Center Offset) and PCV (Phase Center Variation) are corrected using the International GNSS Service (IGS) products, whereas the smartphone PCO is corrected using a recommended value from the reference [29]. The detailed PPP processing strategy is shown in Table 1.

Table 1. PPP processing strategy.

Items	Processing Strategies
Estimation method	Kalman filter
Constellations	GPS (L1, L5)/GLONASS (G1)/BDS (B1)/Galileo (E1, E5a)
Weighting scheme	Combined elevation angle and C/N0 weighting method
Satellite orbit and clock	Real-time precise satellite orbit and clock products from Centre National d'Etudes Spatiales (CNES) [30]
Ionospheric delay	Estimated as random walk process and GIM products are used as pseudo-observations
Tropospheric delay	Hydrostatic delay uses Saastamoinen model correction [31], and zenith wet delay is estimated as random walk noise process
Receiver position	Estimated as random walk process
Receiver clock offset	Estimated as white noise
Inter-system bias	Estimated as white noise
Ambiguities	Estimated as constants

The carrier phase observation precision is set to 0.01 m, whereas the pseudorange observation precision is set to 3 m [23]. In the quad-constellation PPP, the observation precision for different constellations and frequencies is different. We employ a four-order differential method [8,32] to analyze the observation precision for different constellations and frequencies. On the basis of a comparative analysis, the initial pseudorange observation weight ratio is set to 5:1:7:7 for GPS/GLONASS/BDS/Galileo at the L1/G1/B1/E1 frequencies and 1:1 for GPS/Galileo at the L5/E5a frequencies. Since the carrier phase observation precision is similar, an identical weight strategy is used for the quad-constellation carrier phase observations at both frequencies. For observations at the L1/E1 and L5/E5a frequencies, their weight ratio is set to 1:2 for pseudorange and 2:1 for carrier phase.

4.2. Kinematic Experiment in an Open Sky Environment

A kinematic GNSS experiment was conducted using the Xiaomi MI8 smartphone on an open playground of Central South University, China on 2 December 2021. Its trajectory and experiment equipment are shown in Figure 6. A Trimble NetR9 GNSS receiver with a geodetic antenna is employed to acquire the precise kinematic position for use as a benchmark value. The horizontal distance between the geodetic-type antenna and the smartphone is less than one decimeter. The experimental platform was lifted by a pedestrian over the head. A base station is set up at about one kilometer away from the rover station. Thus, the position of the geodetic receiver at the rover station can be precisely determined by the real-time kinematic (RTK) technique. The entire data collection lasts about 24 min, with a sampling rate of 1 HZ. The elevation mask is set to 10 degrees.

Figure 7 shows the total number of satellites received, the total number of satellites used and the number of satellites for different constellations of GPS, GLONASS, BDS and Galileo in the PPP. The number of GPS and BDS satellites observed is obviously more than GLONASS and Galileo. At the same time, the total number of satellites received is 23.8 on average, which is apparently less than the Trimble NetR9 receiver at an average number of 36.0. In contrast, the total number of satellites used for PPP on the Xiaomi MI8 smartphone is only 17.5 on average with an STD of 2.8 due to a lack of parts of observations.

Taking the GPS G01 satellite at the L5 frequency and the GLONASS R09 satellite at the G1 frequency as examples, Figure 8 depicts the variations in elevation angle and C/N0 in the time domain, which shows that the elevation angle steadily changes for both satellites. In contrast, the C/N0 severely fluctuates with a range exceeding 10 dB-Hz for both satellites, which suggests that the C/N0 is prone to be affected by the outer environment due to a polarized antenna embedded inside the smartphone. Figure 9 shows the observation variances derived from Equations (5), (6) and (8) for the GPS G01 at the L5 frequency and the GLONASS R09 at the G1 frequency. The observation variances are displayed in the top, middle and bottom panels, derived from three different weighting scenarios, namely elevation-angle-only, C/N0-only and combined elevation angle and C/N0. As can be seen, the C/N0-derived observation variances severely vary, whereas the observation variances derived from the combined elevation angle and C/N0 are more stable.

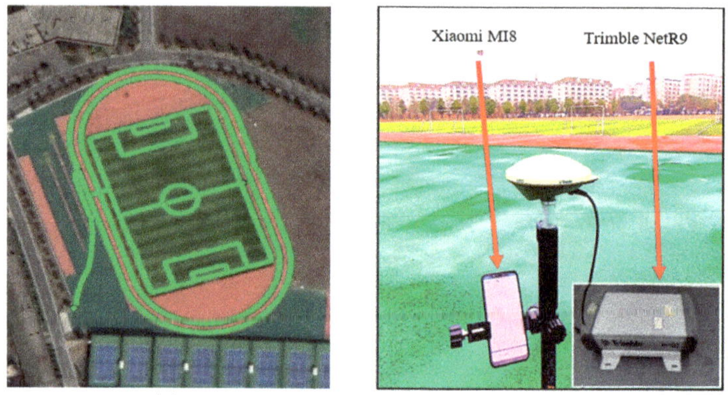

Figure 6. Kinematic experimental trajectory (**a**) and equipment setup (**b**) on an open-sky playground.

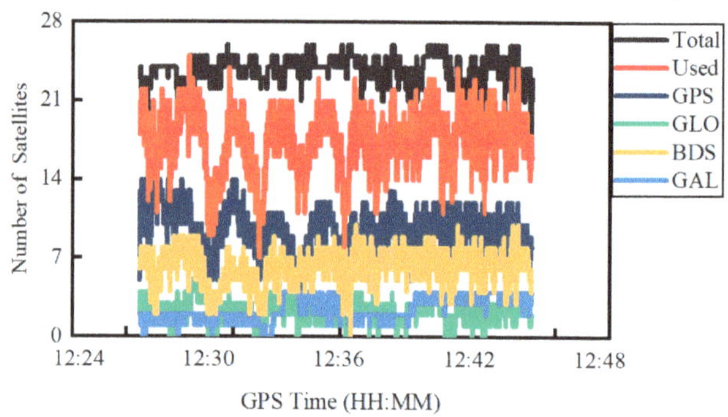

Figure 7. Number of satellites for quad-constellations on 2 December 2021.

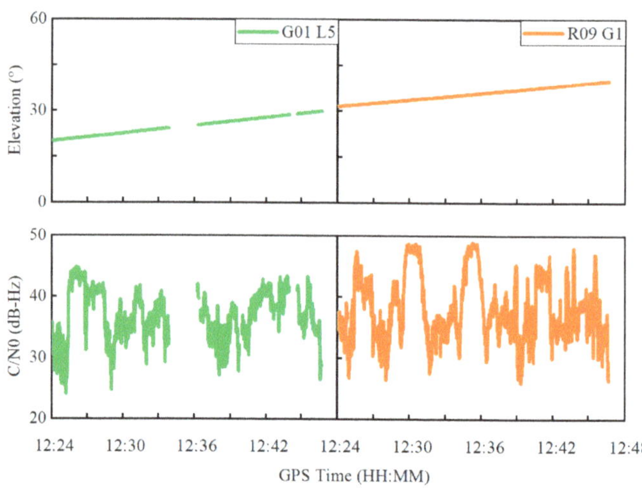

Figure 8. Elevation angle and C/N0 variations for GPS G01 satellite at L5 frequency and GLONASS R09 satellite at G1 frequency.

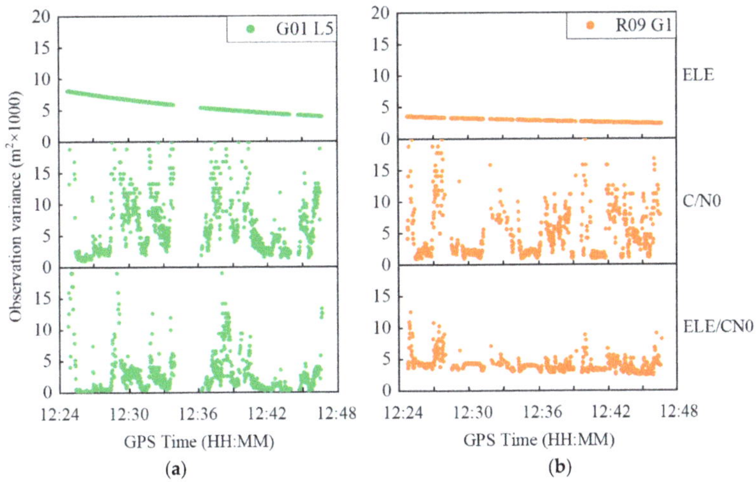

Figure 9. Carrier phase observation variances for GPS G01 satellite at L5 frequency (**a**) and GLONASS R09 satellite at G1 frequency (**b**). ELE, C/N0 and ELE/CN0 represent weighting scenarios of elevation-angle-only, C/N0-only and combined elevation angle and C/N0, respectively.

Figure 10 shows the positioning errors in the east, north and up components using the three weighting scenarios. It can be seen that the positioning error curve derived from the combined weighting method is closer to the zero axis, especially in the east direction. The RMS statistics of the positioning errors are provided in Table 2 in the east, north and up components as well as the three-dimensional (3D) direction. Based on the combined weighting scenario, the RMS values of the position errors are 0.55 m, 0.73 m and 1.35 m in the east, north and up components, respectively, which improve by about 25.7%, 2.7% and 26.2% over the elevation-angle-only weighting scenario, and by about 53.8%, 5.2% and 16.7% over the C/N0-only weighting scenario. For the 3D position, the improvement can reach 22.7% and 24.2% over the elevation-angle-only and C/N0-only weighting scenarios, respectively. The obtained PPP accuracy is comparable to the existing research [12,13].

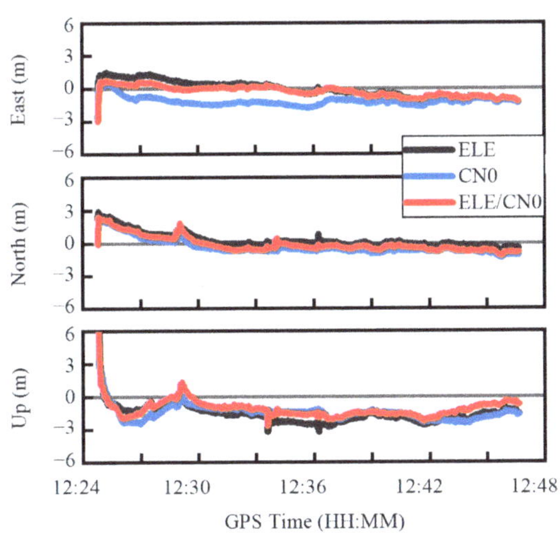

Figure 10. Quad-constellation PPP errors using three different weighting scenarios. ELE, C/N0 and ELE/CN0 represent weighting scenarios of elevation-angle-only, C/N0-only and combined elevation angle and C/N0, respectively.

Table 2. RMS Statistics of the PPP errors in the east, north and up directions.

	ELE	C/N0	ELE/CN0
East (m)	0.74	1.19	0.55
North (m)	0.75	0.77	0.73
Up (m)	1.83	1.62	1.35
3D (m)	2.11	2.15	1.63

4.3. Kinematic Experiment in Constrained Visibility Environment

A kinematic GNSS experiment was conducted using the Xiaomi MI8 smartphone in a constrained satellite visibility environment on 12 October 2021, on the new campus of Central South University, China. The satellite visibility is easily affected by the buildings and trees along its trajectory, as shown in Figure 11. The equipment setup is similar to the previous experiment shown in Figure 5. The experiment equipment was carried by an electric bicycle. The entire data collection lasts about half an hour with a sampling rate of 1 HZ. The elevation mask is set to 10 degrees.

Figure 11. Kinematic experimental trajectory in a constrained satellite visibility environment.

Figure 12 shows the total number of satellites received, total number of satellites used and the number of satellites for different constellations of GPS, GLONASS, BDS and Galileo in the PPP. Similar to the previous experiment, the number of GPS and BDS satellites observed is more than the GLONASS and Galileo satellites. The total number of satellites received for the Trimble NetR9 receiver is less than that in the open area at an average of about six satellites due to the constrained satellite visibility environment. In contrast, the number of satellites used on the Xiaomi MI8 smartphone in the constrained environment is larger than that in the open area by an average of about two satellites, which indicates that the number of GNSS satellites observed on smartphones is unstable in different sessions. Furthermore, due to the observation environment influence, the number of satellites used in PPP significantly fluctuates.

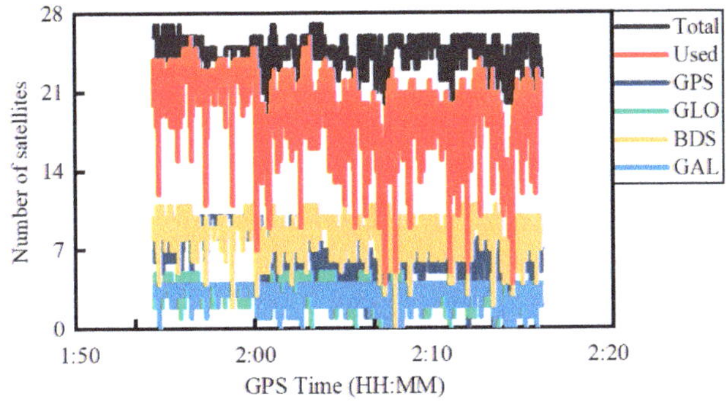

Figure 12. Number of satellites for quad-constellation GNSS on 12 October 2021.

Figure 13 shows the observation variances derived from Equations (5), (6) and (8) for the GPS G06 at the L5 frequency and the GLONASS R04 at the G1 frequency. The observation variances are displayed in the top, middle and bottom panels, derived from three different weighting scenarios, namely elevation-angle-only, C/N0-only and combined elevation angle and C/N0. Similar to the previous experiment, the C/N0-derived observation variances severely vary, while the observation variances derived from the combined elevation angle and C/N0 weighting scenarios are more stable.

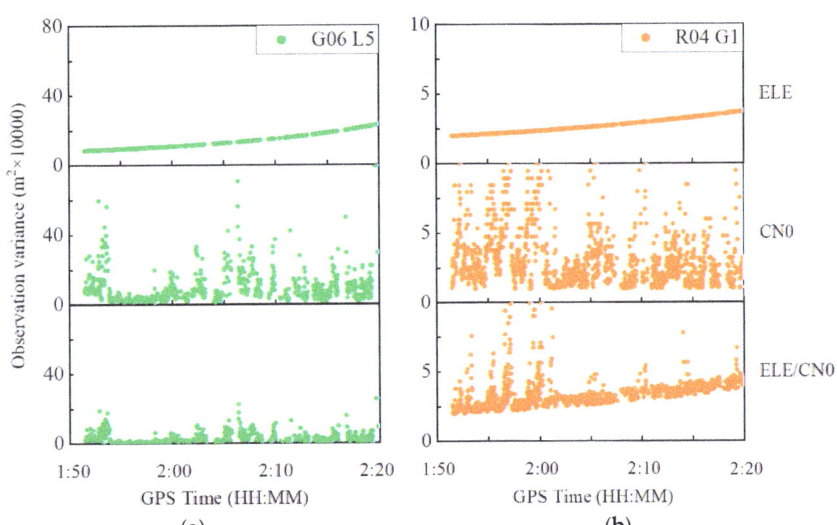

Figure 13. Carrier phase observation variances for GPS G06 satellite at L5 frequency (**a**) and GLONASS R04 satellite at G1 frequency (**b**).

Figure 14 shows the positioning errors in the east, north and up components using the three weighting scenarios. It can be seen that the positioning error curves derived from the combined weighting scenario is much closer to the zero axis, especially in the north and up directions. The RMS statistics of the positioning errors are provided in Table 3. Based on the combined weighting scenario, the RMS values of the PPP errors are 1.00 m, 0.62 m and 2.22 m in the east, north and up components, respectively, which improve by about 19.3%, 12.7% and 55.9% over the elevation-angle-only weighting scenario in the east, north and up directions, respectively and by about 51.6% and 26.7% over the C/N0-only weighting scenario in the north and up components, respectively. For the 3D position, the corresponding improvement can reach 52.0% and 26.0%, respectively. In addition, it is more prone to reconvergence under the constrained satellite visibility condition, which is especially apparent in the up direction. The achieved PPP accuracy is comparable to the existing research [12,13].

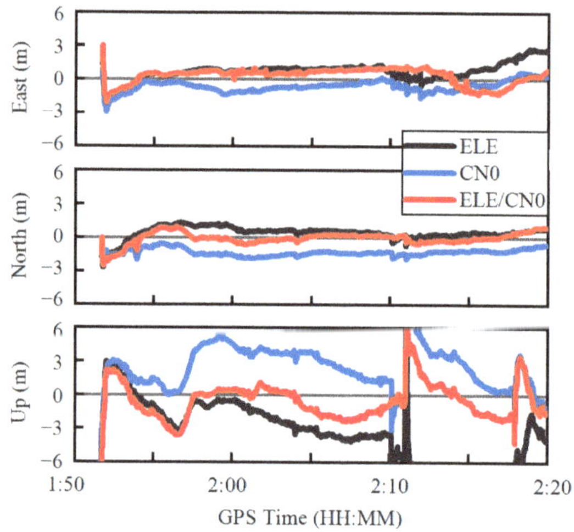

Figure 14. Quad-constellation PPP errors using three different weighting scenarios.

Table 3. RMS statistics of the PPP errors in the east, north and up directions.

	ELE	**CN0**	**ELE/CN0**
East (m)	1.24	0.83	1.00
North (m)	0.71	1.28	0.62
Up (m)	5.03	3.03	2.22
3D (m)	5.23	3.39	2.51

5. Conclusions

The observation weighting method is vital for smartphone-based GNSS PPP. In this study, a combined elevation angle and C/N0 weighting method was proposed by taking advantage of the two indices. The obtained weight values were more stable than the C/N0-only derived weight values. Simultaneously, it also instantaneously responded to the observation accuracy variation. The proposed weighting method was fully evaluated in two GNSS PPP kinematic experiments using a Xiaomi MI8 smartphone.

Two kinematic PPP experiments were carried out in different satellite visibility conditions. Using the combined elevation angle and C/N0 weighting method, the RMS improvement of the 3D position for PPP reaches 22.7% and 24.2% over the elevation-angle-only and C/N0-only weighting methods, respectively in the open sky environment, and 52.0% and 26.0% over the elevation-angle-only and the C/N0-only weighting methods, respectively in the constrained visibility environment. It should be noted that all conclusions were drawn based on the Xiaomi MI8 GNSS kinematic experiments. Further smartphone experiments should be conducted in future research.

Author Contributions: Conceptualization, Y.L. and C.C.; methodology, Y.L.; software, Y.L.; validation, Y.L. and Z.X.; formal analysis, Y.L.; investigation, Y.L. and Z.X.; resources, C.C.; data curation, Y.L. and Z.X.; writing—original draft preparation, Y.L. and C.C.; writing—review and editing, Y.L. and C.C.; visualization, Y.L.; supervision, C.C.; project administration, Y.L. and C.C.; funding acquisition, C.C. All authors have read and agreed to the published version of the manuscript.

Funding: This research was funded by the National Key Research and Development Program of China, grant number 2020YFA0713501; the National Natural Science Foundation of China, grant number 42174040; the Hunan Natural Science Foundation, grant number 2020JJ4111; the Hunan Provincial Innovation Foundation for Postgraduate, grant number CX20200241.

Acknowledgments: The authors wish to acknowledge the Centre National d'Etudes Spatiales (CNES) for providing open access real-time orbit/clock and ionosphere corrections products.

Conflicts of Interest: The authors declare no conflict of interest.

References

1. Li, B.; Shen, Y.; Lou, L. Efficient Estimation of Variance and Covariance Components: A Case Study for GPS Stochastic Model Evaluation. *IEEE Trans. Geosci. Remote Sens.* **2010**, *49*, 203–210. [CrossRef]
2. Li, B.; Lou, L.; Shen, Y. GNSS Elevation-Dependent Stochastic Modeling and Its Impacts on the Statistic Testing. *J. Surv. Eng.* **2016**, *142*, 04015012. [CrossRef]
3. Robustelli, U.; Baiocchi, V.; Pugliano, G. Assessment of Dual Frequency GNSS Observations from a Xiaomi Mi 8 Android Smartphone and Positioning Performance Analysis. *Electronics* **2019**, *8*, 91. [CrossRef]
4. Håkansson, M. Characterization of GNSS Observations from a Nexus 9 Android Tablet. *GPS Solut.* **2018**, *23*, 21. [CrossRef]
5. Filho, V.C.L.; Moraes, A. Modeling Multifrequency GPS Multipath Fading in Land Vehicle Environments. *GPS Solut.* **2020**, *25*, 1–14. [CrossRef]
6. Paziewski, J.; Sieradzki, R.; Baryla, R. Signal Characterization and Assessment of Code GNSS Positioning with Low-Power Consumption Smartphones. *GPS Solut.* **2019**, *23*, 1–12. [CrossRef]
7. Zhang, X.; Tao, X.; Zhu, F.; Shi, X.; Wang, F. Quality Assessment of GNSS Observations from an Android N Smartphone and Positioning Performance Analysis using Time-Differenced Filtering Approach. *GPS Solut.* **2018**, *22*, 70. [CrossRef]
8. Li, G.; Geng, J. Characteristics of Raw Multi-GNSS Measurement Error from Google Android Smart Devices. *GPS Solut.* **2019**, *23*, 90. [CrossRef]
9. Chen, B.; Gao, C.; Liu, Y.; Sun, P. Real-time Precise Point Positioning with a Xiaomi MI 8 Android Smartphone. *Sensors* **2019**, *19*, 2835. [CrossRef]
10. Wu, Q.; Sun, M.; Zhou, C.; Zhang, P. Precise Point Positioning Using Dual-Frequency GNSS Observations on Smartphone. *Sensors* **2019**, *19*, 2189. [CrossRef]
11. Shinghal, G.; Bisnath, S. Conditioning and PPP Processing of Smartphone GNSS Measurements in Realistic Environments. *Satell. Navig.* **2021**, *2*, 1–17. [CrossRef] [PubMed]
12. Zhu, H.; Xia, L.; Wu, D.; Xia, J.; Li, Q. Study on Multi-GNSS Precise Point Positioning Performance with Adverse Effects of Satellite Signals on Android Smartphone. *Sensors* **2020**, *20*, 6447. [CrossRef] [PubMed]
13. Zangenehnejad, F.; Gao, Y. Application of UofC Model Based Multi-GNSS PPP to Smartphones GNSS Positioning. In Proceedings of the 34th International Technical Meeting of the Satellite Division of the Institute of Navigation (ION GNSS+ 2021), St. Louis, MO, USA, 20–24 September 2021; pp. 2986–3003.
14. Geo++ RINEX Logger App. Available online: https://play.google.com/store/apps/details?id=de.geopp.rinexlogger (accessed on 10 January 2020).
15. Estey, L.H.; Meertens, C.M. TEQC: The Multi-Purpose Toolkit for GPS/GLONASS Data. *GPS Solut.* **1999**, *3*, 42–49. [CrossRef]
16. Cai, C.; He, C.; Santerre, R.; Pan, L.; Cui, X.; Zhu, J. A Comparative Analysis of Measurement Noise and Multipath for Four Constellations: GPS, BeiDou, GLONASS and Galileo. *Surv. Rev.* **2016**, *48*, 287–295. [CrossRef]
17. Vaclavovic, P.; Dousa, J. G-Nut/Anubis: Open-Source Tool for Multi-GNSS Data Monitoring with a Multipath Detection for New Signals, Frequencies and Constellations. In *IAG 150 Years*; Springer: Cham, Switzerland, 2015; pp. 775–782. [CrossRef]
18. Gill, M.; Bisnath, S.; Aggrey, J.; Seepersad, G. Precise Point Positioning (PPP) using Low-Cost and Ultra-Low-Cost GNSS Receivers. In Proceedings of the 30th International Technical Meeting of The Satellite Division of the Institute of Navigation (ION GNSS+ 2017), Portland, OR, USA, 25–29 September 2017; pp. 226–236.
19. Takasu, T.; Yasuda, A. Development of the Low-Cost RTK-GPS Receiver with an Open Source Program Package RTKLIB. In Proceedings of the International Symposium on GPS/GNSS, Jeju, Korea, 4–6 November 2009.
20. Zhang, Z.; Li, B.; Shen, Y.; Gao, Y.; Wang, M. Site-Specific Unmodeled Error Mitigation for GNSS Positioning in Urban Environments Using a Real-Time Adaptive Weighting Model. *Remote Sens.* **2018**, *10*, 1157. [CrossRef]
21. King, R.; Bock, Y. *Documentation for the GAMIT GPS Analysis Software*; Massachusetts Institute of Technology: Cambridge, MA, USA, 2000; Available online: http://www-gpsg.mit.edu/~{}simon/gtgk/GAMIT.pdf (accessed on 19 January 2022).
22. Dach, R.; Hugentobler, U.; Fridez, P.; Meindl, M. *Bernese GPS Software, Version 5.0*; Astronomical Institute, University of Bern: Bern, Switzerland, 2007.
23. Wang, L.; Li, Z.; Wang, N.; Wang, Z. Real-time GNSS Precise Point Positioning for Low-Cost Smart Devices. *GPS Solut.* **2021**, *25*, 69. [CrossRef]
24. Carcanague, S. Low Cost GPS/GLONASS Precise Positioning Algorithm in Constrained Environment. Ph.D. Thesis, Institut National Polytechnique de Toulouse-INP, Toulouse, France, 25 February 2014.

25. Li, B.; Zang, N.; Ge, H.; Shen, Y. Single-frequency PPP Models: Analytical and Numerical Comparison. *J. Geod.* **2019**, *93*, 2499–2514. [CrossRef]
26. Zhou, F.; Dong, D.; Li, P.; Li, X.; Schuh, H. Influence of Stochastic Modeling for Inter-System Biases on Multi-GNSS Undifferenced and Uncombined Precise Point Positioning. *GPS Solut.* **2019**, *23*, 59. [CrossRef]
27. Liu, Q.; Hernández-Pajares, M.; Yang, H.; Monte-Moreno, E.; Roma-Dollase, D.; García-Rigo, A.; Li, Z.; Wang, N.; Laurichesse, D.; Blot, A.; et al. The Cooperative IGS RT-GIMs: A Reliable Estimation of the Global Ionospheric Electron Content Distribution in Real Time. *Earth Syst. Sci. Data* **2021**, *13*, 4567–4582. [CrossRef]
28. Wang, G.; Bo, Y.; Yu, Q.; Li, M.; Yin, Z.; Chen, Y. Ionosphere-Constrained Single-Frequency PPP with an Android Smartphone and Assessment of GNSS Observations. *Sensors* **2020**, *20*, 5917. [CrossRef]
29. Netthonglang, C.; Thongtan, T.; Satirapod, C. GNSS Precise Positioning Determinations using Smartphones. In Proceedings of the 2019 IEEE Asia Pacific Conference on Circuits and Systems (APCCAS), Bangkok, Thailand, 11–14 November 2019; pp. 401–404.
30. Kazmierski, K.; Sośnica, K.; Hadas, T. Quality Assessment of Multi-GNSS Orbits and Clocks for Real-Time Precise Point Positioning. *GPS Solut.* **2017**, *22*, 11. [CrossRef]
31. Saastamoinen, J. Atmospheric correction for the troposphere and stratosphere in radio ranging satellites. In *The Use of Artificial Satellites for Geodesy*; Geophysics Monograph Series; American Geophysical Union: Washington, DC, USA, 1972; pp. 247–251.
32. Pirazzi, G.; Mazzoni, A.; Biagi, L.; Crespi, M. Preliminary Performance Analysis with a GPS+Galileo Enabled Chipset Embedded in a Smartphone. In Proceedings of the 30th International Technical Meeting of the Satellite Division of the Institute of Navigation (ION GNSS+ 2017), Portland, Oregon, 25–29 September 2017; pp. 101–115.

sensors

MDPI

Article

Instantaneous Best Integer Equivariant Position Estimation Using Google Pixel 4 Smartphones for Single- and Dual-Frequency, Multi-GNSS Short-Baseline RTK

Chien Zheng Yong [1,2], Ken Harima [3], Eldar Rubinov [4], Simon McClusky [3,5] and Robert Odolinski [1,*]

1 National School of Surveying, University of Otago, 310 Castle Street, Dunedin 9016, New Zealand; chienzheng.yong@utm.my
2 Geomatic Innovation Research Group, Faculty of Built Environment and Surveying, Universiti Teknologi Malaysia, Johor Bahru 81310, Johor, Malaysia
3 Geoscience Australia, GPO Box 378, Canberra, ACT 2601, Australia; ken.harima@ga.gov.au (K.H.); simon.mcclusky@ga.gov.au (S.M.)
4 FrontierSI, Goods Shed, Village Street, Docklands, VIC 3008, Australia; erubinov@frontiersi.com.au
5 Research School of Earth Sciences, Australian National University, Canberra, ACT 2601, Australia
* Correspondence: robert.odolinski@otago.ac.nz

Abstract: High-precision global navigation satellite system (GNSS) positioning and navigation can be achieved with carrier-phase ambiguity resolution when the integer least squares (ILS) success rate (SR) is high. The users typically prefer the float solution under the scenario of having a low SR, and the ILS solution when the SR is high. The best integer equivariant (BIE) estimator is an alternative solution since it minimizes the mean squared errors (MSEs); hence, it will always be superior to both its float and ILS counterparts. There has been a recent development of GNSSs consisting of the Global Positioning System (GPS), Galileo, Quasi-Zenith Satellite System (QZSS), and the BeiDou Navigation Satellite System (BDS), which has made precise positioning with Android smartphones possible. Since smartphone tracking of GNSS signals is generally of poorer quality than with geodetic grade receivers and antennas, the ILS SR is typically less than one, resulting in the BIE estimator being the preferred carrier phase ambiguity resolution option. Therefore, in this contribution, we compare, for the first time, the BIE estimator to the ILS and float contenders while using GNSS data collected by Google Pixel 4 (GP4) smartphones for short-baseline real-time kinematic (RTK) positioning. It is demonstrated that the BIE estimator will always give a better RTK positioning performance than that of the ILS and float solutions while using both single- and dual-frequency smartphone GNSS observations. Lastly, with the same smartphone data, we show that BIE will always be superior to the float and ILS solutions in terms of the MSEs, regardless of whether the SR is at high, medium, or low levels.

Keywords: best integer equivariant (BIE); smartphone positioning; mean squared error (MSE); real-time kinematic (RTK); multi-GNSS

Citation: Yong, C.Z.; Harima, K.; Rubinov, E.; McClusky, S.; Odolinski, R. Instantaneous Best Integer Equivariant Position Estimation Using Google Pixel 4 Smartphones for Single- and Dual-Frequency, Multi-GNSS Short-Baseline RTK. *Sensors* **2022**, *22*, 3772. https://doi.org/10.3390/s22103772

Academic Editors: Ali Khenchaf and Sandra Verhagen

Received: 10 February 2022
Accepted: 9 May 2022
Published: 16 May 2022

1. Introduction

The key to high-precision global navigation satellite system (GNSS) positioning and navigation is carrier-phase integer ambiguity resolution. The ambiguity-fixed GNSS baseline, as obtained by integer least squares (ILS), is commonly expected to be superior to its float counterpart if the integer ambiguity success rate, i.e., the probability of correct integer estimation, is close to the maximum value of one. On the other hand, the float solution usually becomes the positioning preference when the success rate is too low. Alternatively, the best integer equivariant (BIE) estimator, as introduced by Teunissen [1], can be used since it always provides the optimal solution in terms of the mean squared errors (MSEs). Several studies have already investigated the extremely poor multipath suppression and linearly polarized patch antenna in the smartphone that is the foremost hurdle to achieving centimeter-level accurate positioning [2–9]. Since smartphones track GNSS signals with a poorer quality than geodetic-grade receivers and antennas, the ILS SR is typically different from one, and the BIE estimator would then be the preferred option. Laurichesse et al. [10] used undifferenced ambiguity resolution in precise point positioning (PPP) mode with smartphone measurements. Darugna et al. [9] and Warnant et al. [11] compared the positioning performance of different multi-GNSS positioning techniques (i.e., single-point positioning (SPP), differential GPS (DGPS), and real-time kinematic (RTK)) with different smartphone models. A much more recent study from Paziewski et al. [12] assessed the performance of several Android smartphones in relative positioning, whereby centimeter-level precision was achievable with fixed ambiguities. In this contribution, we study, for the first time, the single-baseline RTK positioning performance of the BIE estimator using smartphone GNSS data.

Verhagen and Teunissen [13] proved that this estimator is always optimal in terms of the MSE, while Wen et al. [14] demonstrated the use of the BIE estimator for GNSS precise point positioning (PPP). In Brack et al. [15] and Brack [16], a sequential BIE approach was developed. Subsequently, Teunissen [17] extended the theory of integer equivariant estimation by developing the principle of BIE estimation for the class of elliptically contoured distributions, while Odolinski and Teunissen [18] analyzed the BIE performance for low-cost, single- and dual-frequency, short- to long-baseline multi-GNSS RTK positioning, and they found that the BIE positions reveal a 'star-like' pattern when the ILS SRs are high. Odolinski and Teunissen [19] recently compared the RTK positioning performance of BIE for the multivariate normal and multivariate t-distribution.

The recent development of smartphone GNSS chipsets, such as Broadcom BCM47755 embedded, makes precise code-based positioning [20], PPP [21], and instantaneous, centimeter-level RTK positioning possible with Android-based smartphones [2,3,12,22–24]. In this contribution, we assess the BIE positioning performance using Google Pixel 4 (GP4) smartphones and compare the performance to that of the ILS and float estimators. The performance of the BIE estimator is assessed from the MSE perspective while using multi-GNSS smartphone data for an instantaneous (single-epoch) and single-baseline RTK model, while employing single- and dual-frequency observations.

This contribution is organized as follows: in Section 2, we describe the functional model that is employed for instantaneous single-baseline RTK positioning. In this section, we emphasize the differences among the float, ILS, and BIE estimators. In Section 3, we present the smartphone GNSS data and stochastic model settings. The setup configuration deployed in this study is deemed to be the best configuration by having the smartphones placed in an upright position, as examined and proven in Yong et al. [22]. Then, in Section 4, we provide an analysis of the single- and dual-frequency RTK positioning performance under different model strengths. We further analyze the performance of BIE and compare it to the ILS and float contenders from the MSE perspective. Lastly, a summary with conclusions is given in Section 5.

2. Instantaneous, Single- and Dual-Frequency, Multi-GNSS RTK Using the Float, Integer Least Squares, and Best Integer Equivariant Estimators

In this section, we describe the functional model employed for the single-baseline RTK model while tracking single- and dual-frequency multi-GNSS observations using GP4 smartphones. We also introduce the float, ILS, and BIE estimators.

2.1. Functional Model

We assume that the two GP4 smartphones track GPS, Galileo, QZSS, and BDS code and carrier-phase frequencies on two frequencies. We make use of broadcast ephemerides for satellite orbits and clocks. The relative ionospheric, tropospheric delays and satellite orbit errors can be assumed negligible, since we employ short baselines. The single-epoch (instantaneous) and linearized double-differenced (DD) system of observation equations reads as follows:

$$y = Aa + Bb, \tag{1}$$

where y is the vector of DD carrier-phase and code observations, A is the design matrix of the DD integer ambiguities in vector a, and B corresponds to the design matrix of the real-valued baseline components b. We employ system-specific reference satellites when performing the between-satellite single-differences. We note that, if we would take a common reference satellite on the overlapping frequencies between the systems, it could further strengthen the model [25,26]. For the stochastic model, we use an elevation weighting sine function as employed in RTKlib v2.4.3 [27].

2.2. Float Estimation

To obtain the float solution, denoted with a 'hat' symbol, we estimate the ambiguities and baseline components as real valued parameters in a standard least-squares solution, obtaining

$$[\hat{a},\hat{b}]^T, \begin{bmatrix} Q_{\hat{a}\hat{a}} & Q_{\hat{a}\hat{b}} \\ Q_{\hat{b}\hat{a}} & Q_{\hat{b}\hat{b}} \end{bmatrix}, \tag{2}$$

where $\hat{a}, \hat{b}$ are the vectors of the float ambiguities and baseline components with dimensions n and p, respectively, $Q_{\hat{a}\hat{a}}$, $Q_{\hat{b}\hat{b}}$ denote the corresponding variance covariance (VCV) matrices, and $Q_{\hat{a}\hat{b}} = Q^T_{\hat{b}\hat{a}}$ are the corresponding covariance matrices between the ambiguities and baseline components.

2.3. Integer Least-Squares Estimation

By using the float ambiguities $\hat{a}$ in Equation (2) we can find the integer least-squares solution of the ambiguities, denoted with a 'check' symbol, by solving the following problem:

$$\check{a} = \text{argmin}||\hat{a} - z||^2_{Q_{\hat{a}\hat{a}}}, \tag{3}$$

where $||.||^2_{Q_{\hat{a}\hat{a}}} = (.)^T Q^{-1}_{\hat{a}\hat{a}}(.)$. This ILS problem is efficiently solved using the LAMBDA (least-squares ambiguity decorrelation adjustment) method [28], finally yielding the following fixed baseline solution:

$$\check{b} = \hat{b} - Q_{\hat{b}\hat{a}} Q^{-1}_{\hat{a}\hat{a}}(\hat{a} - \check{a}). \tag{4}$$

The corresponding VCV matrix, provided that the uncertainty in $\check{a}$ can be neglected, reads

$$Q_{\check{b}\check{b}} = Q_{\hat{b}\hat{b}} - Q_{\hat{b}\hat{a}} Q^{-1}_{\hat{a}\hat{a}} Q_{\hat{a}\hat{b}}. \tag{5}$$

The precision of the fixed ILS baseline $\check{b}$ in Equation (4) is driven by the very precise phase data provided that the ILS SR is sufficiently high, while in the single-epoch case the float solution $\hat{b}$ in Equation (2) is driven by the precision of the code data. This means, in the case that the ILS SR is sufficiently high, that the ILS solution is expected to have at least a two-order of magnitude positioning precision improvement compared to that of the float solution.

2.4. Best Integer Equivariant Estimation

When the ILS SR is low, the user will normally prefer the float solution $\hat{b}$ in Equation (2) rather than the ILS solution $\check{b}$ in Equation (4). The alternative is to use the BIE estimator to solve for the ambiguities [1]. Assuming normally distributed GNSS data, the BIE estimator we use is denoted with an 'overline' symbol and reads

$$\bar{a}^{\lambda} = \sum_{z \epsilon \Theta^{\lambda}_{\hat{a}}} z \frac{\exp\left(-\frac{1}{2}||\hat{a} - z||^2_{Q_{\hat{a}\hat{a}}}\right)}{\sum_{z \epsilon \Theta^{\lambda}_{\hat{a}}} \exp\left(-\frac{1}{2}||\hat{a} - z||^2_{Q_{\hat{a}\hat{a}}}\right)}, \tag{6}$$

Note in Equation (6) that the BIE solution is not always necessarily an integer as it is a weighted sum of integers. We also remark that, for the BIE estimator, no ratio test is needed [1].

The BIE baseline solution can then be derived as

$$\breve{b} = \hat{b} - Q_{\hat{b}\hat{a}} Q_{\hat{a}\hat{a}}^{-1} \left(\hat{a} - \overline{a}^{\lambda} \right), \tag{7}$$

where $\check{a}$ in Equation (4) is replaced by $\overline{a}^{\lambda}$ in Equation (6). Note that the true BIE estimator involves an infinite weighted sum over the whole space of integers for z, which is computationally impractical. Hence, in Equation (6) we make use of a finite integer set $\Theta_{\hat{a}}^{\lambda}$ instead [29], which can be defined as follows:

$$||\hat{a} - z||^2_{Q_{\hat{a}\hat{a}}} < \lambda^2, \tag{8}$$

where the threshold λ^2 can be determined from a central chi-squared distribution χ^2 with n degrees of freedom and a small significance level $\alpha = 10^{-9}$. Note that, for very weak models, the number of candidates within this threshold in Equation (8) can reach several tens of thousands of candidates, whereas, for stronger models, at most a few candidates are usually obtained.

3. Google Pixel 4 Smartphone GNSS Data Collection

This section describes the short-baseline setup configurations of the GP4 smartphones while using (1) external antennas and (2) internal smartphone antennas. In this section, we also outline the stochastic model settings.

3.1. Setup Configuration with External and Internal Smartphone Antennas

Figure 1 depicts the external and internal antenna setups to evaluate the positioning performance of the GNSS data observed by GP4 smartphones. The GP4 smartphones are capable of tracking dual-frequency GPS L1 + L5, Galileo E1 + E5a, QZSS L1 + L5, and BDS B1 code and carrier-phase observations. The GP4 smartphones logged the GNSS measurements at a 1-s measurement interval via the Geo++ RINEX Logger vers. 2.1.6.

When evaluating the performance using external antennas, the GP4 smartphones are placed in two separate radiofrequency (RF) shielding boxes to avoid them from receiving the GNSS signals other than from the dedicated reradiating antenna (see Figure 1a–d). The GNSS signals are collected from two distinct active low-cost antennas, Swift GPS500, and then reradiated via a passive antenna inside the RF shielding boxes. The signal amplifier is connected between the rooftop antenna and reradiating antenna to mitigate the effect of signal attenuation over a 30 m LMR-400 flexible low loss communication coaxial cable. A similar setup was validated in Yong et al. [22] that benchmarked the smartphones with survey-grade receivers, to assure that no GNSS signal leakage was experienced in the RF shielding box. The duty-cycling settings of the GP4 smartphones were disabled during the experiment to obtain continuous carrier-phase observations [3].

In addition to the short baseline with external antennas, we assessed the positioning performance of the short baseline while using the smartphone internal antennas (see Figure 1e). Note that the built-in antennas of the GP4 smartphones have been found to be sensitive to poor quality GNSS signals and the surrounding environment [7,22], which means that we can expect larger multipath errors to be present for this setup.

3.2. Stochastic Model Settings

The stochastic model was determined by fitting the empirical 95% confidence interval/ellipses to the formal counterparts, as derived from the corresponding VCV matrices of the positions. The empirical VCV matrix was estimated from the positioning errors obtained by comparing the estimated positions to very precise benchmark coordinates, whereas the formal VCV matrix was obtained by the average of all single-epoch formal

VCV matrices of the entire observation time span [30]. We used independent datasets to analyze and to formulate the stochastic model for the subsequent sections, and the stochastic model settings were determined for different elevation cut-off angles to formulate the most realistic stochastic models possible. By using realistic stochastic models, we could assure that we obtained the best possible ambiguity resolution and positioning performance results. The different elevation cut-off angles were used to mimic situations in urban canyon environments or when low-elevation multipath is present.

Figure 1. GP4 smartphones in a short-baseline setup configuration with external antennas (**a**,**b**) inside an RF shielding box (**c**,**d**), and with the smartphone internal antennas (**e**) on the rooftop of the building of the School of Surveying in Dunedin, New Zealand.

Table 1 depicts the range of the undifferenced and zenith-referenced standard deviations (STDs) utilized in the stochastic models, together with the observation span of the external and internal antenna setup configurations in Figure 1. Note that each GNSS and/or frequency had equal weighting in this article, similar to the RTKlib implementation [27].

Table 1. Stochastic model settings in terms of undifferenced and zenith-referenced STD range (minimum to maximum) of the code and phase observations. In the last column, the time span of the data to be analyzed is also depicted (in universal coordinated time, UTC).

Antennas Used	Range of Phase STDs (m)	Range of Code STDs (m)	DoY (Hours of Day), hh:mm:ss UTC
External antenna	0.001–0.002	1.198–1.413	228–2021 (8 h), 13:35:00–21:34:59
Internal antenna	0.003–0.004	5.985–5.998	344 and 345–2020 (6 h), 21:03:00–02:57:59

Table 1 shows that the code STDs improved by a factor of approximately five when external antennas were used instead of the internal smartphone antennas that were more sensitive to multipath. For example, the code STD increased from a maximum of 1.4 m when using external antennas to approximately 6.0 m when internal antennas were used. The corresponding phase STDs increased from a maximum of 2 mm to 4 mm when using internal antennas. Note that the ambiguity resolution performance in the subsequent sections is driven by the number of satellites and frequencies used, as well as the stochastic model.

4. Instantaneous, Short-Baseline, Single- and Dual-Frequency RTK and BIE Positioning with Google Pixel 4 Smartphones

In this section, we investigate the BIE estimator and compare the performance to that of the commonly used ILS and float estimators. The benefit of using the BIE estimator is that the MSEs are always smaller than or at least as good as the float and ILS counterparts. For instance, in practice, when the ILS success rate is lower than the desired 100%, the user usually opts for the float solution when, in fact, BIE would be the preferred option.

The number of correctly fixed epochs, used below, was determined by the number of epochs where the estimated local east, north, and up coordinate errors were all below or equal to 0.05 m. The ILS success rate was then computed as follows:

$$P_{S_E} = \frac{\text{\# of correctly fixed epochs}}{\text{total \# of epochs}} \times 100\%. \quad (9)$$

In the results below, we investigate the BIE performance for GP4 smartphones when using both external and internal smartphone antennas while collecting single- and dual-frequency multi-GNSS data.

4.1. BIE with External Antennas for Single-Frequency RTK

Figure 2 depicts the float (black), ILS (magenta), and BIE (green dots) horizontal RTK positioning errors using GP4 smartphones in a short-baseline RTK setup, while using low-cost external antennas. The positioning errors were determined by comparing the estimated positions to very precise benchmark coordinates. These benchmark coordinates were determined using geodetic GNSS receivers and antennas, a Kalman filter, and a multi-epoch model while assuming the ambiguities to be time-constant. Any phase center offsets and variations of the smartphones were neglected in this process. From top to bottom rows and left to right columns, we depict the results for various elevation cut-off angles resulting in ILS SRs of 11.5%, 54.9%, 79.7%, 94.3%, 99.8%, and 99.9%, respectively. We depict L1+ E1 + L1 + B1 GPS + Galileo + QZSS + BDS results, where the zoom-in windows show at least a two-order-of-magnitude improvement when going from ambiguity float and incorrectly fixed ILS solutions to that of the correctly fixed ILS positioning errors. Note that the float solutions, as depicted by black dots, become more precise as the model strength increases.

Figure 2 (top row and left column) shows that many of the incorrectly fixed ILS solutions (magenta dots) are at the meter level, and that the BIE solutions (green dots) resemble the float solutions (black dots underneath the green dots). When the ILS SRs increased, however, such as in the right column and second row, we could see BIE solutions starting to outperform their float counterparts, with a much larger density of BIE solutions with millimeter- to centimeter-level positioning precisions, as shown in the zoom-in windows. In the second row and right column, as well as in the third row, we can further see ILS solutions with larger positioning errors than BIE even though the ILS SRs ranged from 94.3% to 99.9%.

To further illustrate the optimal performance of the BIE estimator in terms of the positioning MSEs, we depict in Figure 3 the MSE ratios, with respect to the float MSEs, as a function of the ILS SRs. Note that the MSE is here the sum of the variances of the east, north, and up errors, since our estimated positions are unbiased. The float MSE ratio is equal to one and is depicted as a full blue line, whereas the ILS and BIE counterparts are depicted as dashed magenta and full green lines, respectively. Note that these MSE ratio results resemble those of Odolinski and Teunissen [19], albeit based on completely different datasets (smartphone vs. low-cost RTK receiver data).

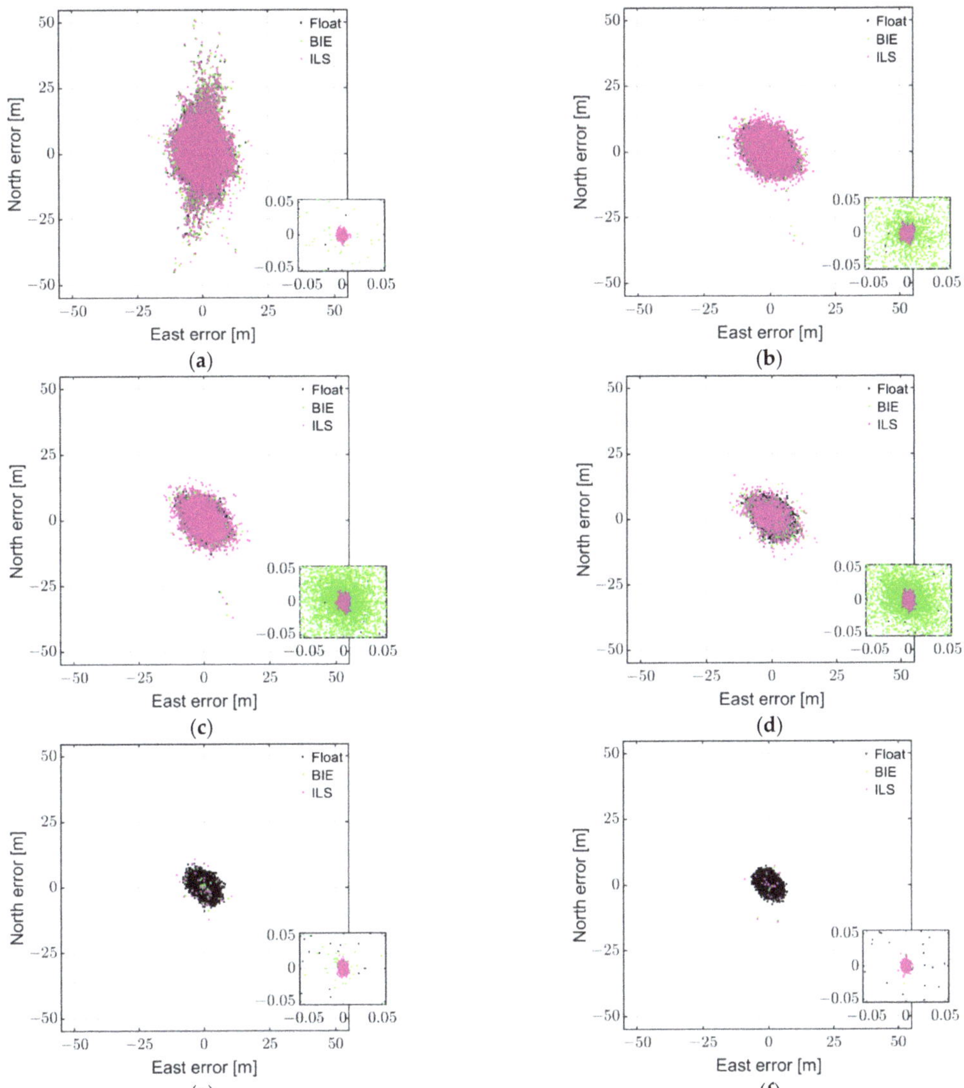

Figure 2. Horizontal (north/east) scatter plot of the multi-GNSS, single-frequency (L1 GPS + E1 Galileo + L1 QZSS + B1 BDS) GP4 data derived with the BIE (*green dots*), ILS (*magenta dots*), and ambiguity float (*black dots*) estimators for instantaneous RTK positioning with external antenna for a short baseline in Dunedin, New Zealand, based on 8 h of data (1 s measurement interval): (**a**) 11.5% ILS SR (38° cut-off); (**b**) 54.9% ILS SR (35° cut-off); (**c**) 79.7% ILS SR (32° cut-off); (**d**) 94.3% ILS SR (30° cut-off); (**e**) 99.8% ILS SR (20° cut-off); (**f**) 99.9% ILS SR (10° cut-off).

Figure 3 shows, as expected, that the BIE MSE ratio is equal to that of the float solutions when the ILS SR is close to 0%, and that BIE is equal to ILS when the ILS SR is close to 100%. Most importantly and for all other cases, we can see that the BIE MSE ratios are smaller than those of the float and ILS solutions, respectively. This shows that using the BIE estimator on smartphone data for RTK positioning will give the optimal positioning performance, as measured by the MSEs.

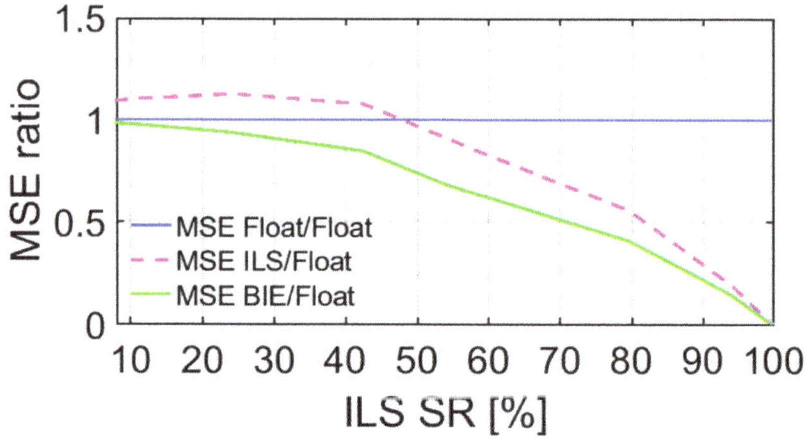

Figure 3. MSE ratios of GP4 single-frequency, short-baseline, and instantaneous RTK positioning errors using external antennas, with BIE (green line), ILS (dashed magenta line), and ambiguity float (blue line), all versus the float solutions.

Table 2 shows the percentage of the 3D position errors within a range of 0.05 m, 1 m, 2 m, and 4 m. The given percentages can provide a practical understanding of the distribution of the RTK positioning errors. The largest percentages for each scenario are shown in bold to distinguish the values from each other when rounded to two decimal places. The percentage of the ILS for the weakest model with the lowest ILS SR of 11.52% shows that the BIE solutions are similar to the float solutions. For the strongest model with the highest ILS SR of 99.92%, the results show that BIE resembled the ILS solutions. For all other scenarios, it becomes clear that BIE outperformed both the ILS and the float solutions in terms of not obtaining very large positioning errors, while also having a smaller likelihood than ILS of very small positioning errors (unless the ILS SR is very high). These results are similar to the cumulative distribution functions (CDFs) as discussed in Odolinski and Teunissen ([18]; Figure 4) and Verhagen and Teunissen ([13]; Figure 1).

To also illustrate the corresponding positioning precisions of the different estimators, Table 2 depicts the positioning standard deviations (STDs), the mean number (#) of satellites, and the employed elevation cut-off angles. We can see in Table 2 that the BIE and float solutions have similar STDs for the east, north, and up components when the ILS SR is 11.5% (with a slightly better performance for the BIE estimator), and that both solutions have STDs that are much better than their ILS counterparts. When the ILS SR increased to 54.9%, we can further see that the BIE estimator start to significantly outperform both the float and the ILS solutions, with STDs in east, north, and up of 2.344 m, 2.217 m, and 7.797 m, respectively. The corresponding STDs for the float and ILS solutions are 2.791 m, 2.605 m, and 9.455 m, and 2.661 m, 2.642 m, and 8.990 m in east, north and up, respectively. In other words, the BIE solutions have STDs that are up to more than 1 m smaller (in the up component) than the ILS counterparts.

Lastly, we can see in Table 2 that the BIE solutions have a better performance than their ILS counterparts when the ILS SRs reached values of 99.8% and 99.9%, with smaller positioning STDs by up to several centimeters in east, north, and up, as well as better performance by even several meters than the float solutions. This implies that when the ILS SR is different from the desirable 100%, the BIE estimator will indeed outperform the float and ILS estimators, and this is true even when smartphone GNSS data are used.

Table 2. Empirical STDs of the ILS, BIE, and float solutions for single-frequency and instantaneous RTK using GP4 smartphones, based on 8 h observations for a short-baseline external antenna data experiment (see Figure 2). Comparisons of the percentage of the 3D position errors within a range of 0.05 m, 1.0 m, 2.0 m, and 4.0 m are also given. Bold values show the largest percentages for each scenario and estimator.

Mode	ILS Success Rate (%)	Percentage of the 3D Position Errors (%)				STD (m)			Mean # of Satellites	Elevation Cut-Off Angle
		≤0.05 m	<1.0 m	<2.0 m	<4.0 m	E	N	U		
ILS	11.52	**11.52**	46.12	69.27	91.18	3.509	5.624	15.044	8.6	38
BIE		0.09	**46.90**	**72.71**	**93.47**	3.238	5.283	14.170		
Float		0.00	44.70	71.34	93.34	3.282	5.332	14.358		
ILS	54.92	**54.92**	**72.79**	85.08	96.16	2.661	2.642	8.990	10.6	35
BIE		5.49	70.92	**88.25**	**97.89**	2.344	2.217	7.797		
Float		0.00	55.65	82.10	97.55	2.791	2.605	9.455		
ILS	79.75	**79.75**	**87.32**	92.86	98.01	1.849	1.883	6.035	11.7	32
BIE		29.05	85.97	**94.54**	**98.92**	1.634	1.596	5.172		
Float		0.00	58.45	84.82	98.17	2.507	2.462	8.096		
ILS	94.29	**94.29**	**96.38**	97.85	99.35	1.126	1.024	3.073	13.5	30
BIE		67.81	95.86	**98.27**	**99.56**	1.014	0.887	2.667		
Float		0.00	60.59	86.41	98.64	2.370	2.258	7.147		
ILS	99.85	**99.85**	**99.90**	99.94	99.99	0.183	0.192	0.425	17.9	20
BIE		99.65	99.90	**99.95**	**99.99**	0.170	0.179	0.400		
Float		0.00	70.98	93.41	99.78	1.846	1.865	4.762		
ILS	99.92	**99.92**	**99.97**	99.98	99.99	0.095	0.147	0.444	23.6	10
BIE		99.91	99.97	**99.98**	**99.99**	0.093	0.143	0.440		
Float		0.00	78.03	96.81	99.98	1.556	1.633	3.411		

4.2. BIE with Internal Antennas for Dual-Frequency RTK

In this section, we investigate the corresponding BIE performance when the internal antennas of the smartphones are used. Since, with the smartphone internal antennas, the multipath errors are more significant than when external antennas are used [22], in this section, we use dual-frequency L1 + L5 GPS, E1 + E5a Galileo, L1 + L5 QZSS, and B1 BDS observations to further strengthen the model.

Figure 4 depicts, as in Figure 2, the float (black), ILS (magenta), and BIE (green dots) horizontal RTK positioning errors using GP4 smartphones in a short-baseline RTK setup, but while using the internal antennas of the smartphones. From top to bottom rows and left to right columns, we depict the results for various elevation cut-off angles resulting in ILS SRs of 9.6%, 53.9%, 72.1%, 84.0%, 91.9%, and 95.4%, respectively. The zoom-in windows show at least a two-order-of-magnitude improvement when going from ambiguity float and incorrectly fixed ILS solutions to that of the correctly fixed ILS positioning errors. Note again that, as the model become stronger, the float solutions, as depicted by black dots, become more precise. We also remark here that it is evident that the float and incorrectly fixed ILS solutions have a much poorer precision than in Figure 2. This degradation in precision when internal antennas are used is indeed due to their sensitive to multipath effects, where the code observations, which dominate the precision of the single-epoch float solutions, are more affected [31].

Figure 4 shows, similar to Figure 2, that many of the incorrectly fixed ILS solutions (magenta dots) have errors at the meter level, and that the BIE solutions (green dots) more or less resembles the float solutions (black dots underneath the green dots) when the models are weak (at the top row). When the ILS SRs increases, however, such as in the right column and second row, we can again see that BIE solutions start to outperform their float counterparts, with a much larger density of BIE solutions with millimeter- to centimeter-level positioning precisions as shown in the zoom-in windows. In the second row and right column, as well as in the third row, we can again see ILS solutions with larger positioning errors than BIE despite the fact that the ILS SRs ranged from 84.0% to 95.4%.

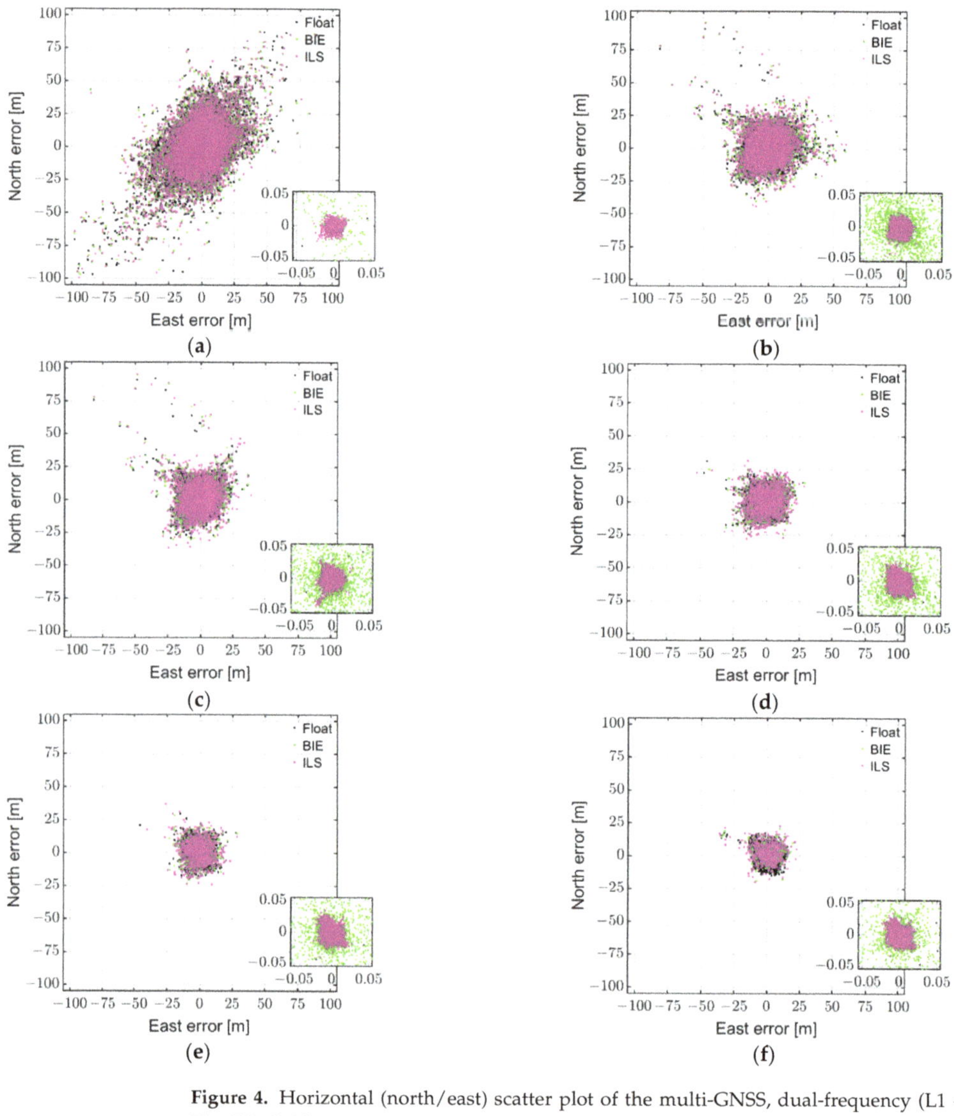

Figure 4. Horizontal (north/east) scatter plot of the multi-GNSS, dual-frequency (L1 + L5 GPS, E1 + E5a Galileo, L1 + L5 QZSS, and B1 BDS) GP4 data derived with the BIE (*green dots*), ILS (*magenta dots*), and ambiguity float (*black dots*) estimators for instantaneous RTK positioning with smartphone internal antennas for a short baseline in Dunedin, New Zealand, based on 6 h of data (1 s measurement interval): (**a**) 9.6% ILS SR (40° cut-off); (**b**) 53.9% ILS SR (30° cut-off); (**c**) 72.1% ILS SR (25° cut-off); (**d**) 84.0% ILS SR (20° cut-off); (**e**) 91.9% ILS SR (15° cut-off); (**f**) 95.4% ILS SR (10° cut-off).

To show the above superior performance of the BIE estimator in a different way, Figure 5 illustrates the 95.4% ILS SR scenario (see Figure 4f), but now with each solution in a separate subplot. The figure shows that the BIE solutions are indeed superior to their float and ILS counterparts, with a better precision than both estimators and fewer large positioning errors than the ILS estimator.

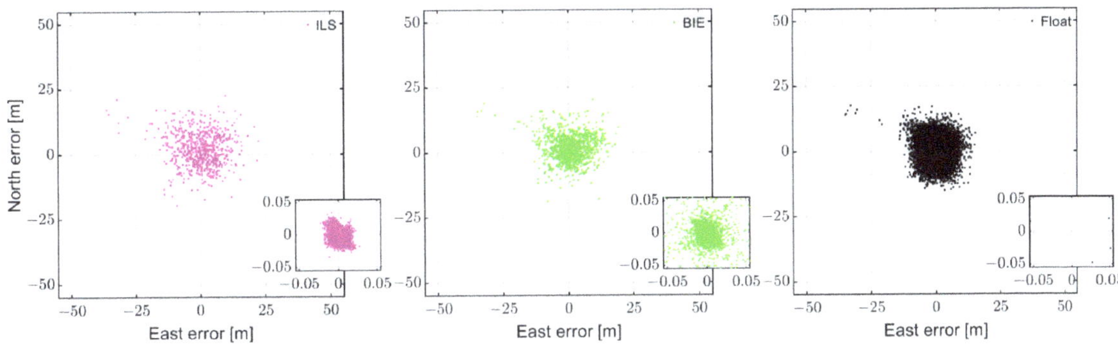

Figure 5. The 95.4% ILS SR scenario (see Figure 4f): horizontal (north/east) scatter plot of the multi-GNSS, dual-frequency (L1 + L5 GPS, E1 + E5a Galileo, L1 + L5 QZSS, and B1 BDS) GP4 data derived with the ILS (left column), BIE (middle column), and ambiguity float (right column) estimators for instantaneous RTK positioning with smartphone internal antennas.

To again illustrate the optimal BIE performance, we depict in Figure 6 the MSE ratios, with respect to the float MSEs, as a function of the ILS SRs. The float MSE ratio is depicted as a full blue line, whereas the ILS and BIE counterparts are depicted as dashed magenta and full green lines, respectively. Figure 6 shows, similar to Figure 3, that the BIE MSE ratio is equal to that of the float solutions when the ILS SR is close to 0%, and that BIE is close to ILS when the ILS SR is also close to 100% (i.e., 95.4%). Most importantly and for all other cases, we can again see that the BIE MSE ratios are smaller than those of the float and ILS solutions. This shows that using the BIE estimator on smartphone data for RTK positioning, even when the internal smartphone antennas are used, will give the optimal positioning performance.

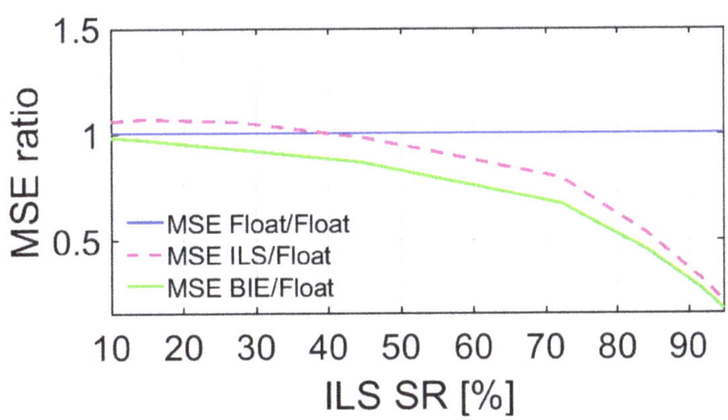

Figure 6. MSE ratios of the GP4 dual-frequency, short-baseline, and instantaneous RTK positioning errors using internal antennas, with BIE (green line), ILS (dashed magenta line), and ambiguity float (blue line), all versus the float solutions.

Table 3 depicts the corresponding percentage of the 3D position errors within a range of 0.05 m, 1.0 m, 2.0 m, and 4.0 m, the positioning standard deviations (STDs), the mean number (#) of satellites, and the employed elevation cut-off angles when the internal smartphone antennas are used. We can again see that the percentages of the position errors are consistent with the CDFs in, e.g., Figure 4 of Odolinski and Teunissen [18]. Similarly, the BIE solutions have always smaller STDs for the east, north, and up components when the ILS SR is between 9.6% and 95.4%, with better STDs by more than 1 m to several tens of centimeters than the ILS solutions in east, north, and up, and better performance by several meters than the float solutions when the ILS SR is high. This implies again that when the ILS SR is different from the desirable 100%, the BIE estimator will indeed outperform the float and ILS estimators, and this is true even when smartphone GNSS data with internal antennas are used.

Table 3. Empirical STDs of the ILS, BIE, and float solutions for dual-frequency and instantaneous RTK using GP4 smartphones, based on 6 h observations for a short-baseline internal antenna experiment (see Figure 4). Comparisons of the percentage of the 3D position errors within a range of 0.05 m, 1.0 m, 2.0 m, and 4.0 m are also given. Bold values show the largest percentages for each scenario and estimator.

Mode	ILS Success Rate (%)	Percentage of the 3D Position Errors (%)				STD (m)			Mean # of Satellites	Elevation Cut-Off Angle
		≤0.05 m	<1.0 m	<2.0 m	<4.0 m	E	N	U		
ILS	9.62	**9.62**	**29.28**	44.44	65.86	10.9938	12.0972	29.7068	7.79	40
BIE		0.52	27.66	**45.26**	**67.95**	10.6723	11.7231	28.5620		
Float		0.00	24.19	41.98	66.46	10.8387	11.8522	28.8412		
ILS	53.89	**53.89**	**64.89**	**72.67**	84.46	6.1821	6.1096	15.9402	11.87	30
BIE		24.98	60.47	72.50	**86.15**	5.8795	5.5962	14.8443		
Float		0.00	31.84	54.61	79.30	6.4436	6.4693	16.6158		
ILS	72.15	**72.15**	**78.87**	**83.63**	90.86	4.6011	4.9372	10.8364	13.74	25
BIE		46.22	73.89	82.53	**91.38**	4.3570	4.5332	9.9186		
Float		0.00	36.32	60.89	84.86	5.2655	5.7106	12.0913		
ILS	84.01	**84.01**	**88.10**	**91.31**	95.32	3.0321	3.0510	7.3630	15.49	20
BIE		65.93	86.18	90.95	**95.72**	2.8652	2.7173	6.8088		
Float		0.00	40.10	65.29	88.75	4.2646	4.5909	9.8794		
ILS	91.90	**91.90**	**94.29**	**96.01**	98.05	1.9998	2.0644	4.1177	17.56	15
BIE		80.03	93.52	95.99	**98.29**	1.8819	1.8900	3.7877		
Float		0.00	45.57	71.17	92.70	3.6434	4.0713	7.1847		
ILS	95.39	**95.39**	**96.64**	97.68	98.76	1.5967	1.3888	2.7278	19.81	10
BIE		88.37	96.35	**97.75**	**99.07**	1.4831	1.2386	2.4483		
Float		0.00	47.81	74.55	94.41	3.3856	3.6980	6.1203		

5. Discussion

In this contribution, we analyzed the best integer equivariant (BIE) estimator for real GNSS data collected by Google Pixel 4 (GP4) smartphones and antennas. We compared the instantaneous (single-epoch) positioning performance of BIE to the float and integer least squares (ILS) estimators that are commonly used when the ILS success rate (SR) is different from one and close to one, respectively. Radiofrequency (RF) shielding boxes and reradiating antennas were used to track GNSS signals from external low-cost antennas, consisting of L1 + L5 GPS, E1 + E5a Galileo, L1 + L5 QZSS, and B1 BDS code and carrier-phase observations. The short-baseline real-time kinematic (RTK) performance was also evaluated while using the GP4 internal smartphone antennas. We investigated the BIE performance both when single-frequency and dual-frequency measurements were employed for the combination of GPS + Galileo + QZSS + BDS. We showed that the BIE positioning performance was superior to that of the ILS and float estimators when the ILS SR is different from one. This was demonstrated to be true on the basis of real multi-GNSS data collected by the GP4 smartphones and antennas.

Our BIE performance evaluation consisted of comparing the estimated positions to very precise benchmark coordinates, and the optimality of the BIE estimator was further evaluated through its position mean squared errors (MSEs) and standard deviations (STDs). It was shown that the BIE performance resembles that of the float estimator when the ILS SR is very low and was similar to that of the ILS when the ILS SR is very high. For all other cases, we demonstrated that BIE outperformed both the float and the ILS estimators even when on the basis of real GP4 smartphone data while using external and internal smartphone antennas. Future studies could involve evaluating the GP4 smartphone BIE RTK positioning performance for longer baselines, when the relative atmospheric delays need to be estimated [19].

Author Contributions: C.Z.Y., conceptualization, methodology, formal analysis, visualization, and writing the manuscript; K.H., software development with BIE implementation into RTKlib; E.R. and S.M., conceptualization; R.O., conceptualization, methodology, formal analysis, investigation, and writing the manuscript. All authors have read and agreed to the published version of the manuscript.

Funding: This research was funded by Geoscience Australia for the project 1002B: "Precise GNSS positioning with smartphones".

Institutional Review Board Statement: Not applicable.

Informed Consent Statement: Not applicable.

Data Availability Statement: The broadcast ephemerides were used for satellite orbits and clocks. The Google Pixel 4 smartphone observation data are stored at the School of Surveying data facilities, the University of Otago, and can be made available upon request by contacting the corresponding author C.Z.Y. via email.

Acknowledgments: The fifth author was the recipient of a Geoscience Australia research grant on the project 1002B: "Precise GNSS positioning with smartphones". We logged the GNSS data using a Geo++ RINEX logger from Android-based smartphones. All this support is gratefully acknowledged.

Conflicts of Interest: The authors declare no conflict of interest.

References

1. Teunissen, P.J.G. Theory of integer equivariant estimation with application to GNSS. *J. Geod.* **2003**, *77*, 402–410. [CrossRef]
2. Pesyna, K.M.; Heath, R.W.; Humphreys, T.E. Centimeter positioning with a smartphone-quality GNSS antenna. In Proceedings of the 27th International Technical Meeting of the Satellite Division of the Institute of Navigation (ION GNSS+ 2014), Tampa, FL, USA, 8–12 September 2014; pp. 1568–1577.
3. Humphreys, T.E.; Murrian, M.; Van Diggelen, F.; Podshivalov, S.; Pesyna, K.M. On the feasibility of cm-accurate positioning via a smartphone's antenna and GNSS chip. In Proceedings of the 2016 IEEE/ION Position, Location and Navigation Symposium, Savannah, GA, USA, 11–16 April 2016; pp. 232–242.
4. Håkansson, M. Characterization of GNSS observations from a Nexus 9 Android tablet. *GPS Solut.* **2019**, *23*, 21. [CrossRef]
5. Tomaštík, J.; Chudá, J.; Tunák, D.; Chudý, F.; Kardoš, M. Advances in smartphone positioning in forests: Dual-frequency receivers and raw GNSS data. *For. Int. J. For. Res.* **2021**, *94*, 292–310. [CrossRef]
6. Aggrey, J.; Bisnath, S.; Naciri, N.; Shinghal, G.; Yang, S. Multi-GNSS precise point positioning with next-generation smartphone measurements. *J. Spat. Sci.* **2019**, *65*, 79–98. [CrossRef]
7. Zhang, X.; Tao, X.; Zhu, F.; Shi, X.; Wang, F. Quality assessment of GNSS observations from an Android N smartphone and positioning performance analysis using time-differenced filtering approach. *GPS Solut.* **2018**, *22*, 70. [CrossRef]
8. Guo, L.; Wang, F.; Sang, J.; Lin, X.; Gong, X.; Zhang, W. Characteristics Analysis of Raw Multi-GNSS Measurement from Xiaomi Mi 8 and Positioning Performance Improvement with L5/E5 Frequency in an Urban Environment. *Remote Sens.* **2020**, *12*, 744. [CrossRef]
9. Darugna, F.; Wübbena, J.; Ito, A.; Wübbena, T.; Wübbena, G.; Schmitz, M. RTK and PPP-RTK Using Smartphones: From Short-Baseline to Long-Baseline Applications. In Proceedings of the 32nd International Technical Meeting of the Satellite Division of The Institute of Navigation (ION GNSS+ 2019), Miami, FL, USA, 16–20 September 2019; pp. 3932–3945.
10. Laurichesse, D.; Rouch, C.; Marmet, F.X.; Pascaud, M. Smartphone applications for precise point positioning. In Proceedings of the 30th International Technical Meeting of the Satellite Division of The Institute of Navigation, Portland, OR, USA, 25–29 September 2017; pp. 171–187.
11. Warnant, R.; Vyvere, D.; Van, L.; Warnant, Q. Positioning with single and dual frequency smartphones running Android 7 or later. In Proceedings of the 31st International Technical Meeting of the Satellite Division of The Institute of Navigation (ION GNSS+ 2018), Miami, FL, USA, 24–28 September 2018; pp. 284–303.
12. Paziewski, J.; Fortunato, M.; Mazzoni, A.; Odolinski, R. An analysis of multi-GNSS observations tracked by recent Android smartphones and smartphone-only relative positioning results. *Measurement* **2021**, *175*, 109162. [CrossRef]
13. Verhagen, S.; Teunissen, P.J.G. Performance comparison of the BIE estimator with the float and fixed GNSS ambiguity estimators. In *A Window Future Geod*; Springer: Berlin, Heidelberg, 2005; pp. 428–433.
14. Wen, Z.; Henkel, P.; Brack, A.; Günther, C. Best integer equivariant estimation for precise point positioning. In Proceedings of the ELMAR-2012, Zadar, Croatia, 12–14 September 2012; pp. 279–282.

15. Brack, A.; Henkel, P.; Günther, C. Sequential best integer-equivariant estimation for GNSS. *Navigation* **2014**, *61*, 149–158. [CrossRef]
16. Brack, A. Partial Carrier-Phase Integer Ambiguity Resolution for High Accuracy GNSS Positioning. Ph.D. Thesis, Lehrstuhl fur Kommunikation und Navigation, Technische Universitat Munchen, München, Germany, 2019.
17. Teunissen, P.J.G. Best integer equivariant estimation for elliptically contoured distributions. *J. Geod.* **2020**, *94*, 82. [CrossRef]
18. Odolinski, R.; Teunissen, P.J.G. Best integer equivariant estimation: Performance analysis using real data collected by low-cost, single- and dual-frequency, multi-GNSS receivers for short- to long-baseline RTK positioning. *J. Geod.* **2020**, *94*, 91. [CrossRef]
19. Odolinski, R.; Teunissen, P.J.G. Best integer equivariant position estimation for multi-GNSS RTK: A multivariate normal and t-distributed performance comparison. *J. Geod.* **2022**, *96*, 3. [CrossRef]
20. Paziewski, J.; Sieradzki, R.; Baryla, R. Signal characterization and assessment of code GNSS positioning with low-power consumption smartphones. *GPS Solut.* **2019**, *23*, 98. [CrossRef]
21. Purfürst, T. Evaluation of Static Autonomous GNSS Positioning Accuracy Using Single-, Dual-, and Tri-Frequency Smartphones in Forest Canopy Environments. *Sensors* **2022**, *22*, 1289. [CrossRef] [PubMed]
22. Yong, C.Z.; Odolinski, R.; Zaminpardaz, S.; Moore, M.; Rubinov, E.; Er, J.; Denham, M. Instantaneous, Dual-Frequency, Multi-GNSS Precise RTK Positioning Using Google Pixel 4 and Samsung Galaxy S20 Smartphones for Zero and Short Baselines. *Sensors* **2021**, *21*, 8318. [CrossRef] [PubMed]
23. Paziewski, J. Recent advances and perspectives for positioning and applications with smartphone GNSS observations. *Meas. Sci. Technol.* **2020**, *31*, 091001. [CrossRef]
24. Hesselbarth, A.; Wanninger, L. Towards centimeter accurate positioning with smartphones. In Proceedings of the 2020 European Navigation Conference (ENC), Dresden, Germany, 23–24 November 2020; pp. 1–8.
25. Odijk, D.; Teunissen, P.J.G. Characterization of between-receiver GPS-Galileo inter-system biases and their effect on mixed ambiguity resolution. *GPS Solut.* **2013**, *17*, 521–533. [CrossRef]
26. Odolinski, R.; Teunissen, P.J.G.; Odijk, D. Combined BDS, Galileo, QZSS and GPS single-frequency RTK. *GPS Solut.* **2015**, *19*, 151–163. [CrossRef]
27. Takasu, T.; Yasuda, A. Development of the low-cost RTK-GPS receiver with an open source program package RTKLIB. In Proceedings of the International Symposium on GPS/GNSS, Jeju, Korea, 4 November 2009; pp. 1–6.
28. Teunissen, P.J.G. The least-squares ambiguity decorrelation adjustment: A method for fast GPS integer ambiguity estimation. *J. Geod.* **1995**, *70*, 65–82. [CrossRef]
29. Teunissen, P.J.G. On the computation of the best integer equivariant estimator. *Artif. Satell.* **2005**, *40*, 161–171.
30. Odolinski, R.; Teunissen, P.J.G.; Odijk, D. First combined COMPASS/BeiDou-2 and GPS positioning results in Australia. Part II: Single- and multiple-frequency single-baseline RTK positioning. *J. Spat. Sci.* **2014**, *59*, 25–46. [CrossRef]
31. Khodabandeh, A.; Zaminpardaz, S.; Nadarajah, N. A study on multi-GNSS phase-only positioning. *Meas. Sci. Technol.* **2021**, *32*, 095005. [CrossRef]

sensors

MDPI

Article

Inherent Limitations of Smartphone GNSS Positioning and Effective Methods to Increase the Accuracy Utilizing Dual-Frequency Measurements

Jeonghyeon Yun, Cheolsoon Lim and Byungwoon Park *

Department of Aerospace Engineering and Convergence Engineering for Intelligent Drone, Sejong University, Seoul 05006, Republic of Korea
* Correspondence: byungwoon@sejong.ac.kr

Abstract: Xiaomi Mi8 with a Broadcom BCM47755 chip, an Android smartphone that supports multi-constellation (GPS, GLONASS, Galileo, BeiDou, and QZSS) and dual-frequency (L1/E1 and L5/E5), was launched in May 2018. Unlike previously released smartphones, it was technically expected to provide robust precise positioning with a fast ambiguity resolution, which led many researchers to be overly optimistic about the applicability of high-accuracy techniques such as real-time kinematic (RTK) systems and precise point positioning (PPP) of smartphones. The global navigation satellite system (GNSS) raw measurement quality of Android smartphones is, however, inherently far lower than that of general GNSS receivers due to their structure, which accordingly makes it difficult for them to be realized. Considering inherent limitations of smartphones such as low-quality antenna, frequent cycle slips, and the duty cycle, a practical strategy including L5 measurements, pseudo-range corrections for L5, and a weighting method is proposed in this paper. The results show that the proposed methods of L5 differential GNSS (DGNSS) and Doppler-based filtering can guarantee a positioning accuracy of 1.75 m horizontally and 4.56 m vertically in an Android device, which is comparable to the performance of commercial low-cost receivers.

Keywords: Android; smartphone; GNSS; L1/L5 dual-frequency; SBAS; Xiaomi Mi8

Citation: Yun, J.; Lim, C.; Park, B. Inherent Limitations of Smartphone GNSS Positioning and Effective Methods to Increase the Accuracy Utilizing Dual-Frequency Measurements. *Sensors* **2022**, *22*, 9879. https://doi.org/10.3390/s22249879

Academic Editors: Yang Gao, Jacek Paziewski, Michael Fu, Augusto Mazzoni and Chris Rizos

Received: 28 October 2022
Accepted: 8 December 2022
Published: 15 December 2022

1. Introduction

In the recent smart mobility era, the use of smartphone location-based services such as remote vehicle driving and mobility location sharing services, as well as navigation and route-finding services, is increasing. For example, Android Auto users can easily use smartphone location-based navigation services in the vehicle by connecting their smartphone to a vehicle that does not have a built-in global navigation satellite system (GNSS) [1]. Because this service relies on the location accuracy of a smartphone installed inside the vehicle rather than a GNSS antenna installed on the vehicle ceiling, incorrect directions often occur. As another example, Tesla recently released a new technology known as Smart Summon, which allows a Tesla vehicle to be called through a smartphone application [2]. It enables a Tesla vehicle to drive itself towards the owner's location based on the relative position between the vehicle and the smartphone. Since the GNSS of a smartphone cannot provide an accurate position, the owner should be cautious of potential malfunctions such as the target point being set incorrectly. However, the GNSS chipset currently used to calculate the location of smartphones has an error of more than 5–10 m in open-sky and 20–100 m in urban areas with many obstacles, such as buildings. Before Google's announcement of providing GNSS raw measurements [3], all the location-based products and services were totally dependent on the location accuracy that the chipsets provide. Smartphone and GNSS chipset vendors previously did not allow general users to feed the correction to the module or access the GNSS pseudo-range [4,5], which resulted in smartphones providing horizontal accuracy up to 10 m with a 95% probability [6,7].

After Android GNSS raw measurements were introduced, many researchers worked on assessing the positioning performance of smartphones. However, the low-cost chips in currently commercialized smartphones do not guarantee reliable high-quality positioning results as general GNSS receivers do. Smartphones use low quality linearly polarized (LP) antenna optimized for voice communication. Since it was not right hand circularly polarized (RHCP) specialized for navigation signals, smartphones are very vulnerable to multipath interference. As a result, multipath errors of smartphones are higher than a few tens of meters [3]. To apply high-precision positioning techniques such as RTK and PPP to smartphones, external equipment or various strong constrains are essential. In particular, continuous carrier-phase measurements are difficult to obtain because of the duty-cycle effect. Moreover, conservation time of smartphone PPP is much longer than geodetic receivers due to unstable carrier-phase measurements [3,6].

The launch of the first dual-frequency GNSS chipset, BCM47755 (designed by Broadcom, San Jose, California, USA), was a big opportunity to reduce the performance gap between the smartphones and general GNSS receivers. The BCM47755 was first installed in the Xiaomi Mi8 (designed by Xiaomi, Haidian District, Beijing, China) in May 2018, and supported multi-constellation (GPS, GLONASS, Galileo, BeiDou and QZSS) and dual-frequency (L1/E1 and L5/E5). Recently, researchers tried to apply high-end technologies, utilized in general GNSS receivers, to Android smartphones. Fortunato applied RTK and PPP methods to the Xiaomi Mi8 and achieved horizontal root mean squares (RMS) of 1.17 m and 2.23 m in dynamic positioning, respectively [7]. Since he utilized the code-minus-carrier phase (CMC) observable method, a separate pretreatment process was required to void the vulnerability by interference, multipath, or cycle slip. Despite preprocessing, ambiguity fix was not enabled. Wu investigated the application of PPP in static and kinematic modes using the Xiaomi Mi8 [6]. The convergence time of dual-frequency smartphones in static experiments was 102 min for 1 m and 116 min for 0.2 m, which is far longer than 40 min of general GNSS receivers. A single-frequency PPP strategy-based clock bias estimation method was studied [8], but the matrix size is too large to be used as real-time high-precision smartphone positioning because the state of the filter had to include the integer ambiguities for all satellites.

Common problems in existing studies include interference, multipath, and cycle slip due to the use of carrier-phase. Additionally, carrier ambiguity fix requires very long convergence times and a complex matrix. To overcome these limitations, this study introduces a method to utilize the L5 measurement, which is strong against noise and multipath error. We also suggest a method to provide the L5 Pseudo Range Correction (PRC) for accurate positioning of an Android smartphone without building a new infrastructure or correction messages. In addition, we present a method to use Doppler measurements instead of carrier phase, and it does not require cycle slip monitor and ambiguity determination processes that caused the complexity and frequent discontinuity due to the instability of carrier-phase measurement. It is expected to be practical and economical to use a smartphone itself aligned with existing infrastructures without adding or modifying the device and current systems.

The remainder of this paper is organized as follows. In Section 2, the obstacles to applying precision positioning method to Android smartphones are described. In Section 3, the strategy for reliable accuracy improvement algorithm for Android is introduced. Further, a field test results after implementing the suggested algorithm on the user side is presented and examined in Section 4. Finally, Section 5 presents the discussion and conclusions.

2. Obstacles to Applying Precision Positioning Technique to an Android Smartphone

2.1. *Low Quality of Android Smartphones Antenna*

RHCP GNSS antennas are used for commercial geodetic receivers, which are optimized to reject reflected signals. However, LP planar inverted F (PIF) antennas are embedded in smartphones and optimized for voice communication with a lower volume. The vulnerability of Android GNSS raw measurements to multipath interference primarily arises due to the antenna configuration.

The PIF antenna significantly affects the quality of the received GNSS signal. First of all, the multipath and noise of smart devices are much larger than the received signals through general GNSS antennas. Noise and multipath error RMS of Android smartphone L1 code was observed up to 10 m under open-sky and reported over 20 m in a multipath environment. It directly affected the smartphone positioning performance, and an error of 11.65 m in vertical RMS was remained even though all the other errors such as ionospheric and tropospheric delay were removed by applying DGNSS [8,9]. Despite the high elevation angle over 60°, weak signal strength of 27 to 33 dB-Hz C/No were observed with frequent loss-of-locks [9], which are very rare cases for RHCP antennas. In addition, uncertainty of the exact phase center was reported to cause the bias of the phase residuals after double-difference [10].

To examine the effect of antenna configuration on the smartphone, we conducted a static test, as shown in Figure 1.

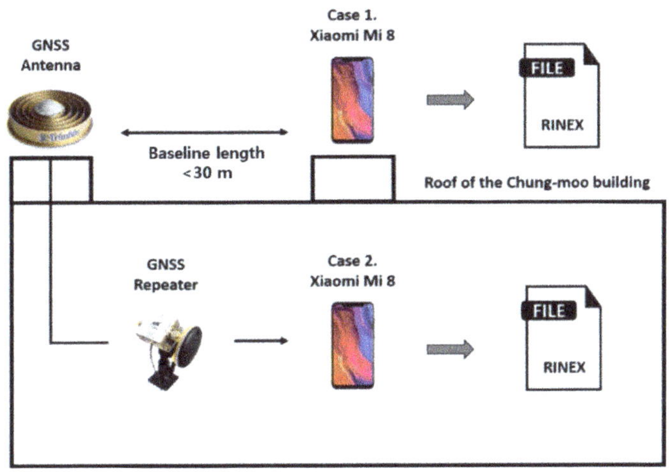

Figure 1. Test configuration for comparison of noise-level with or without GNSS repeater.

Unlike other GNSS receivers, the hardware structure makes it difficult to connect a smartphone directly to a geodetic antenna. GNSS repeaters were used to solve this problem [10,11]. Open-sky GNSS signals were directly re-radiated to one smartphone via a GNSS repeater coupled with a geodetic antenna to eliminate interference from the surrounding environment and amplify the receivable signal strength. We placed a Xiaomi Mi8 Android smartphone to receive signals that were transmitted by a repeater via a Trimble choke ring antenna. The other Mi8 was placed to receive live GNSS signals at the rooftop where the choke ring antenna was implemented. We collected pseudo-range (P) and carrier-phase (Φ) for both cases using the two smartphones at the same period, which are denoted in meters in Equations (1) and (2).

$$P^s_{r,i} = d^s_r - b^s + B_r + T^s_r + I^s_{r,i} + \varepsilon_{P^s_{r,i}}, \tag{1}$$

$$\Phi^s_{r,i} = d^s_r - b^s + B_r + T^s_r - I^s_{r,i} + \lambda^s_i N^s_{r,i} + \varepsilon_{\Phi^s_{r,i}}, \tag{2}$$

where $d, b, B, T, I, \lambda, N$ and ε denote the geometric range from satellite to ground receiver, satellite clock bias, receiver clock bias, tropospheric delay, ionospheric delay, carrier wavelength, integer ambiguity, and measurement noise, respectively. The subscript, r, denotes each test scenario, Open-sky for case 1 and Repeater for case 2. The superscript, s and subscript, i, denote the s-th satellite and type of carrier-frequency, respectively.

The combination of code and carrier described in Equation (3) enables all the geometric and ionospheric terms removed, so that there are only integer ambiguity and noise terms remained in R_r^s.

$$R_r^s = \left(P_{r,L_1}^s - P_{r,L_5}^s\right) + \left(\Phi_{r,L_1}^s - \Phi_{r,L_5}^s\right), \tag{3}$$

In this case, since the integer ambiguity has a constant value that does not change unless a cycle-slip occurs, the integer ambiguity can be estimated by taking the average value. Therefore, it is easy to compare the pseudo-range noise levels of live and re-radiated signals after removing the integer ambiguity from Equation (3), since L1 and L5 pseudo-range noise terms are dominant over carrier-phase noise. Table 1. and Figure 2. shows the measurement noise level with the SNR measured for the two cases. The signal to noise ratios (SNR) of the smartphone live signals were generally lower than re-radiated signal via a choke-ring antenna, and accordingly noise levels of the live signals were far higher.

Table 1. SNR and Noise-level with or without GNSS repeater results.

PRN	Signal Type	SNR [dB-Hz] MEAN	Noise Level [m] RMS	95%
GPS 1	Live	37.8412	2.5963	5.0824
	Repeater	44.9657	1.3853	2.6775
GPS 8	Live	33.1333	4.7425	9.2270
	Repeater	48.9788	0.8384	1.6925
GPS 27	Live	32.9258	6.8558	13.3884
	Repeater	41.7745	2.0198	3.9691
GPS 30	Live	37.0858	6.2621	14.0582
	Repeater	45.8212	1.0534	2.0153

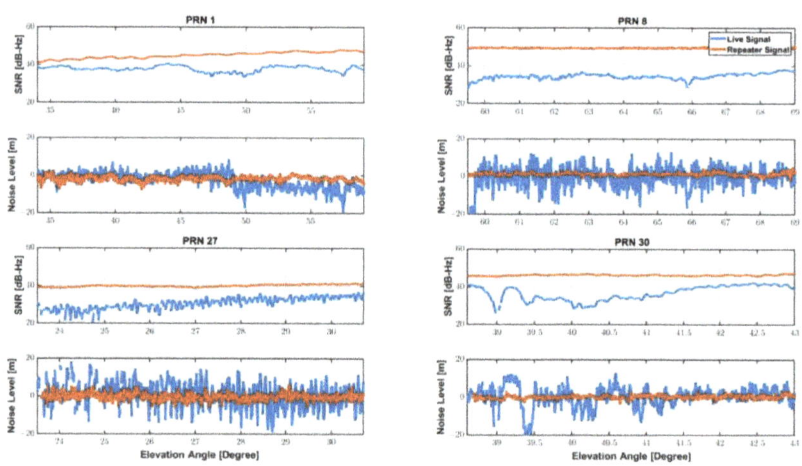

Figure 2. SNR and Noise-level of Live and Re-radiated signal.

The smartphone observed the signals passing through the GNSS repeater with an SNR value of 40 dB-Hz or higher. In contrast, the SNR of the PRN 8 was observed around 30 dB-Hz when a smartphone receiving live signal and did not exceed 40 dB-Hz despite a high elevation angle of 60° or more. Signal SNR observed by a commercial GNSS receiver generally increases as the elevation angle increases. The clear relationship between elevation and SNR, however, is not found in smartphones equipped with low-cost GNSS antenna. To make matters worse, as can be seen from PRNs 27 and 30, SNR and noise fluctuations due to multipath were significantly severe to smartphone measurements.

In summary, the Android GNSS measurement had much a larger noise level and a lower correlation to the satellite elevation than that of a typical GNSS receiver, and is very vulnerable to multipath error [12]. Therefore, we should find ways to reduce the multipath effect on the measurement results and devise a new noise modeling [13] and weighting method appropriate for noise characteristics of smartphone GNSS modules.

2.2. Unstable and Discontinuous Measurements

Smartphone vendors prioritize to maximize power consumption, and accordingly, a duty cycle technique is implemented to maintain a low power consumption because continuously operating the GNSS chipset drains the battery [7,14]. The navigation chip on the smartphone tracks the signal for 200 ms/s and is dormant for 800 ms/s when the duty cycle is on, which limits the continuous acquisition of measurements, as shown in Figure 3. While this feature would not degrade code measurement quality, it does have a significant impact on carrier-phase measurements. Without continuous tracking, several cycle slips may occur between two consecutive measurements, which severely limit the use of advanced phase-based precise positioning techniques such as RTK or PPP. To cope with the hardware limitation for the precise positioning, a function to turn off the duty-cycle, called "Force full GNSS measurements," was added at the Android 9 Pie update in 2018. However, executing the duty-cycle off mode does not completely remove the cycle slips, as shown in Figure 4.

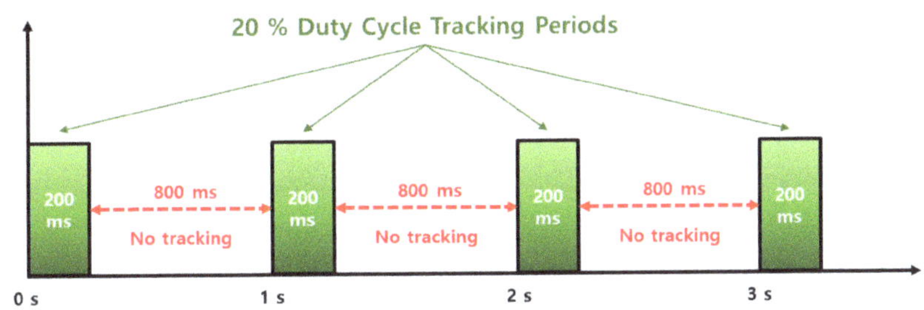

Figure 3. Duty-cycle versus time.

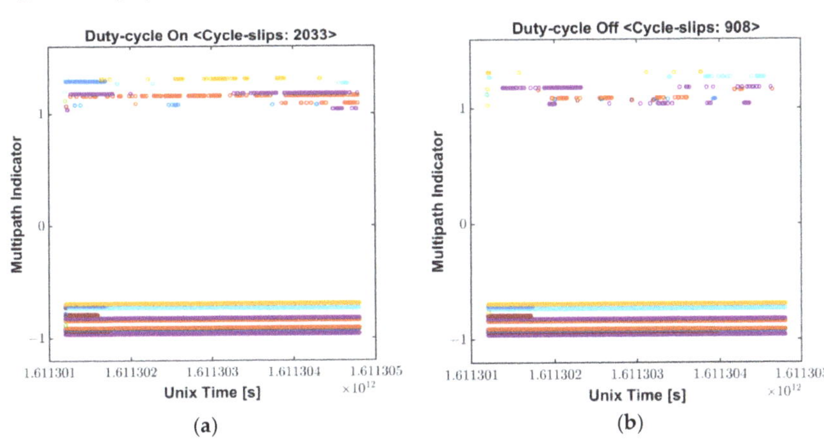

Figure 4. Cycle-slip Flag: (**a**) Duty-cycle On; (**b**) Dufy-cycle Off.

Even though the function of the duty-cycle de-activation was reported to reduce the cycle slips found in the duty-cycle on mode by more than 50%, it is not so perfect as to prevent all the measurements from being slipped. As described in Section 2.1 above, PIF antennas optimized for voice signals are highly affected by multipath errors. Although

"Force full GNSS measurements" may be effective to reduce cycle slips caused by duty-cycle function, it is difficult to completely remove all the cycle slips, therefore a substantial number of cycle-slips still remain. Because of this instability and discontinuity of Android carrier-phase, the ambiguity determination process in precise positioning should be often initialized, which results in long convergence time in PPP. Wu et al., reported that the convergence time within 0.2 m of a smartphone was up to 272 min, which was longer than twice of a geodetic receiver converging time, 116 min [6].

In RTK cases, frequent unstable measurements on a smartphone and their misdetection hinder the ambiguity to be fixed or often cause wrong ambiguity determination, resulting in float solutions for a long period or large position errors. We processed GPS L1/L5 measurement results obtained from a Mi8 to examine how much the duty-cycle off function actually contributes to making the carrier-based positioning robust. GPS observables for the two modes were logged from two Mi8 devices, and their results, processed by a RTKLIB software package [15], were compared in Figure 5. The commonly used RTK algorithm, which could provide cm-level accurate position to commercial GPS receivers, is not valid to Android measurements for both duty-cycle modes. Integer ambiguities of the carrier-phase measurements were rarely determined through all the test periods. Subsequently, position errors in each direction were barely found within 1 m. Frequent initialization and wrong position fixes, which are typical results caused by undetected cycle slips, were found in both modes. In conclusion, the "Force full GNSS measurements" function can effectively eliminate the cycle-slip caused by the duty-cycle, however, a lot of cycle-slip remains due to the structural limitations of the Android smartphone. Even with the duty-cycle turned off, errors of more than 10 m at five points and non-continuous carrier-phase measurements still do not achieve centimeter-level accuracy.

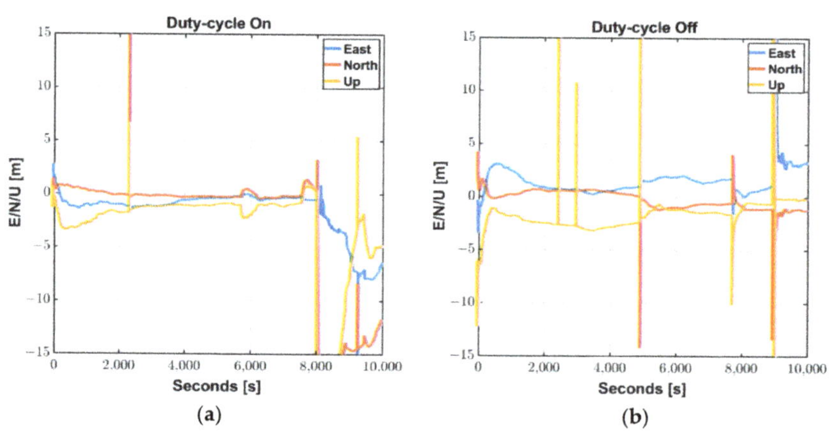

Figure 5. RTKLIB Processing Results of Mi8 L1/L5 GPS Measurements: (**a**) Duty-cycle On; (**b**) Duty-cycle Off.

2.3. *Carrier-Phase Is Not Available on All Android Smartphones*

Smartphones with Android Nougat operating systems (OS) and above allow access to Google's application program interfaces (APIs) that provide GNSS raw measurements of pseudo-range, carrier-phase, and Doppler measurements. Although Android OS is up-to-date and the dual-frequency carrier-phase available chip is included in devices, it is totally up to the manufacturer whether or not each measurement is provided to the public.

Even though code measurements filtering based on carrier-phase is essential to improving the positioning accuracy, recently released smartphones mostly do not support carrier-phase measurements. Galaxy S21 Z-Fold and Z-Flip series, flagship models of Samsung that ranked first with 19.0% market share (Table 2, [16]), do not provide carrier phase measurement to users. Xiaomi, the third ranker of the market and the manufacturer of the dual-frequency carrier-phase enabled smartphone Mi8, stopped providing carriers in its latest release, Mi 9 (see Table 3) [17].

Table 2. Worldwide smartphone sales to end users by vendor in 2021 (Gartner, March 2022 [16]).

Vendor	2021 Units (Thousands of Units)	2021 Market Share (%)
Samsung	272,327.5	19.0%
Apple	239,239.1	16.7%
Xiaomi	189,305.4	13.2%
OPPO	138,242.1	9.6%
Vivo	136,011.3	9.5%
Others	458,733.9	32.0%
Total	1,433,859.4	100.0%

Table 3. Android devices that support raw GNSS measurements (Android Developers [17]).

Model	Android Version	ADR (Carrier-Phase)	L5 Frequency	Global System
Samsung Galaxy S20/S21 (Snapdragon)	12.0	No	Yes	GPS, GLO, GAL, BDS, QZS
Xiaomi Mi9	9.0	No	Yes	GPS, GLO, GAL, BDS, QZS
Samsung Galaxy Note 10	9.0	No	Yes	GPS, GLO, GAL
LG G8 ThinQ	9.0	No	Yes	GPS, GLO, GAL
One Plus 7	9.0	No	Yes	GPS, GLO, GAL
Pixel 3	9.0	No	Yes	GPS, GLO, GAL, BDS

There are few smartphones remaining in the market that carrier-phase based precise positioning methods can be applied to. Instead, dual-frequency, multi-constellation, and Doppler capabilities has become more general. Therefore, it is necessary to find a way to improve the accuracy of smartphones more universally considering the market trends.

3. Strategy for Reliable Accuracy Improvement of Android Smartphone Positioning

3.1. Usage of the L5 Code Measurements

L5 band signals such as GPS L5 and GAL E5 have a high chipping rate, 10 times higher than that of L1 frequency signals [18]. In addition, it includes a pilot channel, which can lead to a total integration time. These new features of L5/E5 signals are less prone to multipath errors; therefore, they provide inherent noise and multipath mitigation capabilities [19].

To compare the multipath error characteristics of L1 and L5 pseudo-range measurements, we constructed a signal reception experiment for 30 min (2019/09/18 04:40:00~05:09:59 UTC) in Children's Grand Park, Seoul, Korea. There were many obstacles such as trees, as shown in Figure 6.

In Equation (4), an ionospheric-free residual is obtained by combining code and phase measurement, which contains multipath error and noise components.

$$MP_{r,i}^{s} + \varepsilon_{P_{r,i}^{s}-\Phi_{r,i}^{s}} = P_{r,i}^{s} - \Phi_{r,i}^{s} - \frac{2 \cdot f_j^2}{f_i^2 - f_j^2}\left(\Phi_{r,i}^{s} - \Phi_{r,j}^{s}\right) \quad + \lambda_i^s N_{r,i}^{s}, \tag{4}$$

We assessed multipath errors of L1 and L5 frequencies by comparing the results when GPS 30 and GAL 13 satellites were employed. The two satellites were observed from the elevation angles smaller than 30° in the southeast, whose signals are heavily affected by nearby obstacles. Figure 7 shows the multipath error and noise for the two satellites. As summarized in Table 4, the RMS and 95% percentile errors of L1 frequency measurements for the two multipath-affected satellites were about 3.4 and 6.5 m, respectively, which is about 2.4 times larger than those of L5 frequency. The maximum multipath of L5 was about 5 m, and it was reduced by 60 to 70% compared to the maximum error of L1.

(a) (b)

Figure 6. Experiment place: Children's Grand Park near the Sejong University: (a) experimental environment; (b) experimental equipment: Android smartphones.

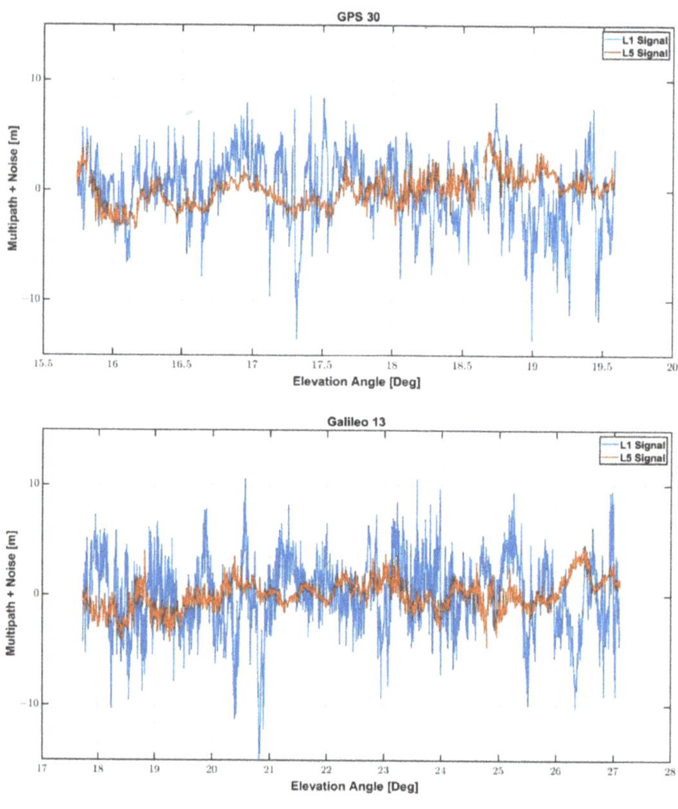

Figure 7. Multipath error and noise at GPS 30 and GAL 13 satellites in L1 and L5 Frequency.

Table 4. Multipath error and noise results.

Results		MAX [m]	STD [m]	RMS [m]	95% [m]
GPS 30	L1 Frequency	13.5913	3.3551	3.3532	6.4155
	L5 Frequency	5.4317	1.5012	1.5004	2.9514
GAL 13	L1 Frequency	15.0005	3.4839	3.4829	6.6878
	L5 Frequency	4.7303	1.4258	1.4254	2.9438

Considering the low noise level and multipath characteristics of the L5 signal, the use of L5 code measurements can achieve higher positioning results than the conventional L1 single-frequency positioning.

3.2. Enhancing Available L5 Signals by Adding Weighted L1 Signals

Noise and multipath performance of L5 measurement is better than that of L1 measurement. In the case of GPS, however, only 17 out of total 31 satellites currently transmit the L5 signal [20]. The number of available L5 GPS signals is frequently observed to be less than four, as shown for about 24 h (1 October 2022 00:00:00~24:00:00 UTC) in Figure 8, which means that smart devices cannot understand their positions for approximately half a day. Although GPS-Galileo multi-constellation might enable the L5 positioning, the increased DOP values due to a bad satellite geometry cannot guarantee high accuracy performance of positioning. At about 23:50, when the L5 satellite's Horizontal DOP (HDOP) was the highest, the L5 DOP was three times larger than that of the L1 DOP, and at about 04:50, when the L5 satellite's Vertical DOP (VDOP) was the highest, the L5 DOP was twice larger than that of the L1 DOP. Moreover, the satellite geometry might be worse than expected because it is reported that fewer Galileo satellites were usually observed by Android smartphones than general commercial receivers, and their signal tracking quality was relatively poor [6]. Therefore, L5-only positioning performance is not guaranteed to provide better results than L1-only because of the satellite geometry, although L5 signal itself has a better performance than L1.

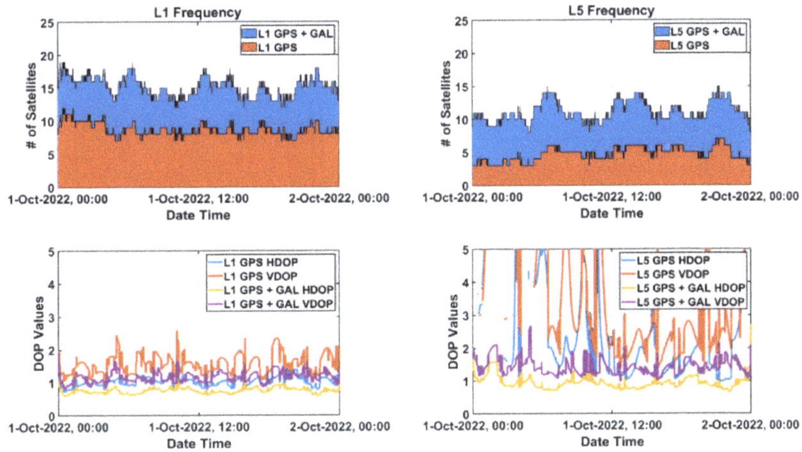

Figure 8. Available satellites status observed at SOUL reference station, Korea on 1 October 2022.

Nevertheless, the excellent L5 signal performance must be utilized, and thus a method to overcome the low number of available L5 signals and poor geometry must be considered. In this study, we suggest adding L1 signals to L5-only positioning and giving different weight by noise-level to each frequency to take advantage of L5 signals.

To implement L1/L5 GPS/Galileo algorithm, observation matrix (H) for each constellation and frequency is defined in Equation (5).

$$\begin{bmatrix} H_{L_1}^{GPS} \\ H_{L_1}^{GAL} \\ H_{L_5}^{GPS} \\ H_{L_5}^{GAL} \end{bmatrix} = \begin{bmatrix} E_{L_1}^{GPS} & \vec{1} & \vec{0} & \vec{0} & \vec{0} \\ E_{L_1}^{GAL} & \vec{0} & \vec{1} & \vec{0} & \vec{0} \\ E_{L_5}^{GPS} & \vec{0} & \vec{0} & \vec{1} & \vec{0} \\ E_{L_5}^{GAL} & \vec{0} & \vec{0} & \vec{0} & \vec{1} \end{bmatrix}, \tag{5}$$

$\vec{e}\left(\vec{e}_x, \vec{e}_y, \vec{e}_z\right)$ is a line-of-sight unit vector from a user to each satellite, E is a group of $\vec{e}$ vectors for each constellation and frequency, which is defined as m-by-3 matrix, $E_{L_1}^{GPS}$ represents the number of GPS L1 signals, m. Since user clock bias of each constellation and frequency is different, H-matrix is defined as $(m+n+o+p)$-by-7 matrix, and n, o, and p stands for the number of Galileo (GAL) L1 signals, GPS L5 signals, and GAL L5 signals, respectively, where $\vec{1}$ and $\vec{0}$ are vectors of ones and zeros, each.

As previously shown, GNSS signal tracking of smart devices is often unstable even at high elevations, and its SNR fluctuation is generally large. Therefore, it is recommended to use a noise modeling function of both elevation angle and signal strength when setting a weighting matrix. One common method for weighting the pseudo-range for the s-th satellite based on its elevation angle (el) and the SNR is Equation (6) [21]. The coefficient k_j for the L1 and L5 frequency has been added to the equation, which need to be assigned based on the noise characteristics for the frequency.

$$(\sigma_j^s)^2 = k_j^2 \cdot \frac{10^{(-0.1 \times SNR^s)}}{\sin^2(el^s)}, \tag{6}$$

According to Section 3.1, noise at L1 frequency was observed to be 3 m-level, and noise at L5 frequency was observed to be 1 m-level. Thus, we set k_1 and k_5 as 3 and 1, respectively. In addition, we assumed that the noise levels of GPS and Galileo were similar. The weighting matrix for both L1 and L5 signals in Equation (7) enables L5 measurements with good signal performance to be used as a primary source of the positioning and several L1 measurements to provide good satellite geometry.

$$R = \begin{bmatrix} R_{L_1|GPS} & 0 & 0 & 0 \\ 0 & R_{L_1|GAL} & 0 & 0 \\ 0 & 0 & R_{L_5|GPS} & 0 \\ 0 & 0 & 0 & R_{L_5|GAL} \end{bmatrix} \tag{7}$$

where

$$R_{j|GPS} = \begin{bmatrix} \left(\frac{1}{\sigma_{j|GPS}^1}\right)^2 & 0 & \cdots & 0 \\ 0 & \left(\frac{1}{\sigma_{j|GPS}^2}\right)^2 & \cdots & 0 \\ \vdots & \vdots & \ddots & \vdots \\ 0 & 0 & \cdots & \left(\frac{1}{\sigma_{j|GPS}^m}\right)^2 \end{bmatrix}, \quad R_{j|GAL} = \begin{bmatrix} \left(\frac{1}{\sigma_{j|GAL}^1}\right)^2 & 0 & \cdots & 0 \\ 0 & \left(\frac{1}{\sigma_{j|GAL}^2}\right)^2 & \cdots & 0 \\ \vdots & \vdots & \ddots & \vdots \\ 0 & 0 & \cdots & \left(\frac{1}{\sigma_{j|GAL}^m}\right)^2 \end{bmatrix}$$

3.3. L5 Pseudo-Range Correction Generation

DGNSS and Wide Area DGNSS (WADGNSS) are typical methods for mitigating code measurement errors [19,20]. Pseudorange received from two GNSS receivers located in close proximity (e.g., within a few hundred kilometers) will contain the same atmospheric errors. To eliminate this, users can mitigate common errors by using base stations with precise and known locations. Since the base receiver knows the actual geometric position between the GNSS satellites and the receiver, it can use this to create corrections from differences in pseudorange measurements received from the satellites. Correction information can be applied in real time on site using radio signals or post-processed using specialized processing software. A similar system that transmits corrections from satellites instead of ground-based transmitters is called a Wide Area Augmentation System (WAAS) or WADGNSS. A Satellite-Based Augmentation System (SBAS), sometimes used synonymously, can include satellite systems implemented in many parts of the world, such as EGNOS, MSAS, QZSS, and GAGAN.

In a similar way for general GNSS receivers, GNSS common errors, such as satellite-related and atmospheric errors, must be mitigated to improve the smartphone position accuracy. The correction messages for the code measurements usually extract or model the error for each satellite itself as defined in Radio Technical Commission for Maritime services (RTCM) Special Committee (SC)-104 version 2 [22] or Radio Technical Commission for Aeronautics (RTCA) DO-229 [23]. On the other hand, Multiple Signal Messages (MSM) that support multi-frequency carrier phase are close to observables [24,25] as defined in RTCM SC-104 version 3. While the bandwidth for the code measurement corrections is bound to several hundred bps [26], the carrier phase corrections cost high-rate transmission data has a bandwidth of at least 9600 bps [27]. However, the overall performance of accuracy and initialization of the smartphones carrier phase positioning has been reported to not be so outstanding as typical commercial receivers. Carrier-phased based positioning employed to smartphones should wait stationary over several or even up to tens of minutes in order to get cm-level results [28]. Furthermore, the long initialization time could not guarantee accurate positioning of cm-level accuracy, which is an added disadvantage [29] and often results in wrong ambiguity resolution or convergence. Therefore, code-based DGNSS positioning is more practical in Android smartphone than phase-based one, and L1/L5 dual frequency for multi-constellation GNSS would enhance the smartphone DGNSS performance.

The errors included in the L5 code measurement should be also mitigated by using correction message since the L5 signal is expected to contribute significantly to the accuracy improvement of the code-based positioning, as discussed in the previous section. Unfortunately, neither DGNSS nor Satellite Based Augmentation System (SBAS) currently provide correction for the frequency other than L1, and this strategy is not able to be applied to smartphones with the existing infrastructures [30]. Moreover, it is unreasonable and impractical to build new infrastructures for L5 signals to improve smartphone positioning accuracy. Therefore, we need to propose a new method to mitigate L5 code measurement errors of smartphones based on the existing infrastructures.

Multi-constellation GNSS errors due to the various sources, i.e., the tropospheric (T) and ionospheric (I) error and satellite-related error (δR) would be effectively mitigated if only the nearby reference station supports as many GNSS constellations as the smart device. PRC and Carrier-Phase Correction (CPC) for DGNSS are generated by Equations (8) and (9), respectively.

$$PRC_{L_1} = d_r^s - P_{r,L_1}^s - b^s + B_r = -T_r^s - I_{r,L_1}^s - \delta R_r^s - \varepsilon_{P_r^s}, \tag{8}$$

$$CPC_{L_1} = d_r^s - \Phi_{r,L_1}^s - b^s + B_r = -T_r^s + I_{r,L_1}^s - \delta R_r^s - \lambda_{L_1}^s N_{r,L_1}^s - \varepsilon_{\Phi_r^s}, \tag{9}$$

where superscript s, subscript L_1, and subscript r represent the satellite id, L1 frequency, and a reference station, respectively. d_r^s is the geometric range between the s-th satellite and the reference station, P_{r,L_1}^s and $\Phi_{r,i}^s$ are L1 frequency pseudo-range and carrier-phase. The satellite clock bias and receiver clock bias are denoted as b^s and B_r, respectively.

Range Rate Correction (RRC) accompanied with PRC are usually calculated by differentiating CPC rather than PRC, as shown in Equation (7) [31], since the time difference of PRC is far noisier than the CPC rate or Doppler [32]. By the time differencing of

Equation (10), the integer ambiguity term $\left(\lambda_{L_1}^{s} N_{r,L_1}^{s}\right)$ has been removed so that only GNSS common error related terms remain in RRC.

$$RRC_{L_1} = C\dot{P}C_{L_1} = -\dot{T}_r^s + \dot{I}_{r,L_1}^s - \delta\dot{R}_r^s - \dot{\varepsilon}_{\Phi_r^s}, \quad (10)$$

The generated PRC and RRC can mitigate the errors in pseudo-range and Doppler $\left(D_{u,L_1}^{s}\right)$ measurements of a rover station, respectively, as Equations (11) and (12) show.

$$P_{u,L_{1correct}}^{s} = P_{u,L_1}^{s} + PRC_{L_1} \approx d_u^s - b^s + B_u + \varepsilon_{P_{ru}^s}, \quad (11)$$

$$D_{u,L_{1correct}}^{s} = D_{u,L_1}^{s} + RRC_{L_1} \approx \dot{d}_u^s - \dot{b}^s + \dot{B}_u + \varepsilon_{D_{ru}^s}, \quad (12)$$

Current DGNSS standard [25] defines PRC and RRC for only L1, and no DGNSS reference station is providing DGNSS services for L2 or L5. Although L5 signals are far more beneficial for smart device positioning than L1, implementing a new L5 DGNSS infrastructure for only a smart device service is not economical. Moreover, it is impossible without revision of standards for L5.

Therefore, we suggest a practical solution for generating L5 PRC correction in a user side using current L1 PRC and SBAS message based on current DGNSS infrastructures as well as standards. The only difference in PRC between L1 and L5 is ionospheric error due to the dispersive signal characteristics [33], and thus we can get Ic_{L_5} for the L5 signals based on the L1 ionospheric correction from the SBAS messages.

$$IC_{L_5} = \gamma \cdot IC_{L_1} \quad (13)$$

where $\gamma = \left(\frac{L_1\ Frequency}{L_5\ Frequency}\right)^2$.

Since the SBAS messages, unlike PRC of DGNSS, enable generating ionospheric correction based on its pierce point, it can compensate for the L5 PRC difference for all the visible satellites as presented in Equation (13). Most GNSS receivers can compensate for the ionospheric error difference by themselves using Equations (14) and (15) because they include built-in SBAS functions.

$$PRC_{L_5} = PRC_{L_1} - (\gamma - 1)IC_{L_1} = -T_r^s - I_{r,L_1}^s - \delta R_r^s - (\gamma - 1)IC_{L_1} - \varepsilon_{P_r^s} \approx -T_r^s - I_{r,L_5}^s - \delta R_r^s - \varepsilon_{P_r^s}, \quad (14)$$

$$RRC_{L_5} = RRC_{L_1} + (\gamma - 1)\dot{I}C_{L_1} = -\dot{T}_r^s + \dot{I}_{r,L_1}^s - \delta\dot{R}_r^s + (\gamma - 1)\dot{I}C_{L_1} - \dot{\varepsilon}_{\Phi_r^s} \approx -\dot{T}_r^s + \dot{I}_{r,L_5}^s - \delta\dot{R}_r^s - \dot{\varepsilon}_{\Phi_r^s}, \quad (15)$$

Finally, the compensated PRC and RRC are fed to the pseudo-range and Doppler measurement as described in Equations (16) and (17). Allocating separate L5 frequency reference stations or related infrastructures and messages is not required by the correction generation.

$$P_{u,L_{5correct}}^{s} = P_{u,L_5}^{s} + PRC_{L_5} \approx d_u^s - b^s + B_u + \varepsilon_{P_{ru}^s}, \quad (16)$$

$$D_{u,L_{5correct}}^{s} = D_{u,L_5}^{s} + RRC_{L_5} \approx \dot{d}_u^s - \dot{b}^s + \dot{B}_u + \dot{\varepsilon}_{D_{ru}^s}, \quad (17)$$

3.4. Code Filtering Using Doppler Measurements

Our last proposal to improve the smart device positioning accuracy is using Doppler measurements instead of carrier-phase to filter the code measurements. Commonly used filtering methods such as Hatch-filter and Kalman-filter usually filter the code measurements using carrier-phase measurements because carrier-phase noises are much smaller than that of code measurements [34,35].

However, many Android smartphones do not support carrier-phases as shown in Table 3. In addition, the ambiguity of carrier-phases on Androids is not kept constant even when the duty-cycle is turned off, which causes frequent cycle slips without any notice or float ambiguity, thus remaining unfixed. The Doppler equivalent of the time derivative of carrier-phase is available from all Android devices on the market and free from the cycle slip.

In commercial geodetic receivers, Doppler measurements have a higher noise level than that of the carrier-phase time differenced measurements [9], however, the noise levels of both measurements are so similar on the Mi8. It is unclear as the chipset vendor does not provide detailed algorithms for measurement generating process, but the smartphone's chipset seems to provide carrier phase measurements generated in a different manner than commercial receivers. Figure 9 compares the noise level of the 2nd order time derivative of carrier-phase measurements against 1st order time derivative of Doppler measurements obtained from Xiaomi Mi8. Excluding the abnormal jumps due to carrier-phase discontinuity, the discrepancy is small with a magnitude of approximately 6.4×10^{-8} m.

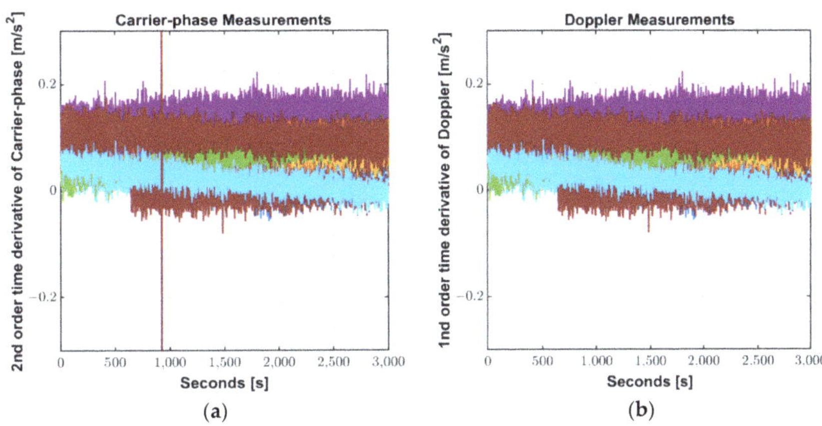

Figure 9. Noise level of measurements: (**a**) 2nd order time derivative of Carrier-phase; (**b**) 1st order time derivative of Doppler.

Sharp jumps due to unexpected cycle-slips of the phase shown in the green dashed line box in Figure 9 can cause filter initialization and divergence, but this can be prevented for Doppler without cycle slip. In addition, as mentioned in Section 2.3, not all smartphones support carrier-phase measurements, but all Doppler measurements are supported. Therefore, a practical filtering method to smart devices L1/L5 code measurements should be performed with Doppler rather than phase difference.

4. Smartphone Positioning Algorithm Implementation and Field Test Results

4.1. Configuration of Field Test

A static field test was carried out for 50 min (1 April 2019 08:00:00~08:50:00 UTC) on the rooftop of the Chung-moo building, Sejong University, Seoul, Korea to obtain raw measurements of dual-frequency smartphone Xiaomi Mi8. Figure 10 shows the actual signal-receiving environment of the static test site. Two Mi8 were installed 30 m apart from the reference station of Trimble NetR9 for DGNSS.

During the test, eight GPS satellites (PRN 1, 7, 8, 11, 16, 18, 27, 30) and three Galileo satellites (PRN 15, 27, 30) were observed under the open-sky. Among them, four GPS (PRN 1, 8, 27, 30) and three Galileo (PRN 15, 27, 30) satellites provided L5 frequency signals. To use SBAS message for the ionospheric compensation, MTSAT Satellite Augmentation System (MSAS) PRN 129 message in #RAWSBASFRAMEA format was transmitted from a Novatel flexpak 6 receiver and decoded. Corrections for L1/L5 DGNSS, PRC_{L_1}, and RRC_{L_1} were obtained by the transmitted RTCM v2 message from the Trimble NetR9 receiver. PRC_{L_5} and RRC_{L_5} were generated after compensating the ionospheric error by the transmitted #RAWSBASFRAMEA from the flexpak 6 receiver as described in Equations (14) and (15).

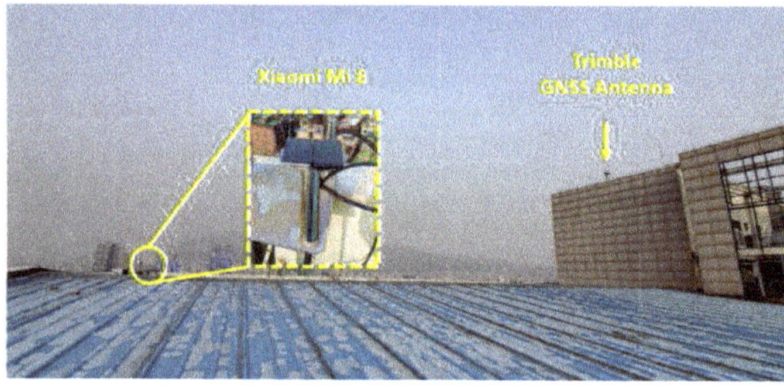

Figure 10. Test Environment at the Roof of the Chung-moo building in Sejong University.

4.2. L1/L5 DGNSS Results

Figure 11 and Table 5 show the L1 and L1/L5 DGNSS positioning results of Mi8 smartphone. The masks of elevation angle and SNR were set to 20° and 20 dB-Hz, respectively. L1 DGNSS used the L1 PRC and RRC generated for all available GNSS by the existing in infrastructure. Then, we estimated ionospheric error for each signal of other GNSS using ionospheric vertical delays at the grid points surrounding its pierce point. Subsequently, in L1/L5 DGNSS, the L5 PRC and RRC compensated by the received SBAS message according to Equations (14) and (15) were fed to the Android L5 GNSS measurements.

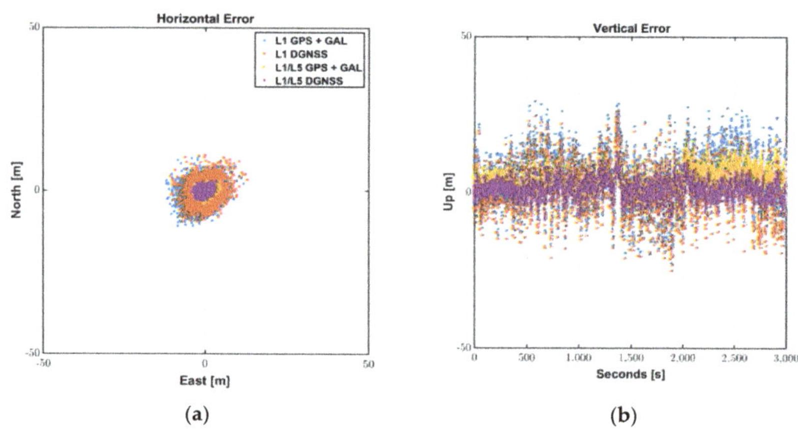

Figure 11. L1/L5 DGNSS positioning results: (**a**) Horizontal Error; (**b**) Vertical Error.

Table 5. Statistics of L1/L5 DGNSS positioning results.

Positioning Results		MEAN [m]	STD [m]	RMS [m]	95% [m]
L1 GPS + GAL	Horizontal	1.7001	4.4629	4.7750	9.3116
	Vertical	5.0637	8.3905	9.7988	19.0114
L1 DGNSS	Horizontal	0.2916	4.3820	4.3909	8.6137
	Vertical	1.0586	8.0881	8.1558	15.8754
L1/L5 GPS + GAL	Horizontal	0.7607	1.4872	1.6703	3.3183
	Vertical	4.6150	3.9226	6.0564	11.1689
L1/L5 DGNSS	Horizontal	0.2991	1.4023	1.4336	2.7209
	Vertical	1.2965	3.1716	3.4259	6.8974

The vertical average error, which was larger than 5 m for L1 GPS + GAL, was reduced to 1.3 m by the compensated L1/L5 DGNSS PRC generation. Moreover, better signal quality by adding L5 satellites made it possible for the Android user get the horizontal position of 2.7 m in 95%. Providing good satellite geometry by L1 multi-constellation GNSS and enabling L5 signals to be available by compensating L5 DGNSS correction would be more beneficial to enhance the position availability and accuracy of users, especially with low satellite visibility in urban or mountainous areas.

Therefore, the Android user can successfully mitigate L5 GNSS error as well as L1 error using existing L1 reference stations and received SBAS messages, without adding infrastructures or modifying the related correction message standards.

4.3. L1/L5 Doppler-Based Kalman-Filter Results

Kalman filter is a recursive filter that estimates the state of a linear dynamics system based on random noise statistical characteristics. A two-step approach to prediction and update can reduce GNSS positioning errors from an optimization point of view. In this study, we defined a state of Kalman filter at time k, 12×1 vector of X_k, which consists of position, velocity, clock bias, clock drift, and Inter-signal Bias, as shown in Equation (18).

$$X_k = \begin{bmatrix} \vec{x} & B_{L_1}^{GPS} & B_{L_1}^{GAL} & \dot{\vec{x}} & \dot{B}_{L_1}^{GPS} & \dot{B}_{L_1}^{GAL} & ISB^{GPS} & ISB^{GAL} \end{bmatrix}^T \tag{18}$$

where $\vec{x}$ is three-dimensional position vector, and $\dot{\vec{x}}$ is its velocity vector. The user clock biases and user clock drifts for GPS and Galileo satellites of L1 frequency are denoted as $B_{L_1}^{GPS}$, $B_{L_1}^{GAL}$, $\dot{B}_{L_1}^{GPS}$ and $\dot{B}_{L_1}^{GAL}$, respectively. Inter-signal Bias, ISB, represents the difference of L1/L5 frequency, which is assumed constant during the session. Dynamics of the state were modeled as the first order state prediction based on the relationship between $\begin{bmatrix} \vec{x} & B \end{bmatrix}$ and $\begin{bmatrix} \dot{\vec{x}} & \dot{B} \end{bmatrix}$, while the ISB was set as a random walk [36]. The noise level ratio of L5 signals to L1, described in Section 3.1, was considered to define covariance matrix of the observables, R_k, for the measurement update. The measurement vector and $(m + n + o + p) \times 1$ vector of Z_k are obtained from observables for L1 and L5. Unlike Kalman filters for general GNSS receivers, carrier-phase observables were replaced with Doppler to reliably filter the Android pseudo-range noise as described in Equation (19).

$$Z_k^s = \begin{bmatrix} \left(P_{u,L_{1correct}}^s - e \cdot R^s + b^s\right) \\ \left(P_{u,L_{5correct}}^s - e \cdot R^s + b^s\right) \\ \left(D_{u,L_{1correct}}^s - e \cdot \dot{R}^s + \dot{b}^s\right) \\ \left(D_{u,L_{5correct}}^s - e \cdot \dot{R}^s + \dot{b}^s\right) \end{bmatrix}, \tag{19}$$

The results of L1/L5 Kalman filter are shown in Figure 12 and summarized in Table 6. The positioning accuracy was improved by 20 to 40% after applying the suggested Doppler-based filter so that the RMS values were reduced to 1.2 m horizontally and 2.3 m vertically. During 95% of the periods of the test sessions, the Android devices were able to provide positions with horizontal accuracy of 2.3 m, which shows that L1/L5 DGNSS was properly filtered by the suggested filter construction.

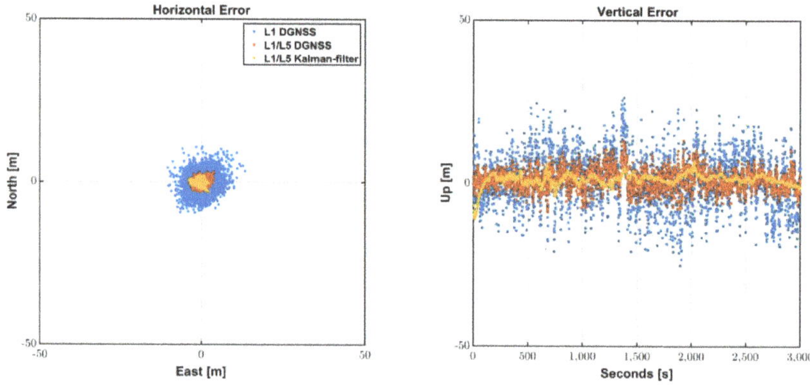

(a) (b)

Figure 12. L1/L5 Kalman-filter positioning results: (**a**) Horizontal Error; (**b**) Vertical Error.

Table 6. Statistics of L1/L5 Kalman-filter positioning results.

Positioning Results		MEAN [m]	STD [m]	RMS [m]	95% [m]
L1 DGNSS	Horizontal	0.2916	4.3820	4.3909	8.6137
	Vertical	1.0586	8.0881	8.1558	15.8754
L1/L5 DGNSS	Horizontal	0.2991	1.4023	1.4336	2.7209
	Vertical	1.2965	3.1716	3.4259	6.8974
L1/L5 KF	Horizontal	0.2449	1.1304	1.1564	2.3188
	Vertical	1.2906	1.9074	2.3028	4.1931

5. Conclusions

Since Google's announcement of providing GNSS raw measurements in 2016, many expected that accurate location information with only several centimeters of error would be provided soon from the palm of the hand. To meet the expectations of the public, many researchers have made efforts to apply PPP or RTK technology to Android devices, and recently published papers have presented the results of feasibility studies. However, considering the smart device manufacturing technologies, market status, and implemented infrastructure for improving positioning accuracy, it might be difficult to realize most of the introduced technologies.

This study discussed why overly optimistic use of high-accurate applications such as RTK or PPP with Android raw measurements alone are difficult to be implemented. LP antenna optimized for voice communication reduces the quality of the GNSS signal and makes the signals be more vulnerable by multipath. Duty-cycle technique for battery saving obstructs measuring continuous carrier-phase, which makes ambiguity estimation of RTK and PPP meaningless. Even when the duty-cycle is turned off, frequent initializations of carrier tracking have been reported. The absence of infrastructures to improve L5 pseudo-range measurements would be a major obstacle when a real service is considered.

To overcome the limitation of smart devices and solve the problems in the real world, practical methods for improving the positioning were proposed as follows:

1. Using weighted L5 code measurements with less-weighted L1 measurements is effective in reducing positioning errors due to the noise and multipath.
2. Feeding L5 PRC after compensating L1 PRC currently in service is efficient way to mitigate the GNSS measurement errors without implementing new infrastructures for L5 service.
3. In the case of the filtering method using Doppler measurements, it can be used even in smartphones that do not support carrier-phase measurements. It also has the advantage of not having to detect an ambiguity cycle sip.

The L5 weighting and Doppler-based filtering method can guarantee that the Android devices provide the positions with an accuracy of 2.32 m horizontally and 4.19 m vertically for most of the time (95%). In addition, the Doppler-based filtering was helpful to increase the reliability of the location information so that the maximum error was bounded within 5 m.

The methods proposed in this paper are of great practical significance as they can be applied to any Android device on the market using the existing infrastructure or service; moreover, they could guarantee assured achievable performance. Therefore, we expect our proposed methods to be a useful and practical solution for not only founding the basis of future Android location-related 4th industrial technologies, but also for improving the location performance of smart devices themselves.

Author Contributions: All the authors have contributed to the presented work. The first author, J.Y. conducted smartphone algorithm proposals and experiments and processed experimental research datasets. C.L. implemented a baseline user SW and correction generating module. B.P. suggested the original concept of the system and supervised its development and the direction of the research. All authors participated in formulating the idea and in discussing the proposed approach and results. All authors have read and agreed to the published version of the manuscript.

Funding: This research was supported by a grant from National R&D Project of "Development of Ground-based Centimeter level High Accuracy Augmentation Technologies for Maritime PNT" of Korea institute of Marine Science & Technology Promotion (KIMST) funded by Ministry of Oceans and Fisheries, Korea (20200451).

Institutional Review Board Statement: Not applicable.

Informed Consent Statement: Not applicable.

Conflicts of Interest: The authors declare no conflict of interest. The funders had no role in the design of the study; in the collection, analyses, or interpretation of data; in the writing of the manuscript, or in the decision to publish the results.

References

1. Khan, Z. Your Guide to Location Services in Android 12. Available online: https://www.androidauthority.com/android-location-services-3219012/ (accessed on 14 November 2022).
2. Merano, M. Tesla FSD "Actual" Smart Summon Is Almost Done. Available online: https://www.teslarati.com/tesla-fsd-actual-smart-summon/ (accessed on 14 November 2022).
3. GSARMT Force. Using GNSS Raw Measurements on Android Devices: White Paper. *Eur. GNSS Agency* **2017**, *10*, 449581. [CrossRef]
4. Park, B.; Lee, J.; Kim, Y.; Yun, H.; Kee, C. DGPS Enhancement to GPS NMEA Output Data: DGPS by Correction Projection to Position-Domain. *J. Navig.* **2013**, *66*, 249–264. [CrossRef]
5. Yoon, D.; Kee, C.; Seo, J.; Park, B. Position Accuracy Improvement by Implementing the DGNSS-CP Algorithm in Smartphones. *Sensors* **2016**, *16*, 910. [CrossRef]
6. Wu, Q.; Sun, M.; Zhou, C.; Zhang, P. Precise Point Positioning Using Dual-Frequency GNSS Observations on Smartphone. *Sensors* **2019**, *19*, 2189. [CrossRef]
7. Fortunato, M.; Critchley-Marrows, J.; Siutkowska, M.; Ivanovici, M.; Benedetti, E.; Roberts, W. Enabling High Accuracy Dynamic Applications in Urban Environments Using PPP and RTK on Android Multi-Frequency and Multi-GNSS Smartphones. In Proceedings of the European Navigation Conference, ENC 2019, Warsaw, Poland, 9–12 April 2019; pp. 1–9.
8. Chen, B.; Gao, C.; Liu, Y.; Sun, P. Real-Time Precise Point Positioning with a Xiaomi MI 8 Android Smartphone. *Sensors* **2019**, *19*, 2835. [CrossRef]
9. Yun, J.; Park, B.; Lim, C.; Yoon, H.; Lee, Y.; Lee, Y. Performance Comparison of L1 and L5 GNSS Measurements Using Android Smartphone. In Proceedings of the ISGNSS 2019 in Conjunction with IPNT Conference, Jeju, Korea, 29 October–1 November 2019; pp. 423–425.
10. Darugna, F.; Wübbena, J.; Ito, A.; Wübbena, T.; Wübbena, G.; Schmitz, M. RTK and PPP-RTK Using Smartphones: From Short-Baseline to Long-Baseline Applications. In Proceedings of the 32nd International Technical Meeting of the Satellite Division of the Institute of Navigation, ION GNSS+, Miami, FL, USA, 16–20 September 2019.
11. Gao, R.; Xu, L.; Zhang, B.; Liu, T. Raw GNSS Observations from Android Smartphones: Characteristics and Short-Baseline RTK Positioning Performance. *Meas. Sci. Technol.* **2021**, *32*, 084012. [CrossRef]
12. Park, K.; Seo, J. Single-Antenna-Based GPS Antijamming Method Exploiting Polarization Diversity. *IEEE Trans. Aerosp. Electron. Syst.* **2021**, *57*, 919–934. [CrossRef]

13. Lee, Y.; Park, B. Nonlinear Regression-Based GNSS Multipath Modelling in Deep Urban Area. *Mathematics* **2022**, *10*, 412. [CrossRef]
14. Warnant, R.; de Vyvere, L.; Warnant, Q. Positioning with Single and Dual Frequency Smartphones Running Android 7 or Later. In Proceedings of the 31st International Technical Meeting of the Satellite Division of the Institute of Navigation, ION GNSS+, Miami, FL, USA, 24—28 September 2018; pp. 284–303.
15. Takasu, T. Open-Source Program Package for RTK-GPS. Available online: https://www.rtklib.com/rtklib_reference.htm (accessed on 26 October 2022).
16. Gartner Inc. Gartner Says Global Smartphone Sales Grew 6% in 2021. Available online: https://www.gartner.com/en/newsroom/press-releases/2022-03-01-4q21-smartphone-market-share (accessed on 26 October 2022).
17. Google Android Developers Raw GNSS Measurements—Android Devices That Support Raw GNSS Measurements. Available online: https://developer.android.com/guide/topics/sensors/gnss?hl=en (accessed on 26 October 2022).
18. Leclère, J.; Landry, R., Jr.; Botteron, C. Comparison of L1 and L5 Bands GNSS Signals Acquisition. *Sensors* **2018**, *18*, 2779. [CrossRef]
19. Circiu, M.; Felux, M.; Thölert, S.; Antreich, F.; Vergara, M.; Sgammini, M.; Enneking, C.; Pullen, S. Evaluation of GPS L5 and Galileo E1 and E5a Performance for Future Multi Frequency and Multi Constellation GBAS. In Proceedings of the Institute of Navigation International Technical Meeting 2015 (ITM 2015), Dana Point, CA, USA, 26 28 January 2015.
20. NOAA GPS.Gov. Available online: https://www.gps.gov/systems/gps/space/ (accessed on 26 October 2022).
21. Tay, S.; Marais, J. Weighting Models for GPS Pseudorange Observations for Land Transportation in Urban Canyons. In Proceedings of the 6th European Workshop on GNSS Signals and Signal Processing, Noordwijk, The Netherlands, 5–7 December 2012; p. 4.
22. Yoon, H.; Seok, H.; Lim, C.; Park, B. An Online SBAS Service to Improve Drone Navigation Performance in High-Elevation Masked Areas. *Sensors* **2020**, *20*, 3047. [CrossRef]
23. RTCA DO-229; Minimum Operational Performance Standards for Global Positioning System. Wide Area Augmentation System Airborne Equipment: Washington DC, USA, 2006.
24. Park, B.; Kee, C. The Compact Network RTK Method: An Effective Solution to Reduce GNSS Temporal and Spatial Decorrelation Error. *J. Navig.* **2010**, *63*, 343–362. [CrossRef]
25. RTCM Standard 10402.3; Differential GNSS (Global Navigation Satellite Systems) Service: Version 3, RTCM 10403.2 RTCM Paper 104-2013. Radio Technical Commission for Maritime Services: Washington, DC, USA, 2010.
26. Kim, J.; Song, J.; No, H.; Han, D.; Kim, D.; Park, B.; Kee, C. Accuracy Improvement of DGPS for Low-Cost Single-Frequency Receiver Using Modified Flächen Korrektur Parameter Correction. *ISPRS Int. J. Geoinf.* **2017**, *6*, 222. [CrossRef]
27. Song, J.; Park, B.; Kee, C. Comparative Analysis of Height-Related Multiple Correction Interpolation Methods with Constraints for Network RTK in Mountainous Areas. *J. Navig.* **2016**, *69*, 991–1010. [CrossRef]
28. Wanninger, L.; Heßelbarth, A. GNSS Code and Carrier Phase Observations of a Huawei P30 Smartphone: Quality Assessment and Centimeter-Accurate Positioning. *GPS Solut.* **2020**, *24*, 64. [CrossRef]
29. Lachapelle, G.; Gratton, P.; Horrelt, J.; Lemieux, E.; Broumandan, A. Evaluation of a Low Cost Hand Held Unit with GNSS Raw Data Capability and Comparison with an Android Smartphone. *Sensors* **2018**, *18*, 4185. [CrossRef]
30. Lim, C.; Shin, D.; Park, B.; Kee, C.; Seo, S.; Park, J.; Cho, A. L1 SFMC SBAS System to Improve the Position Accuracy of Android Device. In Proceedings of the 31st International Technical Meeting of the Satellite Division of the Institute of Navigation, ION GNSS+, Miami, FL, USA, 24—28 September 2018; pp. 455–481.
31. Park, B.; Kim, J.; Kee, C.; Cleveland, A.; Parsons, M.; Wolfe, D.; Kalafus, R. RRC Unnecessary for DGPS Messages. *IEEE Trans. Aerosp. Electron. Syst.* **2006**, *42*, 1149–1160. [CrossRef]
32. Kee, C.; Park, B.; Kim, J.; Cleveland, A.; Parsons, M.; Wolfe, D. A Guideline to Establish DGPS Reference Station Requirements. *J. Navig.* **2008**, *61*, 99–114. [CrossRef]
33. Park, B.; Lim, C.; Wang, J.; Morton, Y.T.J. Horizontal Drift Velocity and Dimensions of Ionospheric Irregularities Using ROT from a GNSS Receiver Array. *IEEE Trans. Geosci. Remote Sens.* **2022**, *60*, 1–14. [CrossRef]
34. Park, B.; Sohn, K.; Kee, C. Optimal Hatch Filter with an Adaptive Smoothing Window Width. *J. Navig.* **2008**, *61*, 435–454. [CrossRef]
35. Park, B.; Lim, C.; Yun, Y.; Kim, E.; Kee, C. Optimal Divergence-Free Hatch Filter for GNSS Single-Frequency Measurement. *Sensors* **2017**, *17*, 448. [CrossRef]
36. Zhou, F.; Dong, D.; Li, P.; Li, X.; Schuh, H. Influence of Stochastic Modeling for Inter-System Biases on Multi-GNSS Undifferenced and Uncombined Precise Point Positioning. *GPS Solut.* **2019**, *23*, 59. [CrossRef]

sensors

MDPI

Article

Optimizing the Use of RTKLIB for Smartphone-Based GNSS Measurements

Tim Everett [1], Trey Taylor [2], Dong-Kyeong Lee [2,*] and Dennis M. Akos [2]

[1] RTK Consultants LLC, Niwot, CO 80503, USA; tim.everett3@gmail.com
[2] Aerospace Engineering Sciences, University of Colorado Boulder, Boulder, CO 80309, USA; fred.tayloriii@colorado.edu (T.T.); dma@colorado.edu (D.M.A.)
* Correspondence: dongkyeong.lee@colorado.edu

Abstract: The Google Smartphone Decimeter Challenge (GSDC) was a competition held in 2021, where data from a variety of instruments useful for determining a phone's position (signals from GPS satellites, accelerometer readings, gyroscope readings, etc.) using Android smartphones were provided to be processed/assessed in regard to the most accurate determination of the longitude and latitude of user positions. One of the tools that can be utilized to process the GNSS measurements is RTKLIB. RTKLIB is an open-source GNSS processing software tool that can be used with the GNSS measurements, including code, carrier, and doppler measurements, to provide real-time kinematic (RTK), precise point positioning (PPP), and post-processed kinematic (PPK) solutions. In the GSDC, we focused on the PPK capabilities of RTKLIB, as the challenge only required post-processing of past data. Although PPK positioning is expected to provide sub-meter level accuracies, the lower quality of the Android measurements compared to geodetic receivers makes this performance difficult to achieve consistently. Another latent issue is that the original RTKLIB created by Tomoji Takasu is aimed at commercial GNSS receivers rather than smartphones. Therefore, the performance of the original RTKLIB for the GSDC is limited. Consequently, adjustments to both the code-base and the default settings are suggested. When implemented, these changes allowed RTKLIB processing to score 5th place, based on the performance submissions of the prior GSDC competition. Detailed information on what was changed, and the steps to replicate the final results, are presented in the paper. Moreover, the updated code-base, with all the implemented changes, is provided in the public repository. This paper outlines a procedure to optimize the use of RTKLIB for Android smartphone measurements, highlighting the changes needed given the low-quality measurements from the mobile phone platform (relative to the survey grade GNSS receiver), which can be used as a basis point for further optimization for future GSDC competitions.

Keywords: Android; smartphone; RTK; PPK; GNSS; RTKLIB; decimeter; Google; carrier; DGNSS

Citation: Everett, T.; Taylor, T.; Lee, D.-K.; Akos, D.M. Optimizing the Use of RTKLIB for Smartphone-Based GNSS Measurements. *Sensors* **2022**, *22*, 3825. https://doi.org/10.3390/s22103825

Academic Editors: Jacek Paziewski, Augusto Mazzoni, Yang Gao and Michael Fu

Received: 14 April 2022
Accepted: 16 May 2022
Published: 18 May 2022

1. Introduction

1.1. Real-Time Kinematic (RTK)

Multiple global navigation satellite system (GNSS) measurements exist that can be used for position, velocity, and time (PVT) computations. These include code, carrier, and doppler measurements. RTK uses carrier phase measurements to provide more precise PVT than code-based positioning. Theoretically, RTK is able to provide centimeter-level positioning to its users [1]. Multiple GNSS receivers exist that provide RTK solutions, but commercial software packages are expensive [2]. A list of the available open-source and closed-source GNSS packages with RTK capabilities is provided by the National Geodetic Survey (NGS) [3]. RTKLIB is one of the open-source packages available to the public.

1.2. What Is RTKLIB?

RTKLIB was first developed in April 2006 by Tomoji Takasu, first released to the public as an open source in January 2009, and is currently distributed under a BSD 2-clause license [4]. The latest version from Takasu's branch is 2.4.3 b34, which was released on 29 December 2020, but multiple forks exist, including the one from the author of this paper, Tim Everett. For user interface, RTKLIB offers graphic user interface (GUI) access points (AP) on Windows and Console apps on both Windows and Linux environments. RTKLIB is a powerful GNSS data analysis tool that can process various types of GNSS, data including receiver independent exchange format (RINEX) [5] and radio technical commission for maritime services (RTCM) formats [6]. Using the input data, it is able to carry out various position computations, including single point positioning (SPP), differential GNSS (DGNSS), real-time kinematic GNSS (RTK-GNSS), post-processing kinematic GNSS (PPK-GNSS), and precise point positioning (PPP). Over the years, numerous updates have been made to the RTKLIB for the support of multi-constellation and multi-frequencies. For this paper, we look at the PPK solution, since it does not require any real-time navigation solution computations, and PPK is able to provide higher accuracy position solutions than SPP and PPP [7]. Furthermore, all the GNSS observables and ephemeris information from nearby base stations required for PPK are available from various public servers.

1.3. *Google Smartphone Decimeter Challenge (GSDC)*

GSDC was a Kaggle competition held by Google in 2021 (with a second expected offering in 2022). The objective was to generate the most accurate position solutions for a large number of raw observation data sets collected using Android devices inside a moving vehicle in the San Francisco Bay area [8]. In 2021, two sets of data, training and test, were provided by Google on Kaggle [9]. There were 29 routes and 73 smartphone data in the train set, and 19 routes and 45 data in the test set. There were multiple data for each route because multiple Android devices, spaced approximately 20 cm apart, were logging the data at the same time. During the drive, truth data was also collected using a NovAtel SPAN ISA-100C unit. This is because the SPAN unit is able to provide a horizontal accuracy of up to 0.01 m RMS [10]. As the smartphones were not in the same location as the antenna, their relative displacements with respect to the NovAtel antenna were taken into account for the smartphone truth determination. The training set was provided with the corresponding truth positions, but the truth for the test set was used for the Kaggle leaderboard determination. The data collected was provided in both RINEX and raw GnssLogger format. Although the raw format is what is logged by the smartphone, Google also provided a translation of the raw measurements into the RINEX format for those who preferred the more commonly used format. The measurements were filtered by Google during the conversion, and more details on this are provided in the Data Conversion section.

The drive routes can be classified into highway, street, and downtown, depending on the amount of expected multipath in the signals [11]. The highway is mostly open-sky, the street is open-sky with some attenuation from trees and nearby buildings, and downtown is heavily affected by attenuation and multipath from high-rise buildings. For the phone setup, the devices were located on the dashboard of the vehicles with no ground plane, which means that they were exposed to significant amounts of additional signal attenuation, and multipath at least at the level as to what would be expected if the antennas were placed on the roof of the vehicle.

The scores for the competition were determined by averaging the 50th percentile and 95th percentile errors of the computed latitude and longitude positions at each time epoch. The errors were computed using truth provided by the NovAtel SPAN unit. As the truth was not released for the test set, the scores were available upon submitting a list of the computed positions to the Kaggle website. In order to allow the participants to test the Kaggle website interface, a baseline results file was provided by Google as well. The algorithms used by Google to obtain the positions in the baseline file were proprietary, but some details about how the results were processed were provided to all participants [12].

1.4. *Performance of RTKLIB and GSDC Participants*

Coupled with commercial receivers such as u-blox, RTKLIB has been able to provide centimeter level of accuracy [13]. However, as RTKLIB was originally designed for commercial GNSS receivers with survey type measurements, multiple adaptations are required for its use in GSDC, where only the GNSS measurements from smartphones are available. The smartphones used in GSDC (Google Pixel4 variants, Samsung S series variants, and Xiaomi Mi8) have several differences from commercial GNSS receivers, particularly with respect to antenna design/performance [14]. Consequently, the quality of the measurements is worse, leading to greater signal noise, increased number of carrier phase cycle slips, and increased multipath. This issue is illustrated by Qiong [15], when raw GNSS measurements from a Xiaomi Mi8 smartphone and a geodetic receiver collected at the same time were processed using RTKLIB. When GPS L1 measurements from both devices were processed using the same kinematic PPP mode, the geodetic receiver had horizontal position errors of approximately 1 m, while the smartphone's errors were approximately 3–5 m, due to fewer observed satellites and lower carrier to noise ratio density (C/N_0) for the tracked satellites.

In the 2021 GSDC, the 1st-place winner used factor graph optimization, coupled with Takasu's version of the RTKLIB, to obtain a score of 1.62 m [11]. Other teams used Google's baseline Android navigation engine [16] or their own proprietary navigation engines [17]. In the Results section of the paper, the performance of the updated RTKLIB against other navigation engines will be provided using the official Kaggle scoring system. The comparison candidates are Google's baseline solution and the scores of other participants. The goal will be to assess where the performance of the suggested RTKLIB adaptation stands among other available navigation engines used in the competition.

2. Strategies

2.1. *PPK Solution*

Post-processed kinematics, or PPK, is a popular use case for RTKLIB, as it allows for high precision position results for previously recorded kinematic data. The PPK solution can utilize the provided GNSS observation file, an additional observation file from a nearby base station, and multi-constellation ephemeris files in order to perform precise positioning at each epoch. In addition to pseudorange measurements, PPK also utilizes the available carrier phase in the navigation solution. PPK in RTKLIB also allows the user to implement a noncausal Kalman filter computation, operating both forwards and backwards in time. By doing this, RTKLIB is able to detect directional anomalies, such as cycle slips, and combine both the results to produce results with better accuracy and integrity.

2.2. *Kinematic Solution Algorithm*

The navigation engine of RTKLIB is based on extended Kalman filter (EKF) and double differencing with respect to a nearby base station. Although the different modes and algorithms implemented for the RTKLIB are available in the user manual [18], a summary of the relevant algorithms utilized for GSDC is provided below. Moreover, although the manual explains the algorithms for estimating the position and velocity states of the receiver, the latest RTKLIB also includes the added acceleration states as well, so the equations below address this update. The EKF is used to compute the estimated state vector $\hat{x}_k$, and its covariance matrix $\boldsymbol{P}_k$, for epoch t_k, using a measurement vector $\boldsymbol{y}_k$.

$$\hat{x}_k^+ = \hat{x}_k^- + \boldsymbol{K}_k(\boldsymbol{y}_k - \boldsymbol{h}(\hat{x}_k^-)) \tag{1}$$

$$\boldsymbol{P}_k^+ = (\boldsymbol{I} - \boldsymbol{K}_k\boldsymbol{H}(\hat{x}_k^-))\boldsymbol{P}_k^- \tag{2}$$

$$\boldsymbol{P}_k^+ = (\boldsymbol{I} - \boldsymbol{K}_k\boldsymbol{H}(\hat{x}_k^-))\boldsymbol{P}_k^- \tag{3}$$

$$\hat{x}_{k+1}^- = \boldsymbol{F}_k^{k+1}\hat{x}_k^+ \tag{4}$$

$$\boldsymbol{P}_{k+1}^- = \boldsymbol{F}_k^{k+1}\boldsymbol{P}_k^+\boldsymbol{F}_k^{k+1T} + \boldsymbol{Q}_k^{k+1} \tag{5}$$

where the $+/-$ signs indicate whether it is before or after the measurement update. $\boldsymbol{F}$ and $\boldsymbol{Q}$ are the state transition matrix and the system noise covariance matrix, respectively, and m is the number of satellites observed and used in the computation. For a given receiver r, stationary base station b, and satellites j and k, we can take the double difference of the pseudorange P, and carrier phase measurements ϕ.

$$P^{jk}_{rb,i} = \rho^{jk}_{rb} + \epsilon_P \tag{6}$$

$$\Phi^{jk}_{rb,i} = \rho^{jk}_{rb} + \lambda\left(B^{j}_{rb,i} - B^{k}_{rb,i}\right) + \epsilon_\phi \tag{7}$$

where ρ is the geometric range, ϵ is the noise, λ and i are the wavelength and corresponding index of the GNSS signal which is, in this case, L1 and L5, and $\boldsymbol{B}$ is the single differenced integer ambiguity. The double difference effectively removes the atmospheric error effects from the measurements. The state vector $\boldsymbol{x}$ we are solving for would include the position $\boldsymbol{r_r}$, velocity $\boldsymbol{v_r}$, and acceleration $\boldsymbol{a_r}$ of the receiver.

$$\boldsymbol{x} = \begin{bmatrix} \boldsymbol{r_r} \\ \boldsymbol{v_r} \\ \boldsymbol{a_r} \\ \boldsymbol{B_1} \\ \boldsymbol{B_5} \end{bmatrix} \quad \boldsymbol{B_i} = \begin{bmatrix} \boldsymbol{B^1_{rb,i}} \\ \boldsymbol{B^2_{rb,i}} \\ \cdots \\ \boldsymbol{B^m_{rb,i}} \end{bmatrix} \tag{8}$$

The measurement vector $\boldsymbol{y}$ is defined with respect to a single satellite with the highest elevation indexed as 1. Although this is effective for a high elevation satellite in open-sky conditions, it is important to note that if this satellite has anomalies, this could negatively affect the quality of the final solution.

$$\boldsymbol{y} = \begin{bmatrix} \boldsymbol{\Phi_1} \\ \boldsymbol{\Phi_5} \\ \boldsymbol{P_1} \\ \boldsymbol{P_5} \end{bmatrix} \quad \boldsymbol{\phi_i} = \begin{bmatrix} \phi^{12}_{rb,i} \\ \phi^{13}_{rb,i} \\ \cdots \\ \phi^{1m}_{rb,i} \end{bmatrix} \quad \boldsymbol{P_i} = \begin{bmatrix} P^{12}_{rb,i} \\ P^{13}_{rb,i} \\ \cdots \\ P^{1m}_{rb,i} \end{bmatrix} \tag{9}$$

For the measurement update, we write the measurement model vector $\boldsymbol{h}(\boldsymbol{x})$, the matrix of partial derivatives $\boldsymbol{H}(\boldsymbol{x})$, and the covariance matrix of measurement errors $\boldsymbol{R}$, as follows:

$$\boldsymbol{h}(\boldsymbol{x}) = \begin{bmatrix} \boldsymbol{h_{\Phi,1}} \\ \boldsymbol{h_{\Phi,5}} \\ \boldsymbol{h_{P,1}} \\ \boldsymbol{h_{P,5}} \end{bmatrix} \quad \boldsymbol{H}(\boldsymbol{x}) = \begin{bmatrix} -\boldsymbol{DE} & 0 & \lambda_1\boldsymbol{D} & 0 \\ -\boldsymbol{DE} & 0 & 0 & \lambda_5\boldsymbol{D} \\ -\boldsymbol{DE} & 0 & 0 & 0 \\ -\boldsymbol{DE} & 0 & 0 & 0 \end{bmatrix} \tag{10}$$

$$\boldsymbol{R} = diag\left(\boldsymbol{DR_{\Phi,1}D^T}, \boldsymbol{DR_{\Phi,5}D^T}, \boldsymbol{DR_{P,1}D^T}, \boldsymbol{DR_{P,5}D^T}\right) \tag{11}$$

$$\boldsymbol{h_{\Phi,i}} = \begin{bmatrix} \rho^{12}_{rb} + \lambda_i\left(B^1_{rb} - B^2_{rb}\right) \\ \rho^{13}_{rb} + \lambda_i\left(B^1_{rb} - B^3_{rb}\right) \\ \cdots \\ \rho^{1m}_{rb} + \lambda_i\left(B^1_{rb} - B^m_{rb}\right) \end{bmatrix} \quad \boldsymbol{h_{P,i}} = \begin{bmatrix} \rho^{12}_{rb} \\ \rho^{13}_{rb} \\ \cdots \\ \rho^{1m}_{rb} \end{bmatrix} \tag{12}$$

$$\boldsymbol{D} = \begin{bmatrix} 1 & -1 & 0 & \cdots & 0 \\ 1 & 0 & -1 & & 0 \\ & & \vdots & \ddots & \vdots \\ 1 & 0 & 0 & \cdots & -1 \end{bmatrix} \quad \boldsymbol{E} = \left[\boldsymbol{e^1_r}, \boldsymbol{e^2_r}, \ldots, \boldsymbol{e^m_r}\right]^{\boldsymbol{T}} \tag{13}$$

$$\boldsymbol{R_{\Phi,i}} = diag\left(2{\sigma^1_{\phi,i}}^2, 2{\sigma^2_{\phi,i}}^2, \ldots, 2{\sigma^m_{\phi,i}}^2\right) \boldsymbol{R_{P,i}} = diag\left(2{\sigma^1_{P,i}}^2, 2{\sigma^2_{P,i}}^2, \ldots, 2{\sigma^m_{P,i}}^2\right) \tag{14}$$

where $\boldsymbol{e}$ is the light of sight (LOS) vector between the satellite and the receiver, σ is the standard deviation of the pseudorange and carrier phase measurement error for each satellite. For the time update of the EKF, the positioning mode is kinematic and the receiver dynamics are turned on. Therefore, the state transition matrix and the system noise covariance matrix are defined as below.

$$F_k^{k+1} = \begin{bmatrix} I_{3\times3} & I_{3\times3}\tau_r & I_{3\times3}\frac{\tau_r^2}{2} & \\ 0 & I_{3\times3} & I_{3\times3}\tau_r & \\ 0 & 0 & I_{3\times3} & \\ & & & I_{(2m-2)\times(2m-2)} \end{bmatrix} \quad (15)$$

$$Q_k^{k+1} = diag\left(\mathbf{0}_{3\times3}, \mathbf{0}_{3\times3}, Q_a, \mathbf{0}_{(2m-2)\times(2m-2)}\right) \quad (16)$$

$$Q_a = E_r^T diag\left(\sigma_{ae}^2\tau_r, \sigma_{an}^2\tau_r, \sigma_{au}^2\tau_r\right) E_r \quad (17)$$

$$\tau_r = t_{k+1} - t_k \quad (18)$$

where τ_r is the receiver sampling interval in seconds between epochs k and $k+1$, and $(\sigma_{ae}, \sigma_{an}, \sigma_{au})$ are the east, north, and up components of the receiver acceleration system process noise.

2.3. Satellite and Correction Data Selection

In order to obtain PPK solutions, observation files from nearby base stations and satellite ephemeris files are required. The ephemeris files were pulled from the International GNSS Service (IGS) website. The files chosen were BRDM files that contain navigation parameters for all GNSS constellations over the course of the data collection day. The BRDM files are generated as part of the Multi-GNSS experiment (MGEX), where the signals from multiple constellations, including GPS, GLONASS, BeiDou, Galileo, QZSS, NAVIC, and SBAS are merged into combined broadcast ephemeris files [19]. The base station observation files were pulled from the National Oceanic and Atmospheric Administration (NOAA) National Geodetic Survey (NGS) website for the Stanford Linear Accelerator Center (SLAC) station. SLAC, currently known as the National Accelerator Laboratory, is one of the 17 Department of Energy national laboratories, and is located at an approximate distance of at most 35 km from the route, which is within the maximum recommended range of 50 km for decimeter-level accuracy RTK or PPK [7]. The provision of GPS, GLONASS, and Galileo observation data, along with the SLAC station's relative proximity to the data collection routes, made it a desirable site to utilize when attempting to eliminate atmospheric and clock errors in the PPK solution.

2.4. Data Conversion

The competition data provided by Google included both raw GNSSLogger and RINEX formats. The RINEX format is compatible with RTKLIB, but the method and parameters used to convert the raw phone data to RINEX was not fully provided to the challenge participants and did not appear to fully align with methodologies suggested by Google in the competition discussion pages. As a result, it was beneficial to convert the raw phone data to RINEX using available open-source tools, as opposed to utilizing the pre-processed files, to provide more user control over the pre-processing data conversion process. To accomplish this conversion, the Rokubun Android GnssLogger to RINEX converter python code was chosen as a starting point [20]. This code is split into two sections, gnsslogger.py, which pulls out and organizes phone data, and rinex3.py that takes in the organized data and writes it out in the RINEX version 3 format. Within both sets of code, adjustments were made to improve the resulting RINEX files. The details of the adjustments and improvements are documented below.

In gnsslogger.py, changes were made to expand compatibility to all phone types used in the competition and to more closely follow filtering rules described by Google in a competition discussion post describing their methodology for generating the included baseline solutions [12]. These changes were primarily integrated by including an observation filter function that served to reject invalid or very low-quality observations as defined by:

- C/N_0 was less than 20 dB-Hz;
- Received space vehicle time uncertainty was greater than 500 nanoseconds;
- Maximum pseudorange uncertainty was greater than 150 m;
- Multipath indicator was set greater than 0;
- Maximum carrier phase uncertainty was greater than 0.1 m;
- Status of the code lock was invalid;
- TOW or TOD values were not decoded and set for all constellations;
- Constellation identifier was invalid.

Finally, the script was modified to provide an option to ignore the cycle slip and half-cycle ambiguity flags that were present in the provided raw data files, as investigating these components showed that their inclusion resulted in usable data often being thrown out. This option was enabled for the processing of the raw data for this solution.

The changes to the rinex.py file were much less involved. The core change made to this script was to replace the single character, unused legacy signal to noise ratio (SNR) field of the RINEX file with the pseudorange and carrier phase uncertainty estimates from the receiver. These receiver estimates can be included in RTKLIB's determination of measurement weights, but for this solution, only elevation-based weighting was utilized.

2.5. Code Changes to the RTKLIB Demo5

In order to get high levels of performance from the RTKLIB PPK processing, changes were made to the demo5 b34e version of RTKLIB. The demo5 code is an open-source, publicly available fork from the RTKLIB 2.4.3 code, and it is maintained on GitHub [21]. It is focused on improving solution performance, reliability, and robustness, particularly for low-cost single and dual frequency receivers. It is kept closely synchronized to the 2.4.3 code version, and all updates from that code are generally ported into the demo5 code. The vast majority of the code is common between the two versions, but some of the more significant differences between them are in the ambiguity resolution algorithms, particularly in the approach to partial ambiguity resolution and the options for dealing with the GLONASS inter-channel biases. However, it should be noted that ambiguity resolution was disabled for this solution, so the differences between the two versions of code are less significant in this case.

The changes made to the demo5 b34e code specifically for this experiment fell into one of two categories. The first and most important is the approach to cycle slip detection. The unmodified RTKLIB code has three methods for detecting cycle slips. The cycle slips can be flagged by the receiver, they can be detected as a Kalman filter error larger than a specified "outlier" threshold, or if dual frequency measurements are available, by using geometry-free linear combinations to detect phase jumps. Cycle slips in the smartphone observations are much more frequent than is typical from higher quality receivers. In addition, the reliability of the receiver cycle-slip flags appears to be much lower, both for false-positives and for false-negatives. In order to compensate for the large number of cycle slips and the low confidence in the receiver to accurately flag them, it was required to improve RTKLIB's ability to detect the cycle slips through other means. The method used in this solution leveraged changing the demo5 code to enable the ability to detect cycle slips using doppler measurements. By checking the difference between the doppler measurement and the change in carrier phase measurement, as carrier phase is essentially the accumulated doppler range, the cycle slips can be detected. This method is included in the open-source versions, but is commented out due to a clock jump issue. By making changes to the code that allow it to process all satellites at an epoch in one call as opposed to processing all of them individually, the common-mode effect of the clock jump can be minimized. If there is a jump in all satellite channels, this would be classified as a clock discontinuity, and if it is present in a subset of channels, it would be identified as a cycle slip.

Other changes were made to reduce the number of observations that were unnecessarily discarded, allowing the solution to degrade more gracefully for poor data sets. An example of this was the adjustment of a check for valid observation data. For this check, instead of discarding any data that contained only pseudorange measurements, the code kept this incomplete data. Due to the fact that some of the Google data could be poor and often included incomplete measurement sets that lacked the carrier, changing this observation check was able to prevent good pseudorange data from being discarded. Additionally, another check in the code that sought to identify valid epoch solutions by ensuring that all epochs had four or more valid carrier phase double differences was changed. Almost all the epochs had at least four valid pseudorange double differences, it was considered restrictive to simply reject these; instead, we chose to use the float solution derived from the pseudoranges when the epoch did not have four or more carrier phase double differences. Continuing with the topic of the double differencing, an adjustment was made that allowed the code to avoid using a satellite with a cycle slip as a reference satellite, as previously, the code just utilized the satellite with the highest elevation. Additionally, a bug was identified and fixed in the geometry-free cycle slip detection, where the code was not checking the L1 frequency for cycle slips and thus, was not progressing as it was supposed to, given a case where only L1 measurements encountered cycle slips.

2.6. Changes to RTKLIB Settings

In addition to the RTKLIB code changes, configuration settings can be adjusted to better align with the expected data input, as well as leverage the code changes made [22]. As a starting point, a config file was used from a previous experiment made with some less challenging smartphone observations [23]. This previous experiment worked with data collected by a Xiaomi Mi8 phone and includes additional changes from the f9p_ppk.conf file provided with the demo5 code not discussed in this section, as they are minor and were specifically tuned to that dataset. The important configuration settings used to produce the results presented will be broken down based on whether they were settings for the processing, adjustments to the outputs, or changes made to the statistics. A summary of the changes with respect to the default f9p_ppk.conf demo5 b34e code is provided in Table 1.

Table 1. Summary of changes to the demo5 b34eRTKlib code default settings file f9p_ppk.conf.

Positioning mode: static -> kinematic
GNSS constellations: disable BeiDou
Enable L5/E5 frequency: L1 + L2/E5b -> L1 + L2/E5b + L5/E5a
Filter Type: Combined -> Combined—no phase reset
SNR Mask L1/L5 (dB-Hz): 35/0 -> 24/24
Integer Ambiguity Res: on -> off
Cycle Slip Threshold: Geometry-Free: 0.05 -> 0.10
Cycle Slip Threshold: Doppler: N/A -> 5.0
Innov (m): 2 -> 1
Time Format: hms -> tow
Phase Error Ratio L5: 300 ->100
Carrier Phase Bias: 0.0001 -> 0.01 cycles

Several changes were made to better accommodate the Google data and perform some simple filtering. First, the positioning mode was set to kinematic, as that is the type of data Google provides for the challenge, and only the GPS, GLONASS, and Galileo constellations were selected for use, as the SLAC station observations used only contain these. Leveraging the ability of the code to use GPS L5 and Galileo E5a, these frequencies were enabled, along with the L1 and L2 frequencies. Settings that were retained from the previous smartphone experiment to filter out some of the lowest quality observations included a 15 degree elevation mask, and a 24 dB-Hz C/N_0 filter. Although this level of C/N_0 is low for decent carrier phase measurements, as RTKLIB does not support separate thresholds for carrier and code, this value was selected to preserve the acceptable code measurements. Furthermore, this threshold was not an assessment of absolute signal quality, but a means to exclude the tail of the distribution. In addition, this is a higher threshold compared to the 20 dB-Hz used by Google in their code-only positioning for the baseline solutions. The solution was run in combined no phase reset mode, which includes running the Kalman filter over the data in both the forward direction and the backwards direction, then combining the results without resetting the ambiguity estimates between the two directions. Continuing with more settings, first, the ambiguity resolution was disabled, as errors within the dataset were too large to resolve consistently, resulting in false fixes and degrading overall solution quality. With the doppler slip now enabled in the RTKLIB code, it could be used, along with geometry-free detection and outlier detection, to identify slips in the carrier. The geometry-free slip threshold was 0.05 m by default, but this was deemed too small to account for the larger errors present in the datasets, so the value was increased to 0.1 m. The doppler slip threshold was set to 5 Hz.

The final changes made to the RTKLIB configuration file revolved around the output and the statistics of those outputs. The output configuration showed no changes, other than to change the time format, which was switched to "ww ssss GPST", which is much closer to the format Google expects in the submitted results. The statistical changes to the configuration file were made to increase the weighting given to the L5 pseudorange measurements, as well as to account for lower confidence in the carrier phase bias due to the increased possibility of missed slips. The L5 phase error ratio was reduced from 300 to 100 due to L5′s higher signal strength and longer codes relative to L1, which results in smaller pseudorange errors. The carrier-phase bias was set to 0.01 cycles. Some of these values were decided upon by trial-and-error, using a select number of datasets from the training set that showed improvement to the overall solution using the values defined above.

2.7. Data Merge

Some of the phone data files were of particularly low quality and contained continuous cycle slips on all satellites for large portions of the data. The raw data for these files indicated hardware clock discontinuities for every epoch, which were potentially caused by the phone's duty cycling the GNSS tracking during data collection. An example of this can be seen in Figure 1, which shows simultaneous observations taken from two adjacent phones, where the red ticks indicate the flagged cycle slips. With little to no valid carrier phase observations in the degraded files, these solutions tended to have very large position errors. Fortunately, the poor solutions also had very large error estimates, and there was at least one dataset in each collection drive that produced good results. Since the distance between the phones within the same collection scenario was small relative to the position errors, merging the solutions for datasets within a collection proved to be an effective method to minimize errors, not only in the problematic datasets, but also for all of the scenarios. The merge was carried out in the position domain instead of the measurement domain, because valid carrier phase measurements were available in at least one of the phones for each epoch, and the position domain merge was sufficient to leverage the benefits of carrier-enhanced positions. Furthermore, due to the approximate 20 cm physical separation between the phones, if the relative displacement vectors are not accounted for, a successful merge of the carrier phase measurements is difficult.

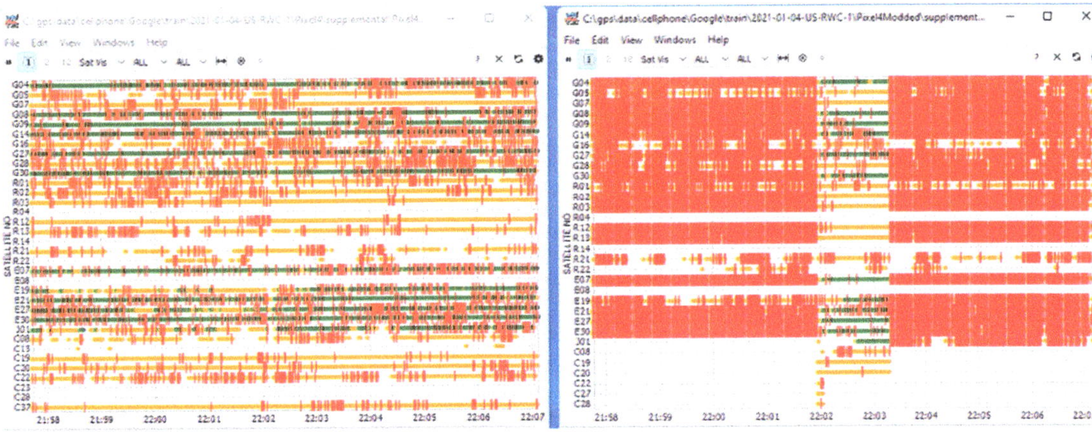

Figure 1. Availability visualization of all satellites present in the observation data for the Pixel 4 and Pixel 4 Modded. The left plot is for Pixel 4, and the right plot is for Pixel 4 Modded. The red markers indicate the flagged potential cycle slips present in the measurements.

The merge was performed by combining the solutions output by RTKLIB for each data collection drive using a weighted approach. The weights were computed as the inverse of the variance of the data naturally computed by RTKLIB. If the weights were approximately zero, they were discarded and the remaining weights which passed this filter were utilized to compute a weighted average merged phone solution. This solution was then distributed back to the individual phones on the drive. The phone merge introduced a minimal error, as the phones were placed approximately 0.2 m from each other on the car dashboard [8], but the improvement to the overall solution that resulted from removing the poor phone data outweighed the small phone distance error.

3. Results

To gain an understanding of the performance of the current version of the RTKLIB with respect to Google's baseline solutions, both the RTKLIB and baseline solutions for the training set were assessed using the provided truth data. Prior to the analysis, training datasets with clock discontinuities were removed, as clock discontinuities suggest no valid carrier phase measurements for that epoch, which will throw off the proposed RTKLIB algorithm. If we look at these problematic datasets, whenever there were discontinuities, the position errors exceeded 100 m. Although the discontinuities exist for the GSDC test set as well, the aforementioned data merge mitigates this issue. We perform the exclusion for the train set analysis to look at the accuracy of the RTKLIB solutions prior to the merge. The excluded data and their number of discontinuities is provided in Table 2.

After the data screening, the training data was separated into three scenarios: highway, street, and urban canyon. Highway indicates all the drives between San Francisco and San Jose, which are open-sky dominant. Urban canyon indicates the drives inside downtown San Jose, which are multipath dominant. Street indicates all the other drives, which have a mixture of both environments. An illustration of the three scenarios are provided in Figure 2. This classification is done to compare the absolute and relative performance of the RTKLIB with respect to the baseline solution.

Table 2. Scenario and device combinations for the datasets that were excluded due to the presence of hardware clock discontinuities.

Scenario–Device	Total Epochs	Total Discontinuities
2020-05-14-US-MTV-2_Pixel4XLModded	129	33
2020-09-04-US-SF-1_Pixel4	258	253
2021-01-04-US-RWC-1_Pixel4Modded	263	222
2021-01-04-US-RWC-1_Pixel4XL	272	272
2021-01-04-US-RWC-2_Pixel4Modded	262	262
2021-01-04-US-RWC-2_Pixel4XL	266	266
2021-01-05-US-SVL-1_Pixel4XL	263	222
2021-01-05-US-SVL-1_Pixel5	269	268
2021-01-05-US-SVL-2_Pixel4XL	257	197
2021-03-10-US-SVL-1_Pixel4XL	242	242
2021-04-29-US-MTV-1_Pixel4	235	121
2021-04-29-US-MTV-1_Pixel5	261	249

Figure 2. Illustration of the different scenarios: Highway (**left**), Street (**middle**), and Urban Canyon (**right**). Each scenario is representative of various GNSS environments: open-sky, mixed, and multipath.

For the accuracy assessment, GSDC requires only the latitude and longitude coordinates. Therefore, in the training set assessment, the altitude for the solutions was set to be the same as the truth. This effectively allows us to attain only the horizontal error. When we look at the cumulative distribution function (CDF) for all the scenarios, as shown in Figure 3, it is clear that the RTKLIB solution was more accurate than Google's provided baseline solution in all scenarios. In addition, by merging the RTKLIB solutions, the accuracy of the final positions improved further. Overall, the baseline solution had 1.94 m for 50th percentile and 10.29 m for the 95th percentile, resulting in a score of 6.11. On the other hand, the merged RTKLIB solution attained 1.21 m for the 50th percentile and 6.5 m for 95th percentile, resulting in a score of 3.86. In terms of relative performance improvements, RTKLIB was the most effective over the baseline for the highway scenario. However, even for the absolute metrics, the unmerged RTKLIB provided almost 0.5 m better 50th percentile and 1.5–5.1 m better 90th percentile performance than the baseline. Furthermore, a limitation exposed by this CDF is that although RTKLIB is effective for open-sky conditions, the difficulty of mitigating multipath residuals remains.

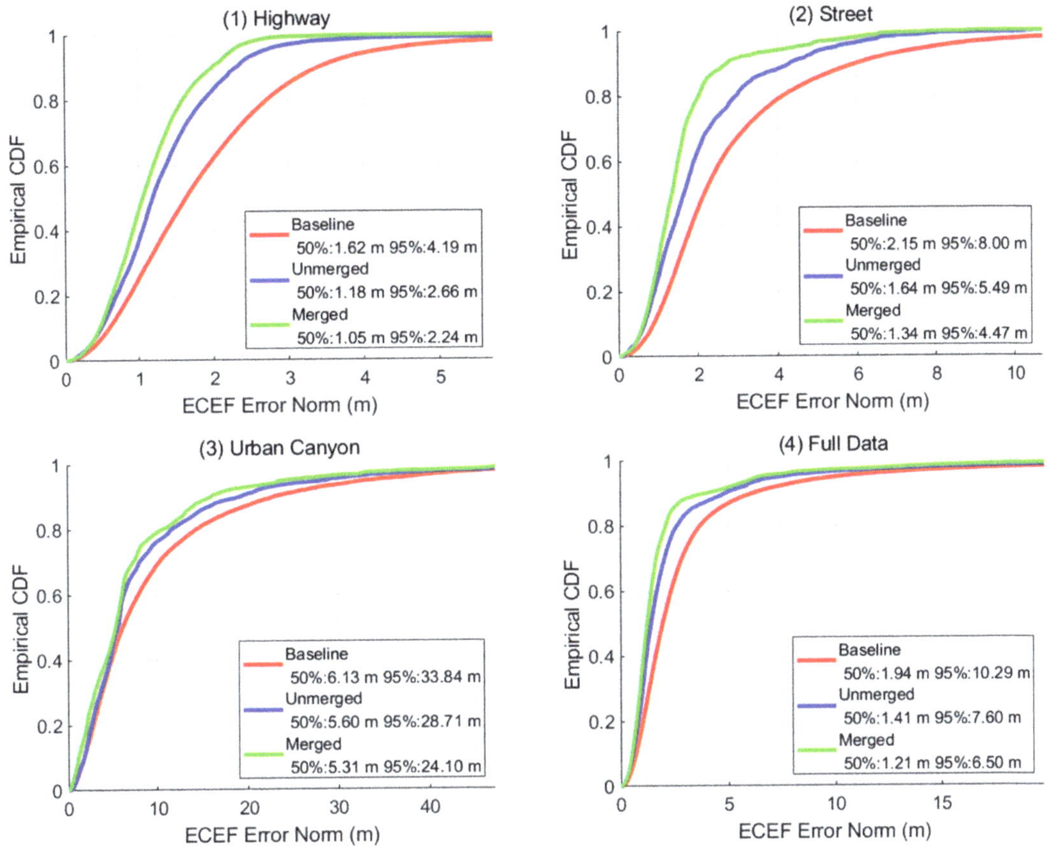

Figure 3. CDF plots of the performance of baseline solutions, unmerged RTKLIB solutions, and final merged solutions for each scenario and the entire training dataset: (**1**) Highway, (**2**) Street, (**3**) Urban Canyon, (**4**) Full Data. The unmerged RTKLIB solution was better than the baseline positions in all scenarios, and the merge further improved the accuracy of the RTKLIB solution.

In order to see how the RTKLIB solution compares to the other participants, the test set results were uploaded to the official Kaggle website. Although the competition is over, the Kaggle site can still be used to assess the test data. For the test data, based on the private leaderboard score, Google's baseline solution scored 5.42 for Kaggle, which is in 692th place out of 810 participants, while the proposed RTKLIB PPK solution scored 2.15 m, which resulted in 5th place, as shown in Figure 4. The private leaderboard score is provided instead of the public leaderboard counterpart, because different subsets of data are used for each scoring scheme, and the private score is used to determine the winners of the contest.

Figure 4. Screenshot of the official Kaggle private leaderboard results for the proposed RTKLIB PPK solution.

The main differences between the baseline solutions that Google provided as an example and the proposed solutions are how the GNSS measurements are computed and how the computed position solutions are processed [10]. For the navigation engine, the Google baseline solution used all satellite constellations, including GPS, GLONASS, BeiDou, Galileo, and QZSS, while the RTKLIB used GPS, GLONASS, and Galileo. For the raw measurements, the baseline solution used only pseudorange, but the RTKLIB also used carrier phase measurements for greater measurement precision. For the measurement processing, the baseline solution used the Klobuchar ionosphere model, EGNOS troposphere model, and ephemeris satellite clock model for error mitigation in single point positioning (SPP), while RTKLIB used double-differencing to remove atmospheric, clock, and orbital errors in its PPK solution. Moreover, while the baseline solution used weighted least squares for data processing, a Kalman filter was used for the RTKLIB in order to alleviate the impacts of measurement noise on the position solution. For the position solution processing, instead of providing the position solution from each device directly, the solutions were merged between devices based on the expected accuracies of the navigation solutions. This is done to mitigate the sections of data where carrier phase measurements from the phones are unreliable or unavailable. This process is difficult for the baseline solution, as no metrics are provided for the expected accuracy of the computed GNSS positions, but in the case of RTKLIB, estimates for the accuracy are provided by the navigation engine.

4. Summary and Conclusions

The latest version of the RTKLIB code, with the aforementioned changes and settings, was able to obtain 5th place ranking, with a score of 2.15. This is a significant improvement compared to the Google baseline score of 5.42, which implies that for the individuals using RTKLIB as the GNSS analysis tool, the provided tools, strategies, and updated code base provide an improved baseline from which to work. The code for this analysis used doppler-carrier comparison, geometry-free linear combination, and outlier detection methods for cycle slip detection, but did not incorporate any other cycle slip detection or mitigation methodologies, measurement error analysis, multipath mitigation, Kalman filter tuning, nor any of the numerous post-processing techniques proposed in various academic publications. Therefore, further performance improvements are expected with their additions. Furthermore, the conversion from the GnssLogger to RINEX can be further investigated, along with the optimization of the receiver parameter settings.

The improvement in the GNSS-derived positions compared to Google's baseline solution is a crucial component of GSDC. This is because the competition has essentially two components: deriving accurate GNSS positions, and post-processing the GNSS solutions. The updated RTKLIB addresses the first component by proposing methods to improve the accuracy of the GNSS solutions and provide the resulting positions. This opens the door to more possibilities in the second component. For those who use Google's baseline solution for post-processing, the RTKLIB GNSS positions will have the potential to provide results with higher accuracy. One of the strategies for the post-processing of the GNSS positions would be map matching with respect to the truth locations of the train data, as the drive trajectories on the road would be similar between collections.

All of the updates for the special GSDC version of RTKLIB are currently in the mainstream demo5 code in the latest b34f release. Moreover, a subset of the demo5 RTKLIB PPK solution code translated to python is available as a base for users who are interested in performing more involved experiments that can be difficult to develop in the original C code environment. Detailed steps required to replicate the results provided in this paper are provided in the blog [22].

Author Contributions: Conceptualization, T.E.; methodology, T.E.; software, T.E.; validation, T.T. and D.-K.L.; formal analysis, D.-K.L.; investigation, T.T. and D.-K.L.; writing-original draft preparation, T.T. and D.-K.L.; writing-review and editing, T.E., T.T., D.-K.L., D.M.A.; visualization, T.E., D.-K.L.; supervision, D.M.A.; project administration, D.M.A.; funding acquisition, T.E. All authors have read and agreed to the published version of the manuscript.

Funding: General financial support for the RTKLIB demo5 code was provided through Open Collective. The donor for the project at the time of this publication was Google Open Source. https://opencollective.com/rtklib-demo5, accessed on 13 April 2022.

Institutional Review Board Statement: Not applicable.

Informed Consent Statement: Not applicable.

Data Availability Statement: The GSDC data can be found on the official Kaggle website: https://www.kaggle.com/c/google-smartphone-decimeter-challenge/data, accessed on 13 April 2022. The observation data for the nearby CORS station can be found at NOAA: https://www.ngs.noaa.gov/UFCORS/, accessed on 13 April 2022. The BRDM files for the satellites can be found at IGS: https://igs.bkg.bund.de/root_ftp/IGS/BRDC/, accessed on 13 April 2022. The original RTKLIB code is found on the official website: http://www.rtklib.com/, accessed on 13 April 2022. The GnssLogger to RINEX conversion script and the updated RTKLIB code can be found on GITHUB: https://github.com/rtklibexplorer/RTKLIB/releases/tag/b34e_smartphone, accessed on 13 April 2022. The python subset of RTKLIB code for PPK solutions can be found at: https://github.com/rtklibexplorer/rtklib-py, accessed on 13 April 2022.

Conflicts of Interest: The authors declare no conflict of interest.

References

1. Wang, J. Stochastic Modeling for Real-Time Kinematic GPS/GLONASS Positioning. *Navigation* **1999**, *45*, 297–305. [CrossRef]
2. Gakstatter, E.; GPS World. RTK GNSS Receivers: A Flooded Market? 21 March 2013. Available online: https://www.gpsworld.com/rtk-gnss-receivers-a-flooded-market/ (accessed on 1 April 2022).
3. Bilich, A. National Geodetic Survey, Last Modified. 14 July 2021. Available online: https://geodesy.noaa.gov/gps-toolbox/exist.htm (accessed on 1 April 2022).
4. Takasu, T. RTKLIB: Open Source Program Package for RTK-GPS. In Proceedings of the FOSS4G 2009, Tokyo, Japan, 20–23 October 2009.
5. IGS/RTCM RINEX Working Group. The Receiver Independent Exchange Format, Version 4.00. 1 December 2021. Available online: https://files.igs.org/pub/data/format/rinex_4.00.pdf (accessed on 1 April 2022).
6. *RTCM Standard 10403.3*; Differential Global Navigation Satellite Systems Services, Version 3. RTCM Special Committee: Washington, DC, USA, 20 May 2021. Available online: https://www.rtcm.org (accessed on 1 April 2022).
7. European Global Navigation Satellite Systems Agency. PPP-RTK Market and Technology Report. 2019. Available online: https://www.euspa.europa.eu/sites/default/files/calls_for_proposals/rd.03_-_ppp-rtk_market_and_technology_report.pdf (accessed on 1 April 2022).
8. Fu, G.; Khider, M.; van Diggelen, F. Android raw GNSS Measurement Datasets for Precise Positioning. In Proceedings of the ION GNSS+ 2020, St. Louis, MI, USA, 22–25 September 2020.
9. Google Smartphone Decimeter Challenge. Available online: https://www.kaggle.com/c/google-smartphone-decimeter-challenge (accessed on 1 April 2022).
10. NovAtel. IMU-ISA-100C Product Sheet, Version 8, Updated 10 May 2016. Available online: https://hexagondownloads.blob.core.windows.net/public/Novatel/assets/Documents/Papers/IMUISA100CD19539v2/IMU-ISA-100C%20Product%20Sheet.pdf (accessed on 1 April 2022).
11. Taro, S. First Place Award Winner of the Smartphone Decimeter Challenge: Global Optimization of Position and Velocity by Factor Graph Optimization. In Proceedings of the ION GNSS+ 2021, Denver, CO, USA, 19–23 September 2021.
12. Sohier, D. Methodology behind the Baseline Location Estimates, Kaggle. Available online: https://www.kaggle.com/competitions/google-smartphone-decimeter-challenge/discussion/238583 (accessed on 28 March 2022).
13. Tomoji, T. Development of the low-cost RTK-GPS receiver with an open source program package RTKLIB. In *International Symposium on GPS/GNSS*; International Convention Center: Jeju, Korea, 2009.
14. Pesyna, K.M., Jr.; Heath, R.W., Jr.; Humphreys, T.E. Centimeter Positioning with a Smartphone-Quality GNSS Antenna. In Proceedings of the ION GNSS+ 2014, Tampa, FL, USA, 8–12 September 2014.
15. Wu, Q.; Sun, M.; Zhou, C.; Zhang, P. Precise Point Positioning Dual-Frequency GNSS Observations on Smartphone. *Sensors* **2019**, *19*, 2189. [CrossRef] [PubMed]
16. Han, K.; Lee, S.; Song, Y.J.; Lee, H.B.; Park, D.H.; Won, J.H. Precise Positioning with Machine Learning based Kalman Filter using GNSS/IMU Measurements from Android Smartphone. In Proceedings of the ION GNSS+2021, Denver, CO, USA, 19–23 September 2021.
17. Fortunato, M.; Tagliaferro, G.; Fernandez-Rodriguez, E.; Critchley-Marrows, J. The Whole Works: A GNSS/IMU Tightly Coupled Filter for Android Raw GNSS Measurements with Local Ground Augmentation Strategies. In Proceedings of the ION GNSS+ 2021, Denver, CO, USA, 19–23 September 2021.
18. Everett, T. RTKLIB Manual: Demo5 Version. 21 December 2021. Available online: https://rtkexplorer.com/pdfs/manual_demo5.pdf (accessed on 1 April 2022).

19. Montenbruck, O.; Steigenberger, P.; Prange, L.; Deng, Z.; Zhao, Q.; Perosanz, F.; Romero, I.; Noll, C.; Stürze, A.; Weber, G.; et al. The Multi-GNSS Experiment (MGEX) of the International GNSS Service (IGS)—Achievements, Prospects and Challenges. *Adv. Space Res.* **2017**, *59*, 1671–1697. [CrossRef]
20. Rokubun. Android GNSS Logger to RINEX Converter. 24 June 2020. Available online: https://github.com/rokubun/android_rinex/ (accessed on 1 April 2022).
21. RTKLIB: Demo5. Available online: https://github.com/rtklibexplorer/RTKLIB/ (accessed on 26 March 2022).
22. Everett, T.; Rtklibexplorer. Google Smartphone Decimeter Challenge. 10 January 2022. Available online: https://rtklibexplorer.wordpress.com/2022/01/10/google-smartphone-decimeter-challenge/ (accessed on 1 April 2022).
23. Everett, T.; Rtklibexplorer. Another Look at L1/L5 Cellphone PPK with RTKlib and an Xiaomi Mi8 Phone. 27 February 2021. Available online: https://rtklibexplorer.wordpress.com/2021/02/27/another-look-at-l1-l5-cellphone-ppk-with-rtklib/ (accessed on 1 April 2022).

sensors

MDPI

Article

Real-Time Loosely Coupled 3DMA GNSS/Doppler Measurements Integration Using a Graph Optimization and Its Performance Assessments in Urban Canyons of New York

Hoi-Fung Ng [1], Li-Ta Hsu [1,*], Max Jwo Lem Lee [1], Junchi Feng [2], Tahereh Naeimi [2], Mahya Beheshti [2] and John-Ross Rizzo [2,*]

1 Department of Aeronautical and Aviation Engineering, The Hong Kong Polytechnic University, Hong Kong 999077, China
2 Grossman School of Medicine, New York University, New York, NY 10016, USA
* Correspondence: lt.hsu@polyu.edu.hk (L.-T.H.); johnrossrizzo@gmail.com (J.-R.R.)

Abstract: Smart health applications have received significant attention in recent years. Novel applications hold significant promise to overcome many of the inconveniences faced by persons with disabilities throughout daily living. For people with blindness and low vision (BLV), environmental perception is compromised, creating myriad difficulties. Precise localization is still a gap in the field and is critical to safe navigation. Conventional GNSS positioning cannot provide satisfactory performance in urban canyons. 3D mapping-aided (3DMA) GNSS may serve as an urban GNSS solution, since the availability of 3D city models has widely increased. As a result, this study developed a real-time 3DMA GNSS-positioning system based on state-of-the-art 3DMA GNSS algorithms. Shadow matching was integrated with likelihood-based ranging 3DMA GNSS, generating positioning hypothesis candidates. To increase robustness, the 3DMA GNSS solution was then optimized with Doppler measurements using factor graph optimization (FGO) in a loosely-coupled fashion. This study also evaluated positioning performance using an advanced wearable system's recorded data in New York City. The real-time forward-processed FGO can provide a root-mean-square error (RMSE) of about 21 m. The RMSE drops to 16 m when the data is post-processed with FGO in a combined direction. Overall results show that the proposed loosely-coupled 3DMA FGO algorithm can provide a better and more robust positioning performance for the multi-sensor integration approach used by this wearable for persons with BLV.

Keywords: localization; navigation; smartphone; GNSS; 3D building models

Citation: Ng, H.-F.; Hsu, L.-T.; Lee, M.J.L.; Feng, J.; Naeimi, T.; Beheshti, M.; Rizzo, J.-R. Real-Time Loosely Coupled 3DMA GNSS/Doppler Measurements Integration Using a Graph Optimization and Its Performance Assessments in Urban Canyons of New York. *Sensors* **2022**, *22*, 6533. https://doi.org/10.3390/s22176533

Academic Editors: Yang Gao, Jacek Paziewski, Michael Fu and Augusto Mazzoni

Received: 3 August 2022
Accepted: 25 August 2022
Published: 30 August 2022

1. Introduction

Mobility and wayfinding are significant obstacles faced by people with BLV, specifically in urban areas. Degradation of the visual system can lead to a dramatic reduction in mobility. It has been shown that 80–90% of people with BLV spend the majority of their time inside buildings, and 30% rarely leave home alone [1,2].

Accurate positioning is essential for localization and navigation in urban canyons. Pedestrians with BLV who live in urban areas could benefit significantly from an integrated navigation solution for use during the activities of daily living. Much research has been performed to improve the autonomy of people with BLV, especially their ability to explore their environment. Rizzo et al. developed an advanced wearable in the form of an instrumented backpack equipped with microcomputers and sensors; this wearable incorporates cameras, inertial measurement units (IMUs), and GNSS positioning to provide a more comprehensive and full-featured navigation solution [3].

This study develops a real-time 3DMA GNSS-positioning system based on state-of-the-art 3DMA GNSS algorithms to advance the approach used for localization by the wearable. The integration of shadow matching and likelihood-based ranging 3DMA GNSS methods were selected to evaluate positioning hypothesis candidates' likelihood scores. The position was then integrated with the velocity estimated by Doppler measurements using the FGO as loosely coupled.

Several experiments were designed in New York City (NYC) to acquire GNSS data during urban commuting (traveling between two defined locations, origin and target). A commercial-grade receiver, u-blox ZED-F9P, was connected to an Nvidia microcomputer, Jetson TX2. The performance of the proposed system was evaluated with the data recorded in a series of trips that took place on the lower east side of NYC (specifically Murray Hill, Manhattan).

The remainder of this paper is organized as follows: Section 2 is an introduction to the integrated navigation system for pedestrians and existing studies. Section 3 is an introduction to the proposed positioning algorithm. Section 4 contains the designed experiment results and analysis. Finally, a conclusion and future work are presented in Section 5.

2. Navigation System for Visually Impaired Pedestrians

2.1. Overview of Navigation System for Visually Impaired Pedestrians

The VIS^{4}ION (Visually Impaired Smart Service System for Spatial Intelligence and Navigation), is an advanced wearable consisting of a backpack with wide-angle high-resolution cameras on the support straps; these cameras, with integrated microphones and IMUs, are connected to a light-weight, portable computer for real-time analysis [1–8]. This wearable system may be used by people with BLV during indoor and outdoor navigation, among other activities of daily living.

The platform provides real-time feedback using a binaural bone conduction headset and a haptic interface, allowing obstacle avoidance and situational awareness. More specifically, this mobile platform enables the users to understand their dynamically changing environment, giving them the agency to travel and wayfind independently. Our current **VIS^{4}ION** system can process 720p video at 10 fps (dependent on the scene/task) and is robust without adding significant cognitive load to the end-user. This wearable runs off a laptop-battery with 66 Wh at 0.5 kg, yielding 3 h of run-time function.

2.2. Importance of GNSS Positioning

Precise and reliable positioning is required to support safe navigation services for persons with BLV. GNSS provides continuous positioning in the global frame in absolute coordinates. However, the performance of GNSS navigation in the urban environment is unsatisfactory. GNSS signals can be blocked or reflected over buildings, resulting in non-line-of-sight (NLOS) reception and the multipath effect [9]. These errors become more severe in highly urban cities with many high-rise buildings, such as Hong Kong and New York. As a result, researchers are trying to improve positioning by integrating different aids, such as inertial measurement units (IMU) and cameras.

One of the most frequently used approaches to integrating the GNSS with IMU is pedestrian dead reckoning (PDR) [10]. The integration of GNSS and IMU takes advantage of two approaches. GNSS positioning can provide absolute coordinates in a global frame. However, the availability is limited by the number of visible satellites. The IMU can provide continuous positioning without subjection to external factors. However, it can only offer relative incremental coordinates in the local frame. At the same time, IMUs suffer from a continuous bias that accumulates over time. As a result, GNSS/INS integration can provide a continuous positioning with absolute coordinates.

A camera is another popular aid for an integrated positioning system, which can provide the receiver's orientation in the local frame. Visual odometry (VO) [11] can extract the features from the image and estimate the relative orientation change of the receiver. VO assumes that feature points are static. Matching two consecutive frames can aid positioning by providing relative position. However, VO is sensitive to illumination conditions and feature availability. In addition, because VO assumes that feature points are static, dynamic objects highly affect performance. When the feature points move in parallel with the camera, the system may think it remains static itself. Therefore, research indicates tightly coupled VO with INS improves positioning performance [12].

Research also suggests using a sky-pointing fisheye camera to detect the NLOS signal via image recognition algorithms [13,14]. Therefore, only the predicted healthy satellites are utilized for GNSS positioning. The limitation of excluding unhealthy satellites is that when many high-rise buildings surround the receiver, exclusion can result in bad satellite geometry and degrading performance.

Besides the aid of external sensors, improving the positioning accuracy of GNSS itself can definitely benefit the whole positioning system. Therefore, researchers are trying to identify and mitigate the NLOS error to improve the GNSS positioning alone. The consistency-check method [15] can detect and isolate unhealthy measurements and positioning performance can be improved to a satisfactory level. However, a consistency check will fail if the unhealthy measurements dominate the healthy ones [16].

Dual-frequencies measurements also demonstrate good performance in mitigating the multipath effects and isolating NLOS errors. Researchers used the nature of the higher resolution of L5-band measurements to design a new weighting scheme; the weighted least squares (WLS) method has been demonstrated to improve performance [17].

2.3. Related Works on 3DMA GNSS

One popular approach to improve GNSS positioning in urban canyons is using a 3D building model to identify and even correct the NLOS reception error. Different research has proposed to simulate the propagation path and model the error for code [18] and carrier phase [19] measurements. Research also incorporates the Fresnel zone analysis to predict GNSS multipath, NLOS, and diffraction effects in urban areas [20]. 3D building models demonstrate a massive improvement in positioning, namely 3D mapping-aided (3DMA) GNSS [21]. 3DMA GNSS has greatly impacted urban positioning in recent years, especially since huge improvements in smartphone positioning have been demonstrated [22]. An existing study uses the 3D building model to exclude the NLOS satellites; the weighted least squares (WLS) solution can therefore be enhanced [23]. However, we have to know the exact position to give a correct prediction and correction. As a result, the solution is usually determined as a particle-based approach. Position hypothesis candidates are distributed and measurements are modelled as the prediction at each candidate. The candidate with the highest similarity between modelled and actual received measurements is assumed to be the receiver location. Two basic categories of 3DMA GNSS algorithms are shadow matching and ranging-based 3DMA GNSS.

Shadow matching [24,25] matches the satellite visibility at different locations. The received satellites are assumed to be the LOS, while the non-received ones are assumed to be NLOS. The ephemeris provides the satellites' position and matches the visibility at each candidate to find the highest visibility similarity. There is also research to further improve the shadow matching by particle filter [26].

Another 3DMA GNSS is the ranging-based method. The receiver location is determined by comparing the modelled and received pseudorange. Pseudorange measurements are modelled at each candidate. For the NLOS-predicted satellite, the NLOS reflection delay is also modelled based on a geometrical or statistical approach. The absolute position of the reflecting point has to be determined for the geometrical approach. A popular approach is ray-tracing [27,28]. It tests and validates the reflection path over each potential reflector, creating a high computational load. Therefore, research on using GPU to accelerate the computing process is relevant [29]. Moreover, an effective computational version called skymask 3DMA GNSS [30] was introduced. It determines the reflecting point over an enhanced skymask. Besides reflection delay, ray-tracing simulation can also calculate the GNSS signal strength based on the multipath propagation model [31,32]. A research study was conducted on the use of the ray-tracing technique to identify propagation obstructions and quantity propagation errors [33]. The study proposed measuring the position integrity as a set-based approach to bound the remaining systematic uncertainty. The statistical approach, also known as likelihood-based ranging [34], uses a skew–normal distribution to model the NLOS delay measurements and then remap the errors to the LOS with the normal distribution.

Performance of shadow matching and ranging-based 3DMA GNSS are different due to the healthy satellite and building geometry distribution. Shadow matching usually outperforms in the across-street direction, while ranging-based 3DMA GNSS obtains higher accuracy in the along-street direction. The complementary nature of the two approaches inspired researchers to integrate them. The latest study shows that an integrated solution of 3DMA GNSS can provide positioning accuracy of around 10 m or less in urban canyons with both single-frequency [34] and multi-frequency [35] measurements.

However, most of these approaches only concern positioning in a single epoch. Hence the performance is not robust for continuous positioning. As a result, a temporal connection is required to improve the reliability of urban GNSS positioning. There is research using the Kalman filter (KF) and extended Kalman filter (EKF) to recursively update the recent state through the prediction based on past estimation and error of current measurements [36,37]. Researchers also use particle filters to effectively distribute and sample the candidates [26,38]. Moreover, a grid filter was adopted to distribute positioning candidates evenly [34]. The filtering techniques demonstrate excellent results in improving the smoothness of the positioning solution. Meanwhile, a machine learning approach can intelligently predict the change of signal status and estimate the most likely path on the map as the optimized positioning estimation [39].

Another approach is using factor graph optimization (FGO) [40], which optimizes the states of all epochs with many constraints. The solution will be highly robust as FGO optimizes all temporal constraints as a batch approach. Researchers open-sourced the FGO code that integrates GNSS positioning with Doppler measurements to provide a multi-epoch optimized solution [41]. FGO also applies to centimeter-level accuracy positioning via carrier-phase measurements, such as GNSS PPP [42] and RTK [41]. Furthermore, 3DMA GNSS-based collaborative positioning can benefit from using FGO [43] to optimize the performance of multi-agent collaborative positioning.

Different research demonstrates that FGO can provide an excellent positioning performance. This study integrates 3DMA GNSS with velocity estimated by Doppler measurements as a loosely-coupled solution, and states were optimized via FGO. The integrated solution can provide a more robust trajectory for pedestrian applications, such as this wearable.

3. Proposed Real-Time 3D Mapping-Aided (3DMA) GNSS-Positioning System

This section introduces the proposed loosely-coupled 3DMA GNSS- and velocity-positioning system via FGO. The flowchart is shown in Figure 1.

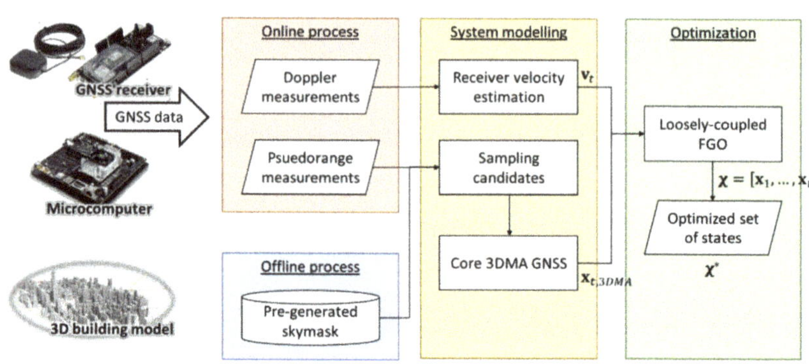

Figure 1. System flowchart on the proposed system.

3.1. *Open-Sourced 3D City Models*

3D models of New York City (NYC) were obtained from open-sourced repositories released by the Department of Information Technology & Telecommunications' (DoITT) 2014 aerial survey [44] with a level of detail (LoD) of 2. The city model was registered to cartesian New York Long Island State Plane FIPS 3104 coordinates. The city model is shown in Figure 2.

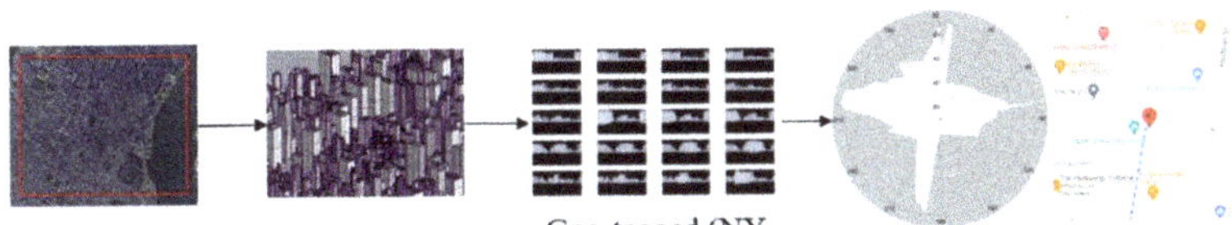

Figure 2. Geo-tagged skymask generation from NYC 3D model.

3.2. *Offline Stage Skymasks Generation*

A skymask is a skyplot with building boundaries for a single location. It is an array with a total of 360 entries that represents the azimuth angle from 0° to 359°. Each entry stores the highest elevation angle of the building blockage in degrees in the corresponding entry (each azimuth angle). Skymasks are generated in an offline stage. The intentional coverage area for the 3DMA GNSS was first selected, and then the 3D city models were downloaded. The models were imported into the Rhino 7 3D-modelling software [45], and then converted to the Unreal engine-supported format for automated skymask generation [46]. The selected area was separated into 4 m catchment areas for each potential location to capture a 360° equirectangular image of the building outline. The skymask generation process was then performed by setting up a virtual camera in the Unreal engine to capture the panorama image at each potential location outside the building and above the terrain. The camera height was set manually based on the covered area, which was 15 m in this study, to best accommodate the elevation variation across the testing area. The elevation of the potential location can be set based on the digital terrain model for mass generation. Saved panorama images were then classified into obstacles and sky. At the cutting edge between obstacles and the sky, pixels were converted to angular position (azimuth and elevation angle) at the skymask. Lastly, each skymask corresponded to one position in the state plane coordinate system, which was converted to WGS84 for real-time 3DMA GNSS positioning. The extracted skymask at each available location was then saved to a specified format for the microcomputer to use during real-time positioning [30].

3.3. *3DMA GNSS Positioning Algorithm*

3DMA GNSS evenly distributes the hypothesis positioning candidates during the online stage around the initial position. After that, the simulated measurements are generated to be compared with the received measurements for each candidate. Due to their computational efficiency, this study integrates shadow matching and likelihood-based ranging 3DMA GNSS. The implementation can be found in [35].

3.3.1. Skymask Context-Based Candidates Sampling

An effective hypothesis positioning candidates sampling is important for 3DMA GNSS. The sampling area must cover the receiver location to achieve the theoretically best performance. Enlarging the sampling radius ensures the receiver location is being covered. However, this creates a massive computational load for the low-end microcomputer, which is not practical for a real-time application. Required computational time is proportional to the number of sampled candidates and received satellites. The computational time is within the necessary output rate. Thus, we proposed to use the surrounding skymasks and principal component analysis (PCA) to determine the street direction and distribute the sampling candidates effectively.

Candidate distribution is based on the weighted least squares (WLS) for the first epoch and using the previous epoch FGO solution as the initial position after it is available. An initial circle with a sampling radius, R, e.g., 50 m, is set up empirically to estimate the surrounding environment by weighted averaging skymask, $\overline{\mathbf{SM}}^{az}$.

$$\overline{\mathbf{SM}}^{az} = \frac{1}{\sum w_k} \sum_{k=1}^{K} w_k \mathbf{SM}_{\mathbf{P}_k}^{az} \text{ where } \mathbf{P}_k \in \{|\mathbf{P} - \mathbf{x}_{init}| < R\} \tag{1}$$

where $\mathbf{SM}_{\mathbf{P}_k}^{az}$ is the skymask of location $\mathbf{P}_k$ where it is within the sampling radius R based on the initial location, $\mathbf{x}_{init}$. az is the array index that represents the azimuth angle from 0° to 359°. $w_k = |\mathbf{P}_k - \mathbf{x}_{init}|^2$ is the weighting of location $\mathbf{P}_k$ based on the distance between the initial location, $\mathbf{x}_{init}$.

The averaged skymask is then converted to vectors in the Earth-Centered-Earth-Fixed (ECEF) frame together with the transformation matrix, $\mathbf{R}$, expressed as,

$$\mathbf{q}^{az} = \mathbf{R} \cdot \left[\sin az \cdot \cos \overline{\mathbf{SM}}^{az}, \cos az \cos \overline{\mathbf{SM}}^{az}\right] \tag{2}$$

where $\mathbf{R}$ is the transformation matrix that converts the vector in the local frame to the world frame in ECEF. Thus, we can form the variance–covariance matrix, $\mathbf{Q}$,

$$\mathbf{Q} = \mathbf{q}^{\mathrm{T}}\mathbf{q} \tag{3}$$

Therefore, we can obtain the eigenvalues, $\boldsymbol{\lambda} = \begin{bmatrix} \lambda_1 & 0 \\ 0 & \lambda_2 \end{bmatrix}$, and eigenvectors, $\mathbf{V} = \begin{bmatrix} \mathbf{v}_1 & \mathbf{v}_2 \end{bmatrix}$, from the variance–covariance matrix, $\mathbf{Q}$. Note that the eigenvalues and eigenvectors are sorted in descending order, e.g., λ_1 and $\mathbf{v}_1$ denotes that they are with the largest eigenvalue. In addition, the eigenvector with a larger eigenvalue can be interpreted as the street's longitudinal direction.

Finally, we can filter the initial circle with the ellipsoid based on the determined eigenvalues and eigenvectors. The length of the semi-major and semi-minor axes are R and $R \cdot \lambda_2/\lambda_1$, respectively. The direction of the semi-major and semi-minor axes are $\mathbf{v}_1$ and $\mathbf{v}_2$, respectively.

$$\mathbf{P}_{j=1\ldots J} = \left\{ \mathbf{P}_k \in \frac{(\mathbf{d}_k \cdot \mathbf{v}_1)^2}{R^2} + \frac{(\mathbf{d}_k \cdot \mathbf{v}_2)^2}{(R \cdot \lambda_2/\lambda_1)^2} < 1 \right\} \tag{4}$$

where $\mathbf{d}_k = \mathbf{P}_k - \mathbf{x}_{init}$ is the vector between the candidate's position, $\mathbf{P}_k$, and initial location, $\mathbf{x}_{init}$.

The distributed candidates are an ellipsoid with a semi-major axis of 50 m. The separation for each candidate is 4 m. The above settings were determined empirically and suitable for real-time processing on the low-end microcomputer used in this study. A semi-major axis of 50 m can cover the position error of the initial position in most cases. In comparison, separation with 4 m can reduce the number of distributed candidates while maintaining an acceptable accuracy level.

The proposed distribution can effectively distribute the hypothesis position candidates based on the surrounding environment. Figure 3 shows two typical cases in urban canyons. In road intersections, two eigenvalues are nearly the same (Figure 3b), such that the candidates' distribution is almost a circle that covers the whole intersection, as shown in Figure 3a. In contrast, when the initial location is in a straight street, the largest eigenvalue is much larger than the other (Figure 3d). The candidates are most likely distributed on the same street but not the next block.

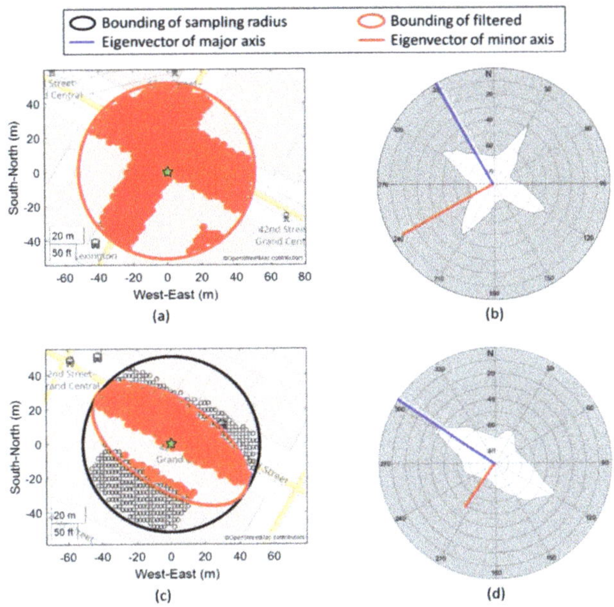

Figure 3. Typical cases of candidate sampling in urban environments on intersection (**a**,**b**) and straight street (**c**,**d**). Note that the eigenvectors (red and blue lines in (**b**,**d**) are projected back to azimuth and elevation angle (local frame) for illustration here.

The prevention of candidate distribution at the next block can potentially mitigate the local minima issue caused by the high similarity of building geometry, as shown in Figure 4. Local minima are located on the next street (the high score part in red near the upper right corner). After applying the proposed skymask context-based candidates sampling strategy, the local minima issue can be mitigated.

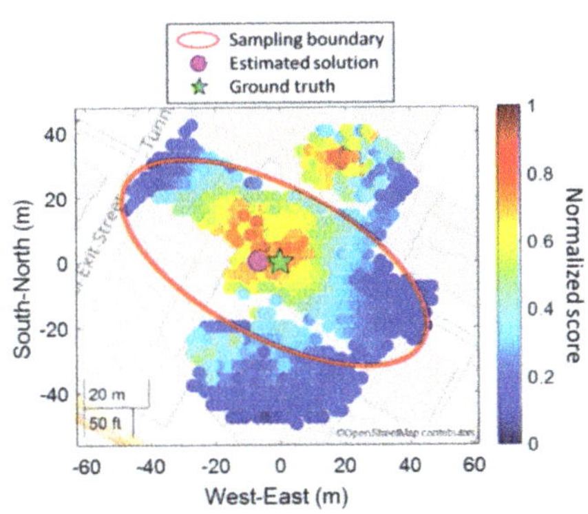

Figure 4. Example of skymask context−based candidates sampling.

To conclude, the skymask context-based candidates sampling can effectively distribute the position candidates. It has two main advantages. The first is to reduce the computational load by reducing the number of distributed candidates based on the surrounding environment. The second advantage is that candidates are most likely distributed on the same street. Therefore, the local minima on the next street can potentially be excluded.

3.3.2. Integrated Solution of 3DMA GNSS

For each candidate $\mathbf{p}_{j=1\ldots J}$, the integrated likelihood score, $S_{j,SDM+LBR}$, will be evaluated,

$$S_{j,SDM+LBR} = \sqrt{S_{j,\mathrm{LBR}} \times S_{j,\mathrm{SDM}}} \tag{5}$$

where $S_{j,\mathrm{LBR}}$ and $S_{j,\mathrm{SDM}}$ are the likelihood score of likelihood-based ranging 3DMA GNSS and shadow matching, respectively. The detail of the calculation of the likelihood scores can be found at [35].

Shadow matching evaluates the visibility consistency between the measured carrier-to-noise ratio (C/N_0) and prediction with skymask. Shadow matching requires all satellites in ephemerides to predict the non-received one. Implementation-wise, we automatically download ephemerides from the day prior. Additionally, we use the same receiving time but a day before to estimate the satellite's angular position for visibility prediction with skymask. If the internet is enabled for the execution platform this can be replaced by assisted GNSS (AGNSS) [47,48] and provide the satellite data to determine position via standard protocol, such as secure user plane location (SUPL). This is more easily achieved with built-in AGNSS devices, such as smartphones [49].

With likelihood-based ranging 3DMA GNSS, we model the pseudorange at each candidate position and compare it with the received pseudorange measurements. The NLOS predicted satellite at a candidate, likelihood-based ranging 3DMA GNSS remaps NLOS pseudorange difference to a LOS one using the distribution model.

The integrated solution of 3DMA GNSS, $\mathbf{x}_{\mathrm{3DMA}}$, is calculated by weighted averaging of the distributed candidates with their likelihood score,

$$\mathbf{x}_{\mathrm{3DMA}} = \frac{\sum_{j=1}^{J} \mathbf{p}_j S_{j,SDM+LBR}}{\sum_{j=1}^{J} S_{j,SDM+LBR}} \tag{6}$$

The receiver location is then optimized via FGO as a loosely-coupled solution.

3.4. Loosely-Coupled Factor Graph Optimization (LC-FGO)

This study also optimized the 3DMA GNSS solution as a batch via forming the graphical optimization. It is associated with FGO, connecting two consecutive epochs' solutions with velocity. The overall structure of the FGO process is shown in Figure 5.

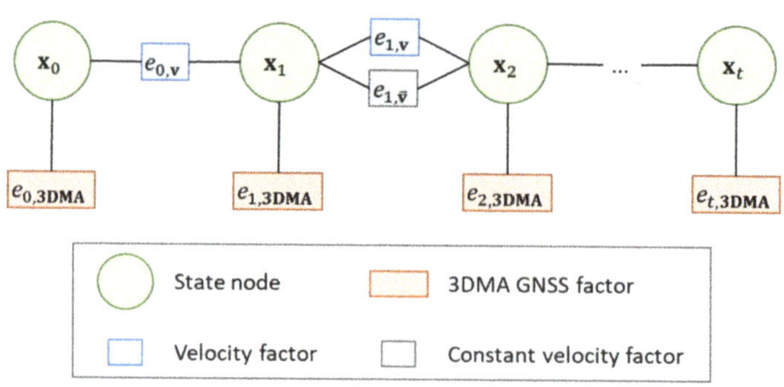

Figure 5. Structure of the proposed loosely-coupled 3DMA GNSS and velocity via FGO.

The error factor between the 3DMA GNSS solution, $\mathbf{x}_{t,3DMA}$, and optimized state, $\mathbf{x}_t$, is given by,

$$\|e_{t,\mathrm{3DMA}}\|^2_{\boldsymbol{\sigma}^2_{\mathrm{3DMA}}} = \|\mathbf{x}_t - \mathbf{x}_{t,\mathrm{3DMA}}\|^2_{\boldsymbol{\sigma}^2_{\mathrm{3DMA}}} \tag{7}$$

where $\boldsymbol{\sigma}^2_{\mathrm{3DMA}} = \alpha \cdot \mathrm{diag}\left(\left[\sigma^2_{\mathrm{3DMA},x}, \sigma^2_{\mathrm{3DMA},y}, \sigma^2_{\mathrm{3DMA},z}\right]\right)$ is a diagonal variance matrix of the 3DMA GNSS. Constant $\alpha = 1$ is an empirically determined tuning factor for 3DMA

GNSS error factor. Variance at each axis is taken by the distance variation between the 3DMA GNSS solution and candidates with the highest 10% likelihood score, divided by the separation of candidates, γ,

$$\sigma^2_{\text{3DMA}} = \frac{1}{\gamma}\text{Var}(|\mathbf{x}_{t,3DMA} - \mathbf{x}_{t,10\%}|) \tag{8}$$

where $\mathbf{x}_{t,10\%}$ represents the candidates' position with the highest 10% likelihood score. $|\cdot|$ denotes the Euclidean distance between two positions.

Receiver velocity, $\mathbf{v}_t$, and clock drift, $c\dot{\delta t}_t$, is estimated by the Doppler measurements of every satellite i at epoch t, $d_t = \left[d_t^1, \ldots, d_t^i\right]$, via the least-squares (LS) method [41]. The error factor between consecutive epochs can be expressed as follows,

$$||e_{t,\mathbf{v}}||^2_{\sigma^2_{\mathbf{v},t}} = ||\mathbf{v}_t - \frac{1}{\Delta t}(\mathbf{x}_{t+1} - \mathbf{x}_t)||^2_{\sigma^2_{\mathbf{v},t}} \tag{9}$$

where Δt is the time difference between epoch t and $t+1$. $\sigma^2_{\mathbf{v}} = \beta\cdot\text{diag}\left(\left[\sigma^2_{v,x}, \sigma^2_{v,y}, \sigma^2_{v,z}\right]\right)$ is a diagonal covariance matrix associated with the velocity $\mathbf{v}_t$ at x-, y-, and z-axis, respectively. And constant $\beta = 5.2$ is an empirically determined tuning factor for velocity error factor. Parameters α and β are determined empirically based on an open-source dataset [50] that covers different typical urban canyon scenarios. All results in this study share the same set of parameters. If tuning factor β increases, the integrated result approaches 3DMA GNSS more. If decreasing the factor below 5.0, the optimized results will be much smoother, but easier to observe a drift if a wrong velocity is estimated.

A constant velocity motion model [51] is included in this graph structure to provide a smoothed trajectory estimation. As this study assumed, users' motions are small with an ignorable acceleration. This factor minimizes the error between the position change between two epochs and the averaged velocity estimated via Doppler measurements, modelled as follows,

$$||e_{t,\bar{\mathbf{v}}}||^2_{\sigma^2_{\bar{\mathbf{v}}}} = ||\frac{1}{2}(\mathbf{v}_t + \mathbf{v}_{t+1}) - \frac{1}{\Delta t}(\mathbf{x}_{t+1} - \mathbf{x}_t)||^2_{\sigma^2_{\bar{\mathbf{v}}}} \tag{10}$$

where $\sigma^2_{\bar{\mathbf{v}}} = \frac{1}{2}\left(\sigma^2_{\mathbf{v},t} + \sigma^2_{\mathbf{v},t+1}\right)$ is the averaged diagonal covariance matrix at time t and $t+1$.

The cost function for the position estimation of the proposed loosely-coupled 3DMA GNSS via FGO is formulated as,

$$\chi^* = \underset{\chi}{\text{argmin}}\sum_t ||e_{t,\mathbf{3DMA}}||^2_{\sigma^2_{\text{3DMA}}} + ||e_{t,\mathbf{v}}||^2_{\sigma^2_{\mathbf{v},t}} + ||e_{t,\bar{\mathbf{v}}}||^2_{\sigma^2_{\bar{\mathbf{v}}}} \tag{11}$$

where $\chi = [\mathbf{x}_1, \ldots, \mathbf{x}_t]$ is the state set of the receiver and χ^* denotes the optimal states set. For computational efficiency, a sliding window for FGO is set as 200 s, which is determined empirically.

4. Experiments and Results

4.1. Experiment Setup

A commercial-grade receiver, u-blox ZED-F9P, was connected to a microcomputer, Nvidia Jetson TX2. A total of four satellite constellations with a single frequency were enabled during the experiments: GPS L1, GLONASS G1, Galileo E1, and Beidou B1. We modified the open-source library RTKLIB [52] for the GNSS-related processes, the main program structure can be found in Appendix B. Google Ceres Solver [53] was used for the nonlinear least squares (NLS) and FGO processes. Several experiments took place on the lower east side of NYC (Murray Hill, Manhattan), map plot can be found in Appendix A. In these experiments, two team members walked fixed navigation routes as if commuting between an origin (NYU Medical Center, New York City) and specific target destinations (storefronts) in a 1 mile radius. A total of 11 trips were made and used for analysis.

The ground-truth reference trajectory was obtained via post-processing. The pedestrian subjects who collected the data walked straight lines and made their best attempts not to veer. Starting and ending locations and locations in-between were labelled manually. We also equipped a smartphone during the experiment and recorded the device location output. We used the smartphone output location to interpolate the longitudinal speed and project the vector between starting and ending location.

4.2. Experiment Results

The evaluation was aimed at comparing the proposed algorithms in both a real-time and post-processing manner, also with several conventional solutions:

1. **NMEA**: receiver output solution.
2. **WLS**: weighted least squares method [52]; uses pseudorange to estimate receiver location.
3. **3DMA GNSS**: snapshot state-of-the-art 3DMA GNSS with positioning hypothesis candidates [35].
4. **LC-FGO (proposed)**: real-time forward (instantaneous) processed loosely-coupled FGO solution with integrated 3DMA GNSS and velocity.
5. **LC-FGO-PP (proposed)**: combined (forward and backward) processed loosely-coupled FGO solution with integrated 3DMA GNSS and velocity.

The optimization frame was under the ECEF coordinate system. The comparison is divided into root-mean-squared error (RMSE) and standard deviation (STD) positioning error in meters. Note that both LC-FGO and LC-FGO-PP share the same graph structure. Only LC-FGO-PP uses historical and future factors and is optimized in a combined direction forward and backwards.

A total of 11 experimental navigation trips were conducted in New York City. The positioning results of different trips are shown in Table 1. In summary, the candidate-based 3DMA GNSS always outperformed the conventional WLS. After integrating the velocity and optimizing it only for forward direction, the positioning accuracy was improved. If constraints optimization is performed in a combined manner, the positioning accuracy becomes higher. Meanwhile, in most cases, FGO outperformed the receiver's output solution (NMEA). From the overall performance of different experiments, the RMSE and STD of 3DMA GNSS are 25.34 m and 19.46 m, respectively. At the same time, LC-FGO is 21.05 m and 14.60 m for RMSE and STD, respectively. LC-FGO-PP have a stronger constraint between epochs and obtained RMSE of 15.97 m while STD is 12.48 m. Two FGO have a smaller RMSE, which means that the overall performance is better than that of 3DMA GNSS, and a lower STD implies that they are more robust. We selected two trips out of eleven (one good and bad case, respectively) to further discuss in this section.

We first present a trip with a good positioning performance (Trip 6). It starts from a relatively open-sky area and travels along straight to a deep urban canyon. The plots are shown in Figure 6.

In this experiment, it can be observed that the 3DMA GNSS outperforms WLS by twofold, and the positioning RMSE are 18.27 m and 39.17 m, respectively. Many solutions for WLS were located on the opposite side or on the wrong street, as shown in Figure 6c. With the aid of 3D models, 3DMA can correct the solution back to the correct street. If further integrated with the Doppler measurements, the positioning error can be suppressed in most cases, especially around epochs 200 s to 600 s. Results in the RMSE of the forward LC-FGO and combined FGO (LC-FGO-PP) are 15.32 m and 14.56 m, respectively.

Trip 6 is followed by a navigation trip with bad positioning results (Trip 2). Trip 2 begins in a deep urban canyon with a walk along the street to a relatively open area which is the opposite to Trip 6. The map and error plot of this experiment are shown in Figure 7.

Table 1. Statistics on positioning results of all experiments.

Navigation Trips	Epochs (s)	Algorithm	RMSE (m)	STD (m)
1	952	1. NMEA	31.09	14.47
		2. WLS	38.30	20.00
		2. 3DMA GNSS	19.70	15.51
		3. LC-FGO	24.66	14.95
		4. LC-FGO-PP	15.54	12.25
2	979	1. NMEA	74.81	31.57
		2. WLS	59.15	26.94
		2. 3DMA GNSS	29.14	16.75
		3. LC-FGO	33.56	17.13
		4. LC-FGO-PP	24.66	13.08
3	574	1. NMEA	19.87	7.35
		2. WLS	62.66	38.62
		2. 3DMA GNSS	27.62	16.40
		3. LC-FGO	22.98	11.04
		4. LC-FGO-PP	21.38	9.51
4	607	1. NMEA	17.20	11.08
		2. WLS	91.98	54.99
		2. 3DMA GNSS	21.08	12.26
		3. LC-FGO	13.01	6.48
		4. LC-FGO-PP	14.09	6.85
5	599	1. NMEA	29.01	7.43
		2. WLS	30.34	10.46
		2. 3DMA GNSS	22.64	13.21
		3. LC-FGO	20.38	10.25
		4. LC-FGO-PP	18.90	11.18
6	934	1. NMEA	36.89	18.01
		2. WLS	39.17	19.54
		2. 3DMA GNSS	18.27	11.28
		3. LC-FGO	15.32	10.44
		4. LC-FGO-PP	14.56	8.73
7	885	1. NMEA	33.36	15.61
		2. WLS	44.25	25.89
		2. 3DMA GNSS	18.64	11.27
		3. LC-FGO	25.17	11.03
		4. LC-FGO-PP	12.17	6.08
8	513	1. NMEA	39.09	11.05
		2. WLS	36.43	15.94
		2. 3DMA GNSS	16.55	9.46
		3. LC-FGO	21.30	8.36
		4. LC-FGO-PP	14.22	7.47
9	878	1. NMEA	24.17	10.38
		2. WLS	40.86	21.21
		2. 3DMA GNSS	41.50	26.91
		3. LC-FGO	44.62	29.99
		4. LC-FGO-PP	37.67	24.91
10	742	1. NMEA	36.01	16.88
		2. WLS	49.43	31.11
		2. 3DMA GNSS	26.72	15.29
		3. LC-FGO	25.49	13.08
		4. LC-FGO-PP	20.46	11.15
11	733	1. NMEA	46.78	18.12
		2. WLS	62.33	37.96
		2. 3DMA GNSS	36.85	28.13
		3. LC-FGO	37.82	27.85
		4. LC-FGO-PP	32.13	26.08

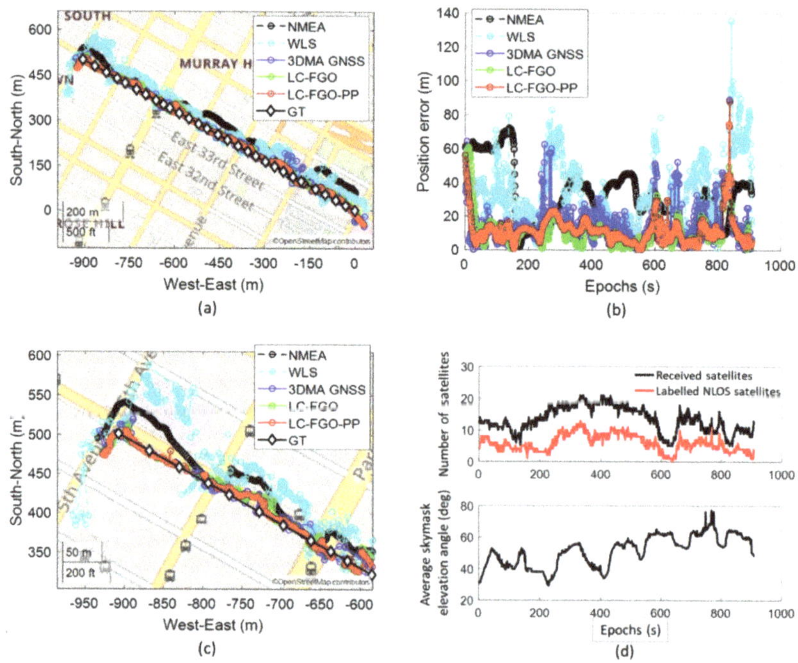

Figure 6. (**a**) map plot, (**b**) positioning errors, (**c**) magnified map plot of last 300 epochs, and (**d**) number of received satellites and average skymask elevation angle of good positioning trip (Trip 6).

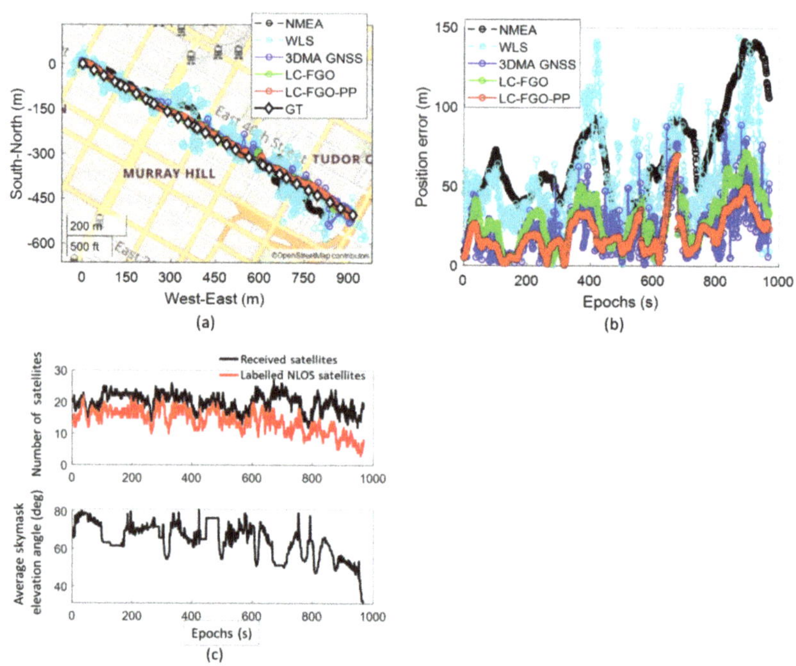

Figure 7. (**a**) map plot, (**b**) positioning errors, and (**c**) number of received satellites and average skymask elevation angle of bad positioning trip (Trip 2).

Similar to Trip 6, both 3DMA GNSS and two LC-FGO algorithms outperform WLS. The RMSE of WLS, 3DMA GNSS, LC-FGO, and LC-FGO-PP are 59.15 m, 29.14 m, 33.56 m, and 24.66 m, respectively. However, the LC-FGO is not outperforming the 3DMA GNSS. The overall positioning error is larger than that in Trip 6 because the environment is more complex. The average skymask elevation angle is higher in Trip 2, resulting in a more severe NLOS reception that mostly occupies a large portion of the total received satellites. The main error comes from the last 200 epochs. The performance of 3DMA GNSS keeps fluctuating during this period. The natural difference between 3DMA GNSS and FGO results in the average performance of this experiment. 3DMA GNSS is a snapshot estimation, and each epoch performance is independent of the others. However, the FGO is different, especially for the forward FGO. The fluctuation of the forward FGO will keep accumulating errors in batch optimization. Therefore, the future estimation is affected. However, the combined FGO, LC-FGO-PP, has a much stronger constraint that tries to optimize the solution in both directions. As a result, the positioning error can be suppressed. Therefore, if the performance of LC-FGO has to be improved, marginalization analysis must be done to find the acceptable error of this graphical problem. And we must adaptively select the existing trustworthy information in the sliding window.

Near the end of the experiment, some 3DMA GNSS solutions wrongly estimated the position of the next block, as shown in Figure 8a. The reason is that the receiver was located in a relatively open area. The PCA result of the average skymask indicates that the two eigenvalues are similar, and there are no clear major or minor axes. Results in the candidate distribute as a full circle, and the solution estimates at the local minima, as shown in Figure 8b. The candidate might have to distribute based on the user's average historical motion to resolve this issue. However, a pedestrian's motion is not as consistent as a vehicle's, therefore distributing candidates based on average motion cannot capture a rapid motion change. Another consideration is the detection of an instant motion change with an inertial measurement unit (IMU) that could be integrated into the camera or platform, more broadly.

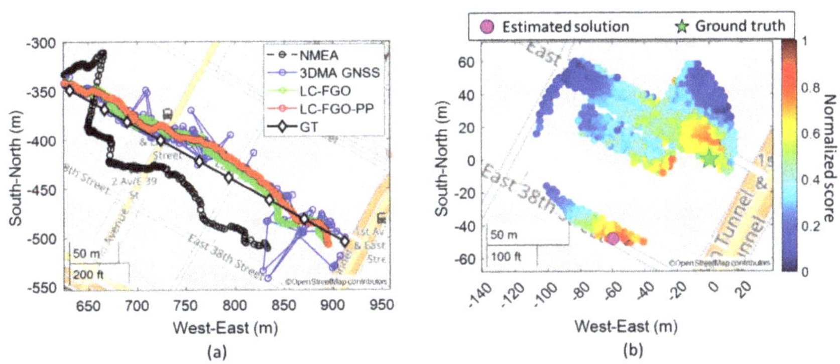

Figure 8. (**a**) zoom−in map plot near the end of Trip 2. (**b**) one of the epochs with large position error due to local minima problem.

Lastly, we also demonstrate a vehicle case in Hong Kong using the same receiver. The experiment covers different scenarios of an urban city, from open-sky areas to deep urban canyons. The data can be found in [50]. The data is collected using the same receiver model, u-blox ZED-F9P, with a patch antenna. The reference trajectory is provided by NovAtel SPAN-CPT [54], a GNSS RTK/INS (fiber-optic gyroscopes, FOG) integrated navigation system. Positioning statistics are shown in Table 2, and the map plot and error plot are shown in Figure 9. The vehicle case covers more scenarios across different environmental complexities and velocities, as shown in Figure 9c,d, respectively. Urban scenarios covering an average skymask elevation angle of around 20 degrees up to nearly 80 degrees are covered. And velocities from 0 m/s up to about 12 m/s are presented in this case.

Table 2. Statistics of vehicle-mounted trip results.

Algorithm	RMSE (m)	STD (m)
1. WLS	14.92	9.20
2. 3DMA GNSS	7.94	4.85
3. LC-FGO	8.09	4.55
4. LC-FGO-PP	5.80	2.95

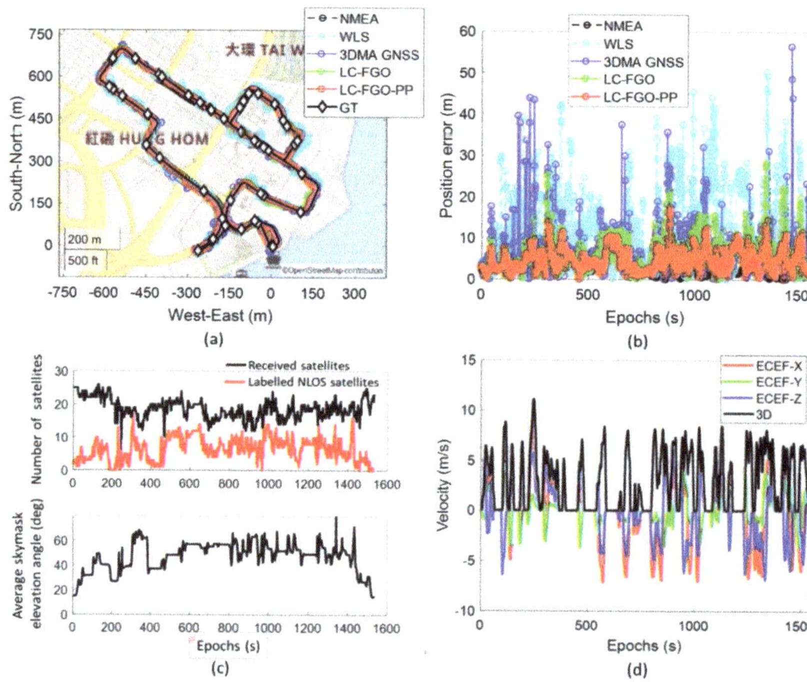

Figure 9. (**a**) map plot, (**b**) positioning errors, (**c**) number of received satellites and average skymask elevation angle, and (**d**) velocities under the ECEF coordinate system provided by NovAtel SPAN−CPT of the vehicle-mounted experiment.

Both 3DMA GNSS and FGO outperformed WLS in this case. RMSE of WLS, 3DMA GNSS, LC-FGO, and LC-FGO-PP are 14.92 m, 7.94 m, 8.09 m, and 5.80 m, respectively. Overall, the positioning outperformed what was noted for New York. There are two main reasons for the excellent positioning performance. Firstly, the local environments in NY and HK were different; the testing areas in NY were more urbanized, i.e., the average skymasks' elevation angle at ground truths in all navigation trips was 58.6° and 46.9° for New York and Hong Kong, respectively. Secondly, measurement noise was notably higher in NY, likely secondary to motion variation [55]. The vehicle (HK) had higher dynamic motion, and measurements suffered less from the multipath effects, therefore better positioning performance could be obtained. We labelled the pseudorange error using the double differencing technique [56] for a good case in the New York dataset (Trip 6) and Hong Kong, as shown in Figure 10. The double difference [56] requires measurements from the reference station. The pseudorange of the commonly received satellites is differenced. The common clock and atmospheric errors are eliminated. Geometric distance, D, is given by the calculated satellite position from the ephemeris, surveyed location of the reference station, and our labelled ground truth. The reference station was set up in an open-sky area where measurement can be assumed to be healthy. Therefore, the remaining value after differencing can be treated as the error caused by the environment of our receiver location. The double-difference-labelled pseudorange error of the *i*-th satellite, $\nabla\Delta\rho^i$, can be calculated by,

$$\begin{aligned} \nabla\Delta\rho^i &= \rho^i_{rcv} - \rho^m_{rcv} - \left(\rho^i_{ref} - \rho^m_{ref}\right) - \nabla\Delta D^i \\ \text{where } \nabla\Delta D^i &= D^i_{rcv} - D^m_{rcv} - \left(D^i_{ref} - D^m_{ref}\right) \end{aligned} \tag{12}$$

where $*_{rcv}$ stands for receiver data while $*_{ref}$ stands for reference station data. $*^m$ stands for the master satellite's data. It is selected in a system-specific pivot satellite manner with the highest elevation angle. ρ and D stand for pseudorange measurement and geometric distance, respectively. Reference station data was retrieved from NYS Spatial Reference Network (NYSNet) for data evaluation in New York. For Hong Kong dataset evaluation, reference station data was retrieved from Hong Kong Satellite Positioning Reference Station Network (SatRef).

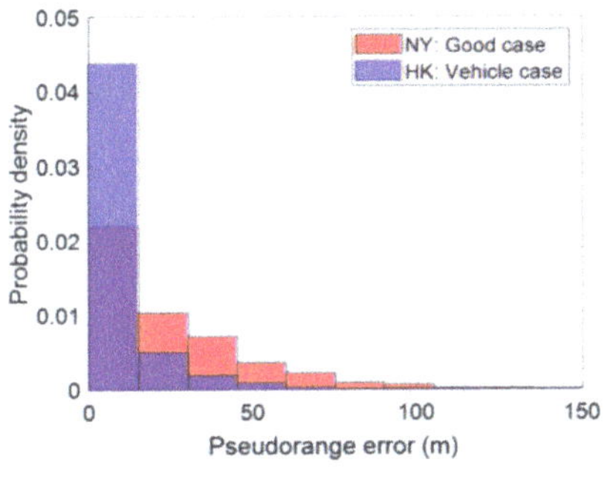

Figure 10. Probability density function plot on pseudorange error labelled by double differencing technique. Note that the master satellite is excluded from pseudorange error labelling, e.g., $i \neq m$ in Equation (12).

The HK dataset reveals a better pseudorange quality, and it is reasonable to expect better positioning performance.

A similar conclusion can be made in this vehicle-mounted experiment based on the results. 3DMA GNSS and LC-FGO obtain similar performance in this data, but comparing the positioning error shown in Figure 9b, LC-FGO can reduce the positioning better than the 3DMA, resulting in a smaller standard deviation on the positioning error. In other words, LC-FGO can provide a smoother and more robust trajectory, which applies to LC-FGO-PP. However, the error of velocity can degrade the integration performance. Figure 11 shows the epoch around 1300 s. Although the 3DMA GNSS performs well, the wrong velocity estimated by Doppler measurements with WLS results in wrong integrated results. As a result, error mitigation or a correction for Doppler measurements have to be explored in the future. Therefore, tightly coupling these approaches with Doppler measurements can potentially address the problem. In doing so, wrong Doppler measurements will be identified and isolated from the state estimation separately. A sophisticated model may be developed to model the Doppler errors [57] so that inaccurate measurements are used in the future. Meanwhile, 3DMA GNSS can be integrated with Doppler measurements more tightly in future work by expressing discrete sampled locations with a continuous mathematical model.

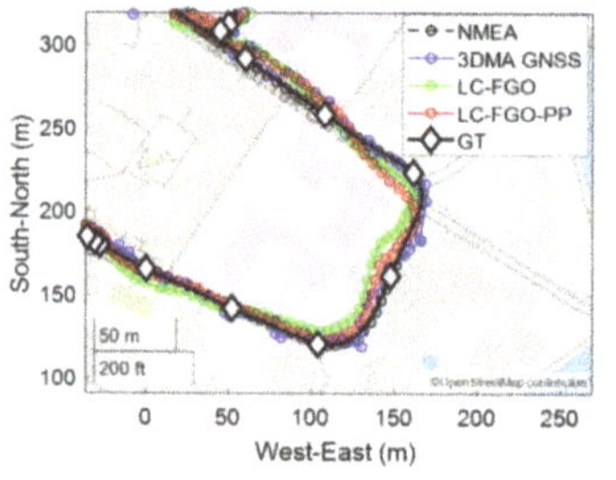

Figure 11. Zoom-in map plot of positioning error caused by badly estimated velocity.

4.3. Computational Load and Storage Requirements

One of the main contributions of this study is to develop a real-time positioning system. Therefore, the processing time of a single epoch solution is important to guidance for a real-time operation that needs to maintain an output rate of 1 Hz. The computational load is directly proportional to the number of distributed candidates (sampling radius) and available satellites. From the result, the average number of total received satellites (including LOS and NLOS) and sampled candidates are 26 and 1143, respectively. The processing time for a single epoch solution is 0.91 s. In other words, the implemented system can provide a real-time operation at a 1 Hz output rate. If a higher output rate is required, using GPU has a huge potential to accelerate the process for real-time applications, such as presented work on using GPU for ray-tracing simulation [29] and correlation-level positioning [58].

Another important point for 3DMA GNSS implementation is the format employed to store the information of 3D building models. It is impossible for a microcomputer to generate skymask online or in real-time. Therefore, the skymask is pre-generated offline and stored in CSV format, as in [50]. If we were to cover the New York downtown area (around 3.6 km by 2.9 km), a total of 812,403 locations (outside the buildings) with 4 m separations for each candidate, the total file size of requisite skymasks would be 1.30 GB. This storage is still manageable for city-scoped applications. If the system has to be extended state-wise, further engineering work must be done to devise a sustainable solution for skymask database implementation.

5. Conclusions and Future Work

This study developed a real-time loosely-coupled 3DMA GNSS with a Doppler measurements positioning system via FGO, and skymask context-based candidate sampling. Our approach distributes the candidates more effectively and mitigates local minima issues. Based on the experimental results, the positioning RMSE of loosely-coupled 3DMA GNSS with Doppler measurements via FGO is around 21 m with STD 15 m (on average). Performance can be further improved when optimizing in a combined direction with RMSE reduced to about 16 m with a STD of 13 m. The FGO can provide a lower standard deviation error than the candidate-based 3DMA GNSS, which means that it can provide a smoother and more robust solution.

However, the performance of LC-FGO still has space to be improved. The results show that candidate-based 3DMA GNSS intermittently outperforms LC-FGO. The reason is the positioning error of 3DMA GNSS keeps contributing to the integration with Doppler measurements. Accumulated error results affect future batch optimization. An adaptive scheme should be developed to select the high confidence information in the sliding window. Moreover, tighter integration of the 3DMA GNSS with Doppler measurements should be effected to improve the performance.

Furthermore, bad Doppler measurements result in wrongly estimated velocity. This will degrade the FGO performance. Doppler measurements error mitigation or correction is the key to improving the positioning. In future research, we will explore how 3DMA

GNSS can more tightly integrate with Doppler measurements to provide a more robust positioning in the urban canyons for smart health applications and beyond.

Author Contributions: Conceptualization, L.-T.H.; Data curation, M.J.L.L., J.F., T.N. and M.B.; Formal analysis, H.-F.N.; Funding acquisition, J.-R.R.; Investigation, L.-T.H., T.N., M.B. and J.-R.R.; Methodology, H.-F.N.; Project administration, L.-T.H., T.N., M.B. and J.-R.R.; Resources, T.N., M.B. and J.-R.R.; Software, H.-F.N., M.J.L.L. and J.F.; Supervision, L.-T.H. and J.-R.R.; Validation, H.-F.N. and J.F.; Visualization, M.J.L.L.; Writing—original draft, H.-F.N.; Writing—review and editing, L.-T.H., M.B. and J.-R.R. All authors have read and agreed to the published version of the manuscript.

Funding: Research reported in this publication was supported in part by the NSF grant 1952180 under the Smart and Connected Community program, as well as by NSF Grant ECCS-1928614, the National Eye Institute of the National Institutes of Health under Award Number R21EY033689, and DoD grant VR200130 under the "Delivering Sensory and Semantic Visual Information via Auditory Feedback on Mobile Technology". The content is solely the responsibility of the authors and does not necessarily represent the official views of the National Institutes of Health and NSF, and DoD.

Institutional Review Board Statement: Not applicable.

Informed Consent Statement: Not applicable.

Data Availability Statement: Not applicable.

Conflicts of Interest: New York University (NYU) and John-Ross Rizzo (J.-R.R.) have financial interests in related intellectual property. NYU owns a patent licensed to Tactile Navigation Tools. NYU, J.-R.R. are equity holders and advisors of said company.

Appendix A

This appendix presents the map plot of a total of 11 experimental navigation trips in New York City, as shown in Figure A1.

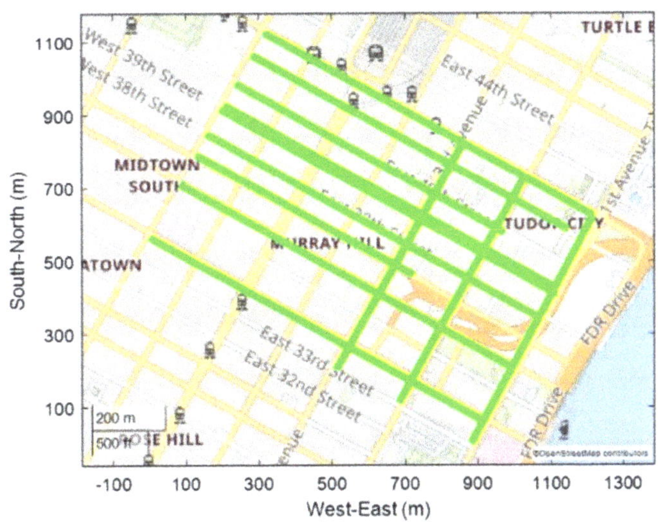

Figure A1. Map plot of a total of 11 experimental navigation trips in New York, each straight green line represents one trip, correspondingly.

Appendix B

This appendix presents the main modified functions and IO from RTKLIB. The development environment is under ubuntu 18.04. RTKLIB 2.4.3 b34 is used.

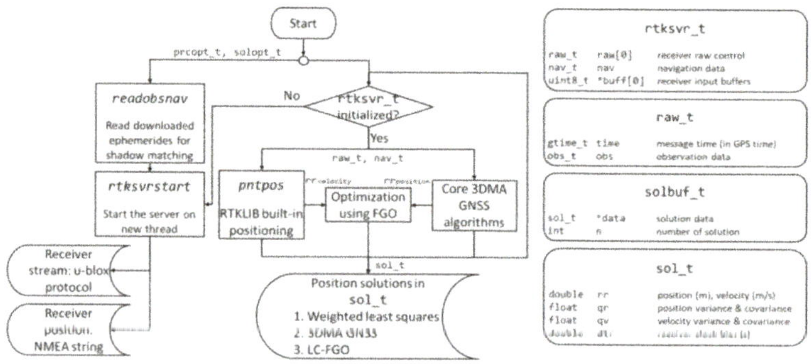

Figure A2. Program flow of this study. The left part is the program structure. The right part is the data dictionary of the primary data structure and members used for implementation.

References

1. Rizzo, J.-R.; Feng, C.; Riewpaiboon, W.; Mongkolwat, P. A Low-Vision Navigation Platform for Economies in Transition Countries. In Proceedings of the 2020 IEEE World Congress on Services (SERVICES), Beijing, China, 18–23 October 2020; IEEE: Piscataway, NJ, USA, 2020; pp. 1–3.
2. Niu, L.; Qian, C.; Rizzo, J.-R.; Hudson, T.; Li, Z.; Enright, S.; Sperling, E.; Conti, K.; Wong, E.; Fang, Y. A Wearable Assistive Technology for the Visually Impaired with Door Knob Detection and Real-Time Feedback for Hand-to-Handle Manipulation. In Proceedings of the IEEE International Conference on Computer Vision Workshops, Venice, Italy, 22–29 October 2017; pp. 1500–1508.
3. Gui, W.; Li, B.; Yuan, S.; Rizzo, J.-R.; Sharma, L.; Feng, C.; Tzes, A.; Fang, Y. An Assistive Low-Vision Platform that Augments Spatial Cognition Through Proprioceptive Guidance: Point-to-Tell-and-Touch. In Proceedings of the 2019 IEEE/RSJ International Conference on Intelligent Robots and Systems (IROS), Macau, China, 4–8 November 2019; IEEE: Piscataway, NJ, USA, 2019; pp. 3817–3822.
4. Boldini, A.; Garcia, A.L.; Sorrentino, M.; Beheshti, M.; Ogedegbe, O.; Fang, Y.; Porfiri, M.; Rizzo, J.-R. An inconspicuous, integrated electronic travel aid for visual impairment. *ASME Lett. Dyn. Syst. Control.* **2021**, *1*, 1–9. [CrossRef]
5. Li, X.; Cui, H.; Rizzo, J.-R.; Wong, E.; Fang, Y. Cross-Safe: A Computer Vision-Based Approach to Make All Intersection-Related Pedestrian Signals Accessible for the Visually Impaired. In Proceedings of the Science and Information Conference, Las Vegas, NV, USA, 25–26 April 2019; Springer: Berlin/Heidelberg, Germany, 2019; pp. 132–146.
6. Rizzo, J.-R.; Pan, Y.; Hudson, T.; Wong, E.K.; Fang, Y. Sensor Fusion for Ecologically Valid Obstacle Identification: Bu ilding A Comprehensive Assistive Technology Platform for the Visually Impaired. In Proceedings of the 2017 7th International Conference on Modeling, Simulation, and Applied Optimization (ICMSAO), Sharjah, United Arab Emirates, 4–6 April 2017; IEEE: Piscataway, NJ, USA, 2017; pp. 1–5.
7. Shoureshi, R.A.; Rizzo, J.R.; Hudson, T.E. *Smart Wearable Systems for Enhanced Monitoring and Mobility. Advances in Science and Technology*; Trans Tech Publications: Bach, Switzerland, 2017; pp. 172–178.
8. Phamduy, P.; Rizzo, J.-R.; Hudson, T.E.; Torre, M.; Levon, K.; Porfiri, M. Communicating through touch: Macro fiber composites for tactile stimulation on the abdomen. *IEEE Trans. Haptics* **2017**, *11*, 174–184. [CrossRef] [PubMed]
9. Groves, P.D. Multipath vs. NLOS signals. *Inside GNSS* **2013**, *8*, 40–42.
10. Asano, S.; Wakuda, Y.; Koshizuka, N.; Sakamura, K. A Robust Pedestrian Dead-Reckoning Positioning Based on Pedestrian Behavior and Sensor Validity. In Proceedings of the 2012 IEEE/ION Position, Location and Navigation Symposium, Myrtle Beach, SC, USA, 23–26 April 2012; IEEE: Piscataway, NJ, USA, 2012; pp. 328–333.
11. Steinbrücker, F.; Sturm, J.; Cremers, D. Real-Time Visual Odometry from Dense RGB-D Images. In Proceedings of the 2011 IEEE International Conference on Computer Vision Workshops (ICCV Workshops), Barcelona, Spain, 6–13 November 2011; IEEE: Piscataway, NJ, USA, 2011; pp. 719–722.
12. Qin, T.; Li, P.; Shen, S. VINS-Mono: A Robust and Versatile Monocular Visual-Inertial State Estimator. *IEEE Trans. Robot.* **2018**, *34*, 1004–1020. [CrossRef]
13. Suzuki, T.; Kubo, N. N-LOS GNSS signal detection using fish-eye camera for vehicle navigation in urban environments. In Proceedings of the 27th International Technical Meeting of the Satellite Division of the Institute of Navigation, ION GNSS 2014, Tampa, FL, USA, 8–12 September 2014; pp. 1897–1906.

14. Moreau, J.; Ambellouis, S.; Ruichek, Y. Fisheye-Based Method for GPS Localization Improvement in Unknown Semi-Obstructed Areas. *Sensors* **2017**, *17*, 119. [CrossRef]
15. Groves, P.D.; Jiang, Z. Height Aiding, C/N0 Weighting and Consistency Checking for GNSS NLOS and Multipath Mitigation in Urban Areas. *J. Navig.* **2013**, *66*, 653–669. [CrossRef]
16. Hsu, L.-T.; Tokura, H.; Kubo, N.; Gu, Y.; Kamijo, S. Multiple Faulty GNSS Measurement Exclusion Based on Consistency Check in Urban Canyons. *IEEE Sens. J.* **2017**, *17*, 1909–1917. [CrossRef]
17. Ng, H.-F.; Zhang, G.; Yang, K.-Y.; Yang, S.-X.; Hsu, L.-T. Improved weighting scheme using consumer-level GNSS L5/E5a/B2a pseudorange measurements in the urban area. *Adv. Space Res.* **2020**, *66*, 1647–1658. [CrossRef]
18. Bradbury, J.; Ziebart, M.; Cross, P.A.; Boulton, P.; Read, A. Code Multipath Modelling in the Urban Environment Using Large Virtual Reality City Models: Determining the Local Environment. *J. Navig.* **2007**, *60*, 95–105. [CrossRef]
19. Lau, L.; Cross, P. Development and testing of a new ray-tracing approach to GNSS carrier-phase multipath modelling. *J. Geod.* **2007**, *81*, 713–732. [CrossRef]
20. Bradbury, J. Prediction of Urban GNSS Availability and Signal Degradation Using Virtual Reality City Models. In Proceedings of the 20th International Technical Meeting of the Satellite Division of The Institute of Navigation (ION GNSS 2007), Fort Worth, TX, USA, 25–28 September 2007; 2007; pp. 2696–2706.
21. Groves, P.D. It's Time for 3D Mapping–Aided GNSS. *Inside GNSS Magazine*, July 2016; pp. 50–56.
22. Diggelen, F.V. End Game for Urban GNSS: Google's Use of 3D Building Models. *Inside GNSS Magazine*, 21 March 2021; pp. 42–49.
23. Obst, M.; Bauer, S.; Wanielik, G. Urban Multipath Detection and Mitigation with Dynamic 3D Maps for Reliable Land Vehicle Localization. In Proceedings of the 2012 IEEE/ION Position, Location and Navigation Symposium, Myrtle Beach, SC, USA, 23–26 April 2012; IEEE: Piscataway, NJ, USA, 2012; pp. 685–691.
24. Wang, L.; Groves, P.D.; Ziebart, M.K. GNSS Shadow Matching: Improving Urban Positioning Accuracy Using a 3D City Model with Optimized Visibility Scoring Scheme. *NAVIGATION J. Inst. Navig.* **2013**, *60*, 195–207. [CrossRef]
25. Wang, L.; Groves, P.D.; Ziebart, M.K. Smartphone Shadow Matching for Better Cross-street GNSS Positioning in Urban Environments. *J. Navig.* **2015**, *68*, 411–433. [CrossRef]
26. Yozevitch, R.; Moshe, B.B. A Robust Shadow Matching Algorithm for GNSS Positioning. *Navig. J. Inst. Navig.* **2015**, *62*, 95–109. [CrossRef]
27. Hsu, L.-T.; Gu, Y.; Kamijo, S. 3D building model-based pedestrian positioning method using GPS/GLONASS/QZSS and its reliability calculation. *GPS Solut.* **2016**, *20*, 413–428. [CrossRef]
28. Miura, S.; Hsu, L.-T.; Chen, F.; Kamijo, S. GPS Error Correction with Pseudorange Evaluation Using Three-Dimensional Maps. *IEEE Trans. Intell. Transp. Syst.* **2015**, *16*, 3104–3115. [CrossRef]
29. Ziedan, N.I. Urban Positioning Accuracy Enhancement Utilizing 3D Buildings Model and Accelerated Ray Tracing Algorithm. In Proceedings of the 30th International Technical Meeting of The Satellite Division of the Institute of Navigation (ION GNSS+ 2017), Portland, OR, USA, 25–29 September 2017; Oregon Convention Center: Portland, OR, USA, 2017; pp. 3253–3268.
30. Ng, H.-F.; Zhang, G.; Hsu, L.-T. A Computation Effective Range-based 3D Mapping Aided GNSS with NLOS Correction Method. *J. Navig.* **2020**, *73*, 1202–1222. [CrossRef]
31. Suzuki, T.; Kubo, N. Correcting GNSS Multipath Errors Using a 3D Surface Model and Particle Filter. In Proceedings of the 26th International Technical Meeting of the Satellite Division of The Institute of Navigation (ION GNSS+ 2013), Nashville, TN, USA, 16–20 September 2013; pp. 1583–1595.
32. Suzuki, T.; Kubo, N. GNSS Positioning with Multipath Simulation Using 3D Surface Model in Urban Canyon. In Proceedings of the 25th International Technical Meeting of the Satellite Division of The Institute of Navigation (ION GNSS 2012), Nashville, TN, USA, 17–21 September 2012; pp. 438–447.
33. Schön, S.; Baasch, K.-N.; Icking, L.; KarimiDoona, A.; Lin, Q.; Ruwisch, F.; Schaper, A.; Su, J. Towards Integrity for GNSS-Based Urban Navigation—Challenges And Lessons Learned. In *Proceedings of the 2022 IEEE Intelligent Vehicles Symposium (IV)*; Aachen, Germany, 4–9 June 2022, IEEE Press: Piscataway, NJ, USA, 2022; pp. 1774–1781.
34. Zhong, Q.; Groves, P.D. Multi-Epoch 3D-Mapping-Aided Positioning using Bayesian Filtering Techniques. *NAVIGATION* **2022**, *69*, navi.515. [CrossRef]
35. Ng, H.-F.; Zhang, G.; Luo, Y.; Hsu, L.-T. Urban positioning: 3D mapping-aided GNSS using dual-frequency pseudorange measurements from smartphones. *NAVIGATION J. Inst. Navig.* **2021**, *68*, 727–749. [CrossRef]
36. Groves, P.D. *Principles of GNSS, Inertial, and Multisensor Integrated Navigation Systems*, 2nd ed.; Artech House: Norwood, MA, USA, 2013.
37. Zhang, G.; Wen, W.; Xu, B.; Hsu, L.-T. Extending Shadow Matching to Tightly-Coupled GNSS/INS Integration System. *IEEE Trans. Veh. Technol.* **2020**, *69*, 4979–4991. [CrossRef]
38. Suzuki, T. Integration of GNSS Positioning and 3D Map using Particle Filter. In Proceedings of the 29th International Technical Meeting of the Satellite Division of The Institute of Navigation (ION GNSS+ 2016), Portland, OR, USA, 12–16 September 2016; Oregon Convention Center: Portland, OR, USA, 2016; pp. 1296–1304.
39. Ziedan, N.I. Optimized Position Estimation in Multipath Environments using Machine Learning. In Proceedings of the 34th International Technical Meeting of the Satellite Division of The Institute of Navigation (ION GNSS+ 2021), St. Louis, MI, USA, 20–24 September 2021; Union Station Hotel: St. Louis, MI, USA, 2021; pp. 3437–3451.

40. Sünderhauf, N.; Protzel, P. Towards Robust Graphical Models for GNSS-Based Localization in Urban Environments. In Proceedings of the International Multi-Conference on Systems, Signals & Devices, Chemnitz, Germany, 20–23 March 2012; IEEE: Piscataway, NJ, USA, 2012; pp. 1–6.
41. Wen, W.; Hsu, L.-T. Towards Robust GNSS Positioning and Real-time Kinematic Using Factor Graph Optimization. In Proceedings of the 2021 IEEE International Conference on Robotics and Automation (ICRA), Xi'an, China, 30 May–5 June 2021; IEEE: Piscataway, NJ, USA, 2021; pp. 5884–5890.
42. Watson, R.M.; Gross, J.N. Evaluation of Kinematic Precise Point Positioning Convergence with an Incremental Graph Optimizer. In Proceedings of the 2018 IEEE/ION Position, Location and Navigation Symposium (PLANS), Monterey, CA, USA, 23 April 2018; pp. 589–596.
43. Zhang, G.; Ng, H.-F.; Wen, W.; Hsu, L.-T. 3D Mapping Database Aided GNSS Based Collaborative Positioning Using Factor Graph Optimization. *IEEE Trans. Intell. Transp. Syst.* **2021**, *22*, 6175–6187. [CrossRef]
44. Department of City Planning NYC 3D Model by Community District. Available online: https://www1.nyc.gov/site/planning/data-maps/open-data/dwn-nyc-3d-model-download.page (accessed on 11 May 2021).
45. Robert McNeel Rhinoceros. Available online: https://www.rhino3d.com/ (accessed on 11 May 2021).
46. Epic Games Unreal Engine. Available online: https://www.unrealengine.com/en-US (accessed on 11 May 2021).
47. van Diggelen, F. *A-GPS: Assisted GPS, GNSS, and SBAS*; Artech: Morristown, NJ, USA, 2009; Volume 33.
48. van Diggelen, F. Assisted GNSS. In *Position, Navigation, and Timing Technologies in the 21st Century*; Wiley-IEEE Press: Piscataway, NJ, USA, 2020; pp. 419–444.
49. Groves, P.D. Shadow Matching: A New GNSS Positioning Technique for Urban Canyons. *J. Navig.* **2011**, *64*, 417–430. [CrossRef]
50. Hsu, L.-T.; Kubo, N.; Wen, W.; Chen, W.; Liu, Z.; Suzuki, T.; Meguro, J. UrbanNav:An Open-Sourced Multisensory Dataset for Benchmarking Positioning Algorithms Designed for Urban Areas. In Proceedings of the 34th International Technical Meeting of the Satellite Division of The Institute of Navigation (ION GNSS+ 2021), St. Louis, MI, USA, 20–24 September 2021; Institute of Navigation: Manassas, VA, USA, 2021; pp. 226–256.
51. Li, W.; Cui, X.; Lu, M. A robust graph optimization realization of tightly coupled GNSS/INS integrated navigation system for urban vehicles. *Tsinghua Sci. Technol.* **2018**, *23*, 724–732. [CrossRef]
52. Takasu, T. RTKLIB: Open Source Program Package for RTK-GPS. In Proceedings of the FOSS4G 2009, Sydney, Australia, 20–23 October 2009.
53. Agarwal, S.; Mierle, K. Ceres solver: Tutorial & reference. *Google Inc.* **2012**, *2*, 8.
54. Kennedy, S.; Hamilton, J.; Martell, H. Architecture and System Performance of SPAN-NovAtel's GPS/INS Solution. In Proceedings of the IEEE/ION PLANS 2006, San Diego, CA, USA, 25–27 April 2006; pp. 266–274.
55. Suzuki, T.; Matsuo, K.; Amano, Y. Rotating GNSS Antennas: Simultaneous LOS and NLOS Multipath Mitigation. *GPS Solut.* **2020**, *24*, 86. [CrossRef]
56. Xu, B.; Jia, Q.; Luo, Y.; Hsu, L.T. Intelligent GPS L1 LOS/multipath/NLOS classifiers based on correlator-, RINEX- and NMEA-level measurements. *Remote Sens.* **2019**, *11*, 1851. [CrossRef]
57. Xie, P.; Petovello, M.G. Measuring GNSS Multipath Distributions in Urban Canyon Environments. *IEEE Trans. Instrum. Meas.* **2015**, *64*, 366–377.
58. Xu, L.; Ziedan, N.I.; Niu, X.; Guo, W. Correlation acceleration in GNSS software receivers using a CUDA-enabled GPU. *GPS Solut.* **2017**, *21*, 225–236. [CrossRef]

sensors

MDPI

Article

An Improved Ambiguity Resolution Algorithm for Smartphone RTK Positioning

Yang Jiang [1,*], Yuting Gao [2], Wei Ding [3], Fei Liu [4] and Yang Gao [1]

1 Department of Geomatics Engineering, University of Calgary, Calgary, AB T2N 1N4, Canada; ygao@ucalgary.ca
2 College of Geomatics, Xi'an University of Science and Technology, Xi'an 710054, China; ygao@xust.edu.cn
3 School of Geomatics, Liaoning Technical University, Fuxin 123000, China; dingwei@lntu.edu.cn
4 Profound Positioning Inc., Calgary, AB T2P 3G3, Canada; fliu@profoundpositioning.com
* Correspondence: yang.jiang1@ucalgary.ca

Abstract: Ambiguity resolution based on smartphone GNSS measurements can enable various potential applications that currently remain difficult due to ambiguity biases, especially under kinematic conditions. This study proposes an improved ambiguity resolution algorithm, which uses the search-and-shrink procedure coupled with the methods of the multi-epoch double-differenced residual test and the ambiguity majority tests for candidate vectors and ambiguities. By performing a static experiment with Xiaomi Mi 8, the AR efficiency of the proposed method is evaluated. Furthermore, a kinematic test with Google Pixel 5 verifies the effectiveness of the proposed method with improved positioning performance. In conclusion, centimeter-level smartphone positioning accuracy is achieved in both experiments, which is greatly improved compared with the float and traditional AR solutions.

Keywords: smartphone positioning; real-time kinematic (RTK); ambiguity resolution (AR); global navigation satellite system (GNSS)

Citation: Jiang, Y.; Gao, Y.; Ding, W.; Liu, F.; Gao, Y. An Improved Ambiguity Resolution Algorithm for Smartphone RTK Positioning. *Sensors* **2023**, *23*, 5292. https://doi.org/10.3390/s23115292

Academic Editor: Robert Odolinski

Received: 16 May 2023
Revised: 30 May 2023
Accepted: 31 May 2023
Published: 2 June 2023

1. Introduction

High-precision smartphone positioning is increasingly demanded to enable potential applications such as lane-level vehicle navigation, augmented reality walking/driving, and precise agriculture via phones [1–3]. Although real-time kinematic (RTK) techniques with ambiguity resolution (AR) have been widely used for high-end Global Navigation Satellite System (GNSS) receivers, advanced positioning algorithms are required to deal with a much-poorer-quality pseudorange and carrier phase measurements from smartphones [4–6]. First, the high and unstable stochastic properties of the measurements, as well as their possible outliers, pose challenges since they often affect the precision of float RTK solutions and reduce the ambiguity-fix success rate [7]. Second, smartphone initial phase bias (IPB) will corrupt the integer property of the carrier ambiguities [8–10]. Third, the biases caused by smartphone antennas and carrier phase multipath effects would also affect the position determination [11,12]. As a result, some smartphone carrier phase ambiguities are contaminated with ambiguity biases, making fixed solutions significantly biased [13].

Many studies have been conducted to explore AR with smartphone GNSS measurements. After assessing the observation quality of smartphone pseudorange and carrier phase measurements, Paziewski et al. [14] found that a stochastic model could be developed based on signal-to-noise ratio measurements and obtained results better than the traditional elevation-dependent model. Gao et al. [15] proposed using the raw measurement's standard deviation and the multipath indicator provided by the Android application programming interface to improve the performance of smartphone AR. In addition, many studies have been undertaken to deal with these issues in realistic smartphone applications, such as multipath [16,17] and antenna offset problems [18,19], and their impacts on carrier phase ambiguities are widely noticed. To resolve the IPBs, Geng et al. [9] proposed a method for post-processing calibration. Yong et al. [20] concluded that the antenna offsets could be minimized by keeping the smartphone in an upright position. Additionally, the implementation of partial AR (PAR) can reduce the impact of the ambiguity biases on the fixed solutions [21–23], as confirmed by [24,25]. Although many studies have been undertaken to assess the performance of smartphone AR, few implementations focus on kinematic experiments in real-time smartphone applications. However, with the user motion, the time-varying multipath effects and antenna offsets increase the difficulty of handling the carrier phase ambiguity biases, which in turn pose challenges for smartphone kinematic AR. The precision of float solutions is also lower in kinematic applications due to the unstable stochastic properties of measurements, which generally decrease the ambiguity–fix success rate.

This study focuses on improving the AR algorithm for smartphones to achieve high-precision smartphone GNSS positioning. Specifically, the integer ambiguities are resolved by coupling the search-and-shrink procedure with testing methods, including the multi-epoch double-differenced (DD) model residual test and the ambiguity majority tests for candidate ambiguities and vectors. To verify the proposed method, two smartphone experiments are conducted, where Xiaomi Mi 8 is used for a static test, and Google Pixel 5 is used for a kinematic test.

Section 2 presents the methodology used in this study. Section 3 introduces the static dataset and its evaluation outcomes of AR efficiency. Section 4 discusses the kinematic positioning performance. The conclusions are summarized in Section 5.

2. Methodology

To realize AR for kinematic applications with single- or dual-frequency GNSS, including smartphones, the GNSS community uses the popular method of LAMBDA [20,26]. A unimodal transformation and a search-and-shrink scheme based on the integer least-squares (ILS) principle find the optimal ambiguity integers with a real-time computational load [27]. Until recently, the method of best integer equivariance (BIE) has proven to be a better replacement [28], where the AR performance is optimized in the sense of minimizing the mean square error [29–31]. However, these methods, including BIE, naturally consider the input float ambiguity estimations to be unbiased integers. This does not fit in the application of smartphone AR since the existence of IPB, carrier phase multipath effects, and antenna offsets lead to non-negligible ambiguity biases. As a result, the unimodal transformation and the search-and-shrink scheme are inaccurate and likely to produce a set of incorrect ambiguity integers [31]. Although PAR coupled with improved ambiguity validation strategies, such as protection-level, are proposed [32,33], they are generally not sufficiently efficient to identify the correct ambiguity integer set for smartphones due to the volume of such ambiguity biases and the significant measurement noises.

Primarily, in this study, the AR scheme using the search-and-shrink procedure is applied without the unimodal transformation, where only one integer ambiguity is resolved at a time. To select the optimal ambiguity candidate, the methods of ambiguity majority tests, as well as the ambiguity validation with the multi-epoch DD residual test are applied based on the candidate ambiguity vectors by LAMBDA with the ILS principle [34]. In the following subsections, the commonly used GNSS mixed-integer model, the AR scheme of search-and-shrink, the ambiguity majority tests, and ambiguity validation with the multi-epoch DD residual test is explained in detail.

2.1. GNSS Mixed-Integer Model

For smartphone RTK based on the DD model with an acceptable baseline length, most GNSS error sources, such as the ionosphere, troposphere, and others, are eliminated. Therefore, the linearized single-epoch observation equation is a mixed-integer model, including parameters of integer-valued carrier phase ambiguities and the real-valued baseline vector [35]:

$$y = Aa + Bb + e, \tag{1}$$

where $y \in \mathbb{R}^m$ refers to the measurement vector contaminated by a zero-mean normal-distributed noise $e \sim N(0, Q_{yy})$; $a \in \mathbb{Z}^n$ and $b \in \mathbb{R}^p$ denote the integer-valued ambiguities and the real-valued baseline vector, respectively, and $A \in \mathbb{R}^{m\times n}$ and $B \in \mathbb{R}^{m\times p}$ represent their respective design matrices. With least-squares or Kalman filtering, the float solution discards the integer nature of ambiguities [35]:

$$\begin{bmatrix} \hat{a} \\ \hat{b} \end{bmatrix} \sim N\left(\begin{bmatrix} a \\ b \end{bmatrix}, \begin{bmatrix} Q_{\hat{a}\hat{a}} & Q_{\hat{a}\hat{b}} \\ Q_{\hat{b}\hat{a}} & Q_{\hat{b}\hat{b}} \end{bmatrix} \right), \tag{2}$$

where $\hat{a}$ and $\hat{b}$ denote the float solutions with respect to ambiguities and coordinates, and $Q_{\hat{a}\hat{a}}$, $Q_{\hat{a}\hat{b}}$, $Q_{\hat{b}\hat{a}}$, and $Q_{\hat{b}\hat{b}}$ represent the variance and covariance components of the estimated parameters. To solve for the integer ambiguities $\check{a} = I(\hat{a})$, many integer equivariance estimators can be used, such as ILS and BIE [28]. In this case, ILS is discussed, which provides $\check{a}_{ILS}$:

$$\check{a}_{ILS} = argmin\|\hat{a} - z\|^2_{Q_{\hat{a}\hat{a}}}, \forall z \in \mathbb{Z}^n; \tag{3}$$

If the integer ambiguities are accepted by ambiguity validation methods, such as a ratio test, a fixed solution can be achieved by readjusting $\hat{b}$, which gives $\check{b}$:

$$\check{b} = \hat{b} - Q_{\hat{b}\hat{a}} Q^{-1}_{\hat{a}\hat{a}} (\hat{a} - \check{a}); \tag{4}$$

Although for smartphones, ambiguities suffer from biases, principles such as ILS are valid for describing the integer components; therefore, they are mostly unbiased. In this way, those biases are absorbed by the carrier phase measurement noises, and the magnitude of the residuals contributes to larger position errors, e.g., 3 cm to 10 cm. Overall, the fixed solutions should remain bias-free and with high precision. However, to solve the ILS principle of (3), the popular search-and-shrink procedure is widely used to reduce the computational complexity of the traditional methods [36,37], which can cause ambiguous candidate vectors to be biased on smartphones. The detailed algorithm and the reasons are explained in the following subsection.

2.2. Search-and-Shrink Procedure for Ambiguity Resolution

For a two-dimensional (2D) example where the ambiguities are a_1 and a_2, the search-and-shrink procedure calculates the conditional ambiguity of a_1 once if $\hat{a}_2$ is rounded to an integer $\hat{z}_2$ by bootstrapping using [25]:

$$\hat{a}_{1|2} = \hat{a}_1 - \sigma_{\hat{a}_1\hat{a}_2} \sigma^{-1}_{\hat{a}_2\hat{a}_2} (\hat{z}_2 - \hat{a}_2), \tag{5}$$

where $\hat{a}_{1|2}$ denotes the conditional ambiguity of a_1, $\sigma_{\hat{a}_1\hat{a}_2}$ represents the covariance of a_1 and a_2, and $\sigma_{\hat{a}_2\hat{a}_2}$ signifies the variance of a_2. Then, with the conditional ambiguities, the boundary value of $\|\hat{a} - z\|^2_{Q_{aa}}$ is reduced so that a smaller search space can be reached. Iteratively, the search space shrinks to reach the k integer ambiguity candidate vector, where k is user specified. In the sequel, $\check{A}$ is used to denote the combination of candidate vectors:

$$\check{A} = \begin{bmatrix} \check{a}^1 & \check{a}^2 & \cdots & \check{a}^k \end{bmatrix}, \check{A} \in \mathbb{Z}^{n\times k}, \tag{6}$$

Commonly, many ILS-based algorithms have applied this procedure [35,38], where the final decision is made as $\check{a}^1$ after the ambiguity validation. For the traditional implementation of the LAMBDA method, vector $\hat{a}$ is sorted by the increasing order of the variances of its elements, meaning $\sigma_{a_1a_1} \leq \sigma_{a_2a_2}$, where the initial search space is defined as follows:

$$F(z) = \|\hat{a} - z\|^2_{Q_{\hat{a}\hat{a}}} \leq \chi^2, \tag{7}$$

where z, with the minimum value of function $F(z)$, is the optimal ILS solution $\check{a}$, and χ^2 can be predetermined and can also shrink during the search [39,40]. LAMBDA is coupled with a unimodular transformation based on the L^TDL factorization of $Q_{\hat{a}\hat{a}}$ so that the search process can be more efficient [39].

However, in the case of smartphone carrier phase measurements, many ambiguities are biased and have lost their integer nature. Therefore, the above-mentioned search-and-shrink procedure is less valid because it assumes an integer grid. Specifically, if a_2 is naturally not an integer, the conditional ambiguity of $\hat{a}_{1|2}$ can be affected by the ambiguity bias of a_2. For higher-dimension problems, those ambiguity biases can accumulate while the search space shrinks, causing the further search space to be incorrect. Furthermore, coupling with the unimodular transformation can expose such bias to even more original ambiguities after the back transformation. In this case, current AR methods with the search-and-shrink procedure, such as LAMBDA, frequently produce an incorrect set of integer ambiguities. Unfortunately, an improved procedure considering such ambiguity biases is currently unavailable. This study primarily focuses on selecting the ambiguities to be conditionally rounded, without unimodular transformation, to mitigate the impact of ambiguity biases.

2.3. *Ambiguity Majority Tests for Candidate Ambiguities and Vectors*

Although LAMBDA with unimodular transformation can spread the impact of ambiguity biases, it is expected that some ambiguities would not vary much among different candidate vectors. This is because their ambiguity variances are relatively small due to better observation conditions; therefore, they are sorted in front of the ambiguity vector, which also makes them less biased by the search-and-shrink procedure. Therefore, these ambiguities are more trustworthy to be integers. Then, we innovatively calculate the modulus of each ambiguity a_i, namely M_{a_i}, from the combined candidate vector $\check{A}$ by the LAMBDA method, these were formulated as follows:

$$M_{a_i} = mod(\check{A}_{i,j})_{j\in\{1,\ldots,k\}}, \tag{8}$$

where $mod(\cdot)$ denotes the modulo operation. Therefore, the number of candidates of ambiguity a_i that equals M_{a_i}, namely N_{a_i}, can be calculated as follows:

$$N_{a_i} = \sum_{j\in\{1,\ldots,k\}} (\check{A}_{i,j} = M_{a_i}); \tag{9}$$

The value of N_{a_i} reflects the confidence in the selected M_{a_i}. Straightforwardly, with a higher value of N_{a_i}, the corresponding ambiguity can be more trusted since it is less variant among different ambiguity vectors. As a drawback, with a k value larger than 2, some candidate vectors by the LAMBDA method can be too abnormal to introduce them to the majority test, which can reduce its accuracy. Here, the value of $\|\hat{a} - z\|^2_{Q_{\hat{a}\hat{a}}}$ is not used to weigh the candidates for the majority test since it is potentially affected by the ambiguity biases and cannot be accurate. To exclude the abnormal candidate ambiguities, we apply the novel majority test to each candidate ambiguity vector, which was formulated as follows:

$$N_{\check{a}^j} = \Sigma_{i=1}^{n}\Sigma_{l=1;l\neq j}^{k}(\check{A}_{i,l} = \check{A}_{i,j}), \tag{10}$$

where $N_{\check{a}^j}$ denotes the majority test for a candidate ambiguity vector $\check{a}^j$. With the larger value of $N_{\check{a}^j}$, $\check{a}^j$ can be expected to be closer to the center of the candidates; therefore, it is a reasonable metric to roughly measure its correctness. It is also noted that this algorithm shown in (8)–(10) performs a partial integer bootstrapping method [20,41]. To further increase its accuracy, the method of multi-epoch residual test is used for ambiguity validation, which excludes those abnormal candidate vectors with large loss values before applying this ambiguity majority test, as shown in the next subsection; After that, (9) can be applied to finally determine one ambiguity to be fixed. After updating the float ambiguities using (5), a new round of majority test can trigger. Iteratively, the full ambiguity set is resolved into integers, and the fixed solutions are obtained.

2.4. Multi-Epoch Residual Test for Ambiguity Validation

For multiple adjacent GNSS epochs of (1), the integer part of a will be identical as long as no cycle slips occur. For kinematic applications of the smartphone, the non-integer part of b is time-dependent. Therefore, we write a multi-epoch combined DD observation model as follows:

$$y_{t_{1,w}} = Aa + B_{t_1,w} b_{t_1,w} + e_{t_{1,w}}, Q_{y_{t_{1,w}}}, \tag{11}$$

$$y_{t_{1,w}} = \begin{bmatrix} y_{t_1} \\ y_{t_2} \\ \vdots \\ y_{t_w} \end{bmatrix}, B_{t_1,w} = \begin{bmatrix} B_{t_1} & & & \\ & B_{t_2} & & \\ & & \ddots & \\ & & & B_{t_w} \end{bmatrix}, b_{t_1,w} = \begin{bmatrix} b_{t_1} \\ b_{t_2} \\ \vdots \\ b_{t_w} \end{bmatrix}, e_{t_{1,w}} = \begin{bmatrix} e_{t_1} \\ e_{t_2} \\ \vdots \\ e_{t_w} \end{bmatrix}, \tag{12}$$

where the epoch times $t_1, t_2, \cdots, t_w$ are adjacent and w defines the window size of the model. Here, with the total number of measurements $m \times w$, the number of estimation parameters are n ambiguities and $p \times w$ user coordinates. Hence, with a larger value of w, the estimation redundancy grows, and the model complexity also increases. For ambiguity validation, the integer part of a is provided from a candidate vector $\check{a}^i$, where the non-integer part can be estimated using least squares, denoted as $\hat{b}^i_{t_{1,w}}$, as well as the measurement residuals, denoted as $\hat{r}^i_{t_{1,w}}$:

$$\hat{b}^i_{t_{1,w}} = \left(B^T_{t_1,w} Q^{-1}_{y_{t_{1,w}}} B_{t_1,w}\right)^{-1} B^T_{t_1,w} Q^{-1}_{yy_{t_{1,w}}} \left(y_{t_{1,w}} - A\check{a}^i\right), \tag{13}$$

$$\hat{r}^i_{t_{1,w}} = y_{t_{1,w}} - A\check{a}^i - B_{t_1,w}\hat{b}^i_{t_{1,w}}, \tag{14}$$

Therefore, the problem of ambiguity validation can be interpreted as an overall residual test based on $\hat{r}^i_{t_{1,w}}$, which can be calculated as follows [42–44]:

$$T_q = \frac{\left\|\hat{r}^i_{t_{1,w}}\right\|^2_{Q_{y_{t_{1,w}}}}}{q}; \tag{15}$$

where T_q denotes the overall test statistics, and q represents the degree of freedom, in this case, $q = (m - p) \times w$. In the sequel, the calculated T_q is used as the residual test statistics of the DD model, called DD residual test statistics. It should be noted that considering the existence of cycle slips, a practical implementation of this method should adaptively reduce the window size for satellites subjected to the cycle slip detection. In addition, to determine a proper window size w, the time correlation of the carrier phase measurements should be considered depending on the GNSS device [45,46]. In the case of the experiment smartphones, Google Pixel 5 and Xiaomi Mi 8, this study used $w = 10$ s, to balance the algorithm efficiency and real-time computational load.

Detailed algorithm implementation parameters are summarized in Table 1. For each iteration, 10 candidate ambiguity vectors are generated using the LAMBDA method, which is reduced to 4 candidates by both the majority test using (10) and residual test using (15). Then, the single fixable ambiguity can be determined using (9).

Table 1. Detailed algorithm implementation parameters for the proposed AR algorithm.

Parameter Descriptions	Values
Number of candidate vectors by LAMBDA	10 candidates
Majority test selection for candidate ambiguity vectors	First 4 candidates
Residual test selection for candidate ambiguity vectors	First 4 candidates
Residual test window size	10 epochs

In the following subsections, a static experiment is presented to show the efficiency of the proposed method in smartphone AR, followed by a kinematic experiment to demonstrate its positioning accuracy.

3. Static Experiment—Smartphone Ambiguity Resolution Efficiency

To evaluate the performance of the proposed method in terms of AR, this subsection focuses on an experiment based on the Xiaomi Mi 8 smartphone. Although previous studies widely use external GNSS antennas and signal repeaters to enhance smartphone observation quality, this study is based on the smartphone as it is, where no external antennas and signal repeaters are applied. To extract the evaluation details such as ambiguity biases, this experiment is static, with the smartphone in the upright position and with known reference coordinates. First, the data show the number of satellites and a skyplot. Second, the positioning solutions are discussed, where the proposed AR method is compared with the LAMBDA method with a full ambiguity resolution (FAR) strategy [47], r-ratio ambiguity validation with a 2.0 ratio threshold [48] due to the generally lower success rate on smartphones, and the float solutions. Later, the efficiencies of the ambiguity majority test and the DD residual test are demonstrated. Last, with the fixed smartphone position and the resolved ambiguity integers, the estimated ambiguity biases are captured and analyzed. In the sequel, the error statistics of root-mean-square (RMS), standard deviation (STD), and mean values are commonly used.

This experiment was conducted in Calgary, Canada, at the local time of 1 PM, 5 March 2021, where the software Geo++ RINEX Logger, version 2.4.3, is used to collect the 1-Hz GNSS data from Xiaomi Mi 8. The base station receiver is a Trimble NetR9, with a baseline length of 9.23 km, which logs at the rate of 1 Hz. Overall, the dataset includes 5261 GNSS epochs, with a total duration of 87.7 min. For data processing, two GNSS constellations, GPS and Galileo, are used, with the signals of GPS L1 C/A and L5 (Q) and that of Galileo E1 (C) and E5a (Q), with an elevation cut-off angle of 4° and a signal-to-noise-ratio threshold of 10 dB-Hz. For the float ambiguity solutions, a Kalman filtering scheme with kinematic parameter settings is used, shown in Table 2, which can be found in Takasu and Yasuda [49]. Additionally, the ionosphere and troposphere model corrections, referring to the Klobuchar model and the Saastamoinen model with the Neill mapping function, respectively, are applied to the measurements of (1) in advance [50–52]. After the double-differencing and these model corrections, ionosphere and troposphere errors can be considered to be eliminated in our processing [53].

Table 2. Filtering parameters for static and kinematic experiments.

Filtering Parameters	Values
Stochastic modelling method	Elevation-dependent model [54], $\sigma_P^2 = 0.5 + \frac{1.5}{\sin^2(E)}$ m^2, $\sigma_\Phi^2 = 10^{-6} + \frac{4 \times 10^{-6}}{\sin^2(E)}$ m^2
Initial state variance	Coordinates $\sigma_{0,crd}^2 = 1000.0$ m^2, ambiguities $\sigma_{0,amb}^2 = 1000.0$ m^2
Process noise	Coordinates $Q_{crd} = 20.0$ m^2, ambiguities $Q_{amb} = 0.00001$ m^2

For AR, the proposed method is used and compared with the LAMBDA method. Here, to avoid the antenna offset problem among signal frequencies, only the signals on the first frequency, that is, GPS L1 C/A and Galileo E1 (C), are used for AR.

Figure 1 reveals that, on average, Xiaomi Mi 8 has 10.1, 1.6, 5.5, and 5.2 satellites on the signals of GPS L1 C/A and L5 (Q) and that of Galileo E1 (C) and E5a (Q), respectively. Overall, the observed satellites are 13 to 18 in total, while GPS and Galileo have 8 to 11 and 5 to 7, respectively, which are sufficient for the AR experiment. As shown in Figure 2, G14, G28, G30, and E21 are mostly under high-elevation conditions.

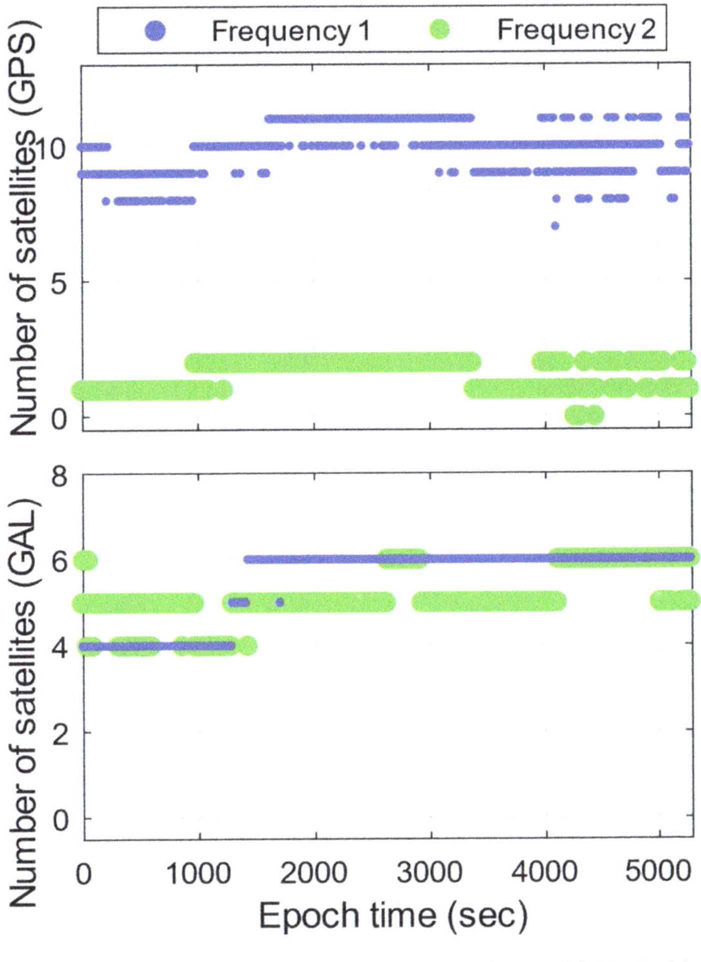

Figure 1. Numbers of GPS and Galileo (GAL) satellites of Xiaomi Mi 8 for dual frequency. For GPS, Frequency 1 is L1 C/A, and Frequency 2 is L5 (Q). For Galileo, Frequency 1 is E1 (C), and Frequency 2 is E5a (Q).

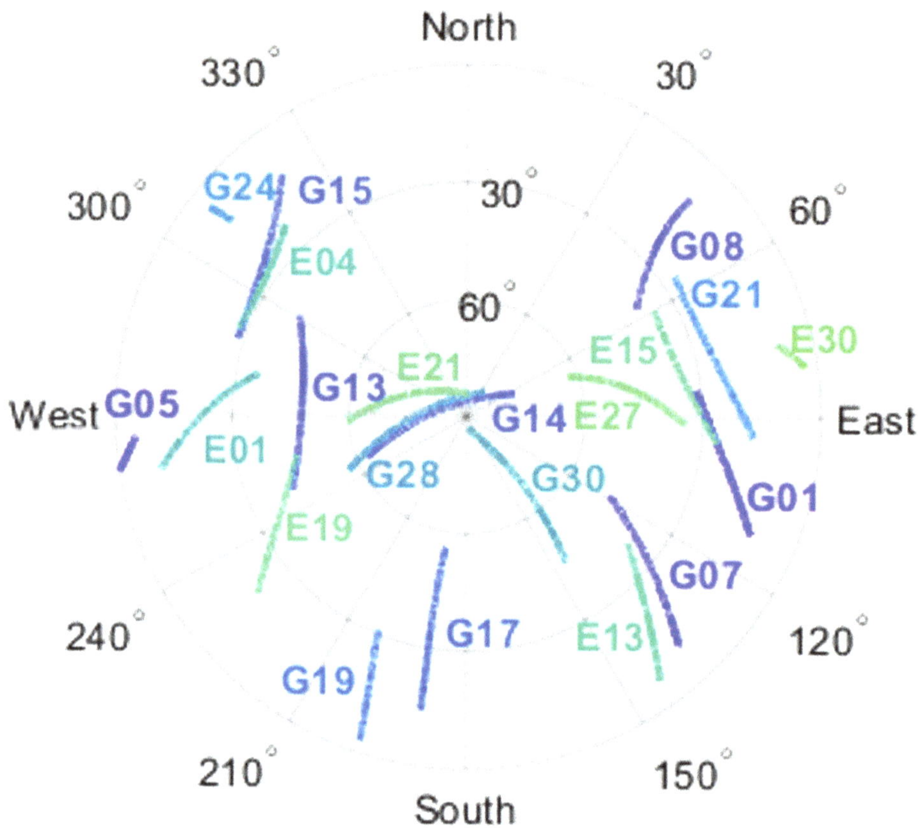

Figure 2. Skyplot of GPS and Galileo satellites of Xiaomi Mi 8.

Figure 3 provides the positioning error time-series of Xiaomi Mi 8 using the proposed method, compared with the LAMBDA method and float solutions. Although it is a static experiment, the data are processed based on kinematic filtering settings; therefore, the results are representative for kinematic applications as well. The static coordinates of the smartphone are calculated by static post-processing with AR, which gives the millimeter-level STD so it can be adopted as a reference. Generally, the proposed method reaches centimeter-level RMS values, which means 100% correct fixation. For comparison purposes, the LAMBDA method is significantly biased because it lacks consideration of the ambiguity biases and often provides incorrect ambiguity vectors. After a period of convergence, it reaches the correct ambiguity set, which took 3000 epochs. As a result, the LAMBDA method reaches an accuracy of 0.6 m, 0.7 m, and 1.3 m, which are worse than that of the float solutions, reaching 0.2 m, 0.2 m, and 0.1 m, respectively. As a primary conclusion, the proposed method provides the correct ambiguity set in most cases.

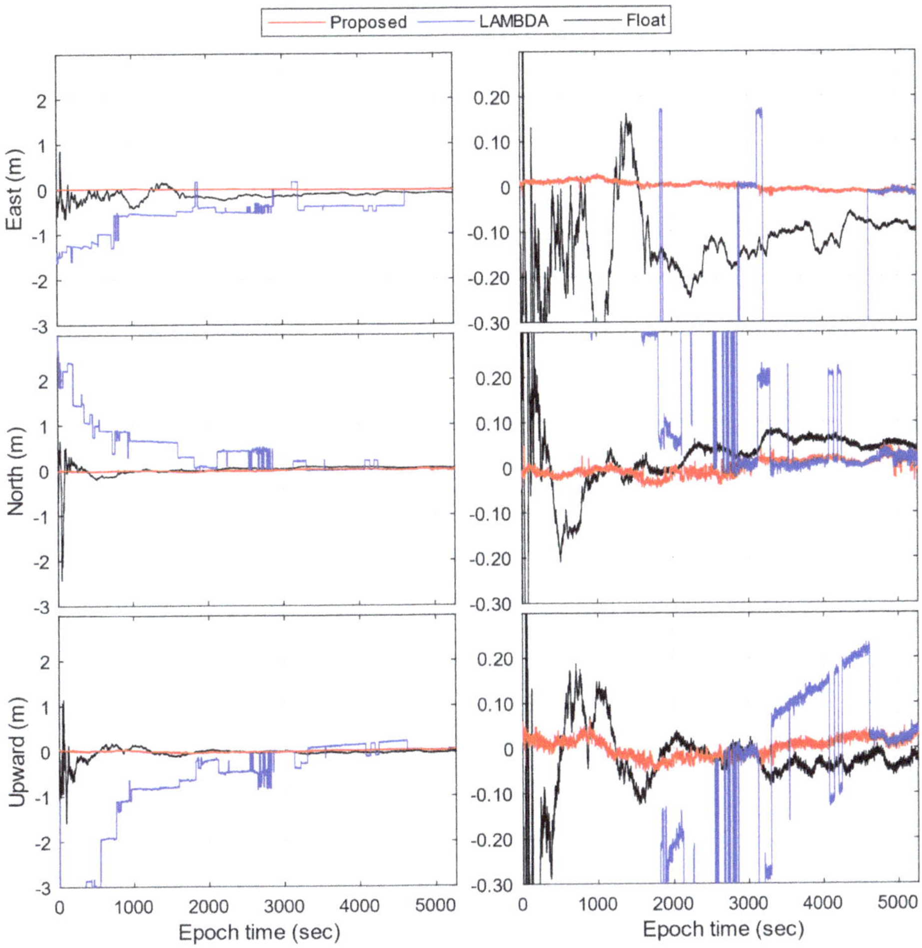

Figure 3. Positioning error time-series of Xiaomi Mi 8 by the proposed method, the LAMBDA method, and float solutions. The figures on the left are zoomed-out plots, while the figures on the right are zoomed-in plots.

In Figure 4, the DD residual test is evaluated in terms of its efficiency. Here, the resolved ambiguity vector for each epoch is obtained by the proposed AR method, which is compared with the other candidate ambiguity vectors from the LAMBDA method in the majority test in terms of the DD residual test statistics. As shown, the test statistics of the candidates are mostly from 0.2 cm to 3.5 cm, while the resolved ambiguity vectors have DD residual test statistics that are mostly from 0.1 cm to 0.5 cm. For 74% of all epochs, the resolved ambiguity vectors have the lowest residual test statistics, which have the second lowest residual test statistics for 2% of epochs. In summary, the DD residual test is a promising way to select trustworthy candidate ambiguity vectors.

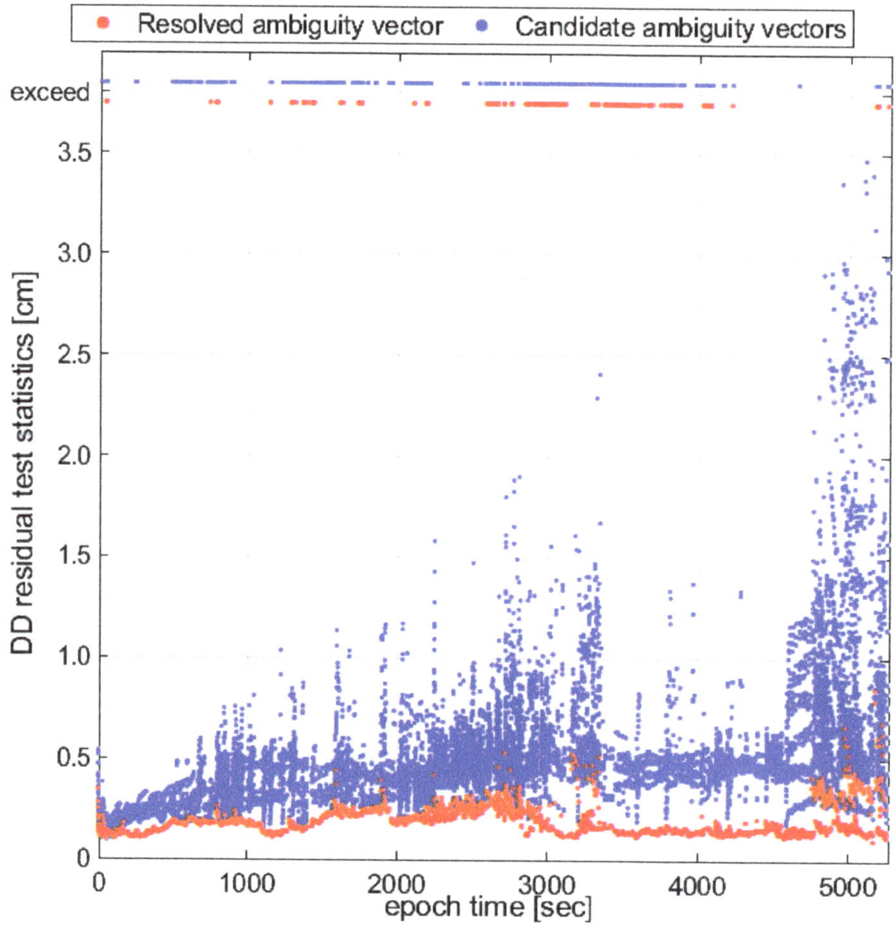

Figure 4. DD residual test statistics of the resolved ambiguity vector compared with candidate ambiguity vectors from the LAMBDA method in the majority test.

For the ambiguity majority tests, Figure 5 illustrates the AR efficiency by comparing the test values from the resolved ambiguity vector and the candidate ambiguity vectors from the LAMBDA method by the majority test, reflected by (9). At a glance, the resolved ambiguity vector gives higher test values than the other candidates in most cases, that is, for 3714 of all epochs (71%). For all epochs, the resolved ambiguity vector mostly ranks first to second among all candidate vectors, which is 1.3 on average. Therefore, it is reasonable to exclude all candidates whose majority test values are lower than the third.

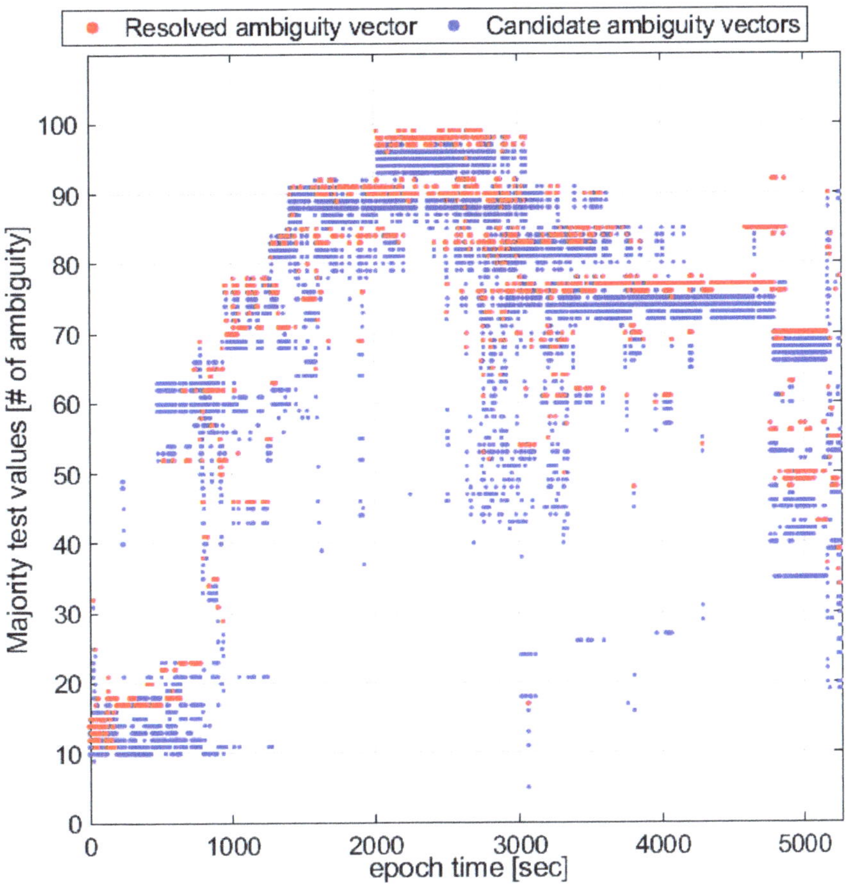

Figure 5. Majority test values of the resolved ambiguity vectors compared with candidate ambiguity vectors from the LAMBDA method in the majority test.

After achieving AR, the ambiguity biases can be captured using the static reference coordinate of the smartphone and resolved ambiguities, as input by (1). Figure 6 provides the time-series of the captured ambiguity biases. Although satellite ambiguities with healthy conditions are witnessed, such as E15 and E19, it is evident that ambiguity biases exist, which can evolve over time, especially for G13, G17, G21, and E01. Moreover, it is observed that GPS satellites suffer half-cycle slips, where sudden jumps of 0.5 cycles can frequently occur; see G07 and G14 as examples. Similar performances can be seen in [9], which agrees with our converged LAMBDA solutions shown in Figure 3 from epoch 4500. However, these contribute to the ambiguity biases, making it less possible to correctly achieve AR in real-time applications. Consequently, it is reasonable that conventional AR methods that have excellent performances on common GNSS receivers, such as LAMBDA, have limited efficiency on smartphone applications because there are frequent ambiguity biases.

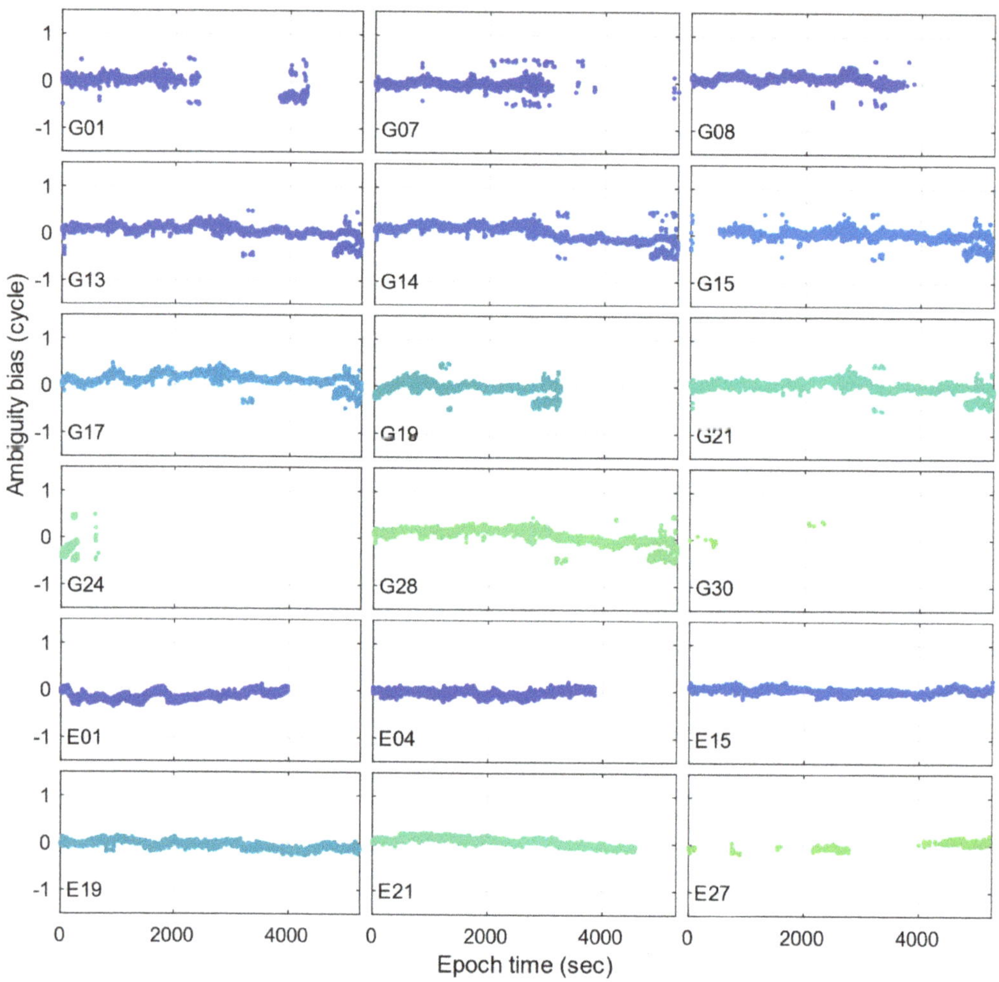

Figure 6. Ambiguity bias time-series for each GPS and Galileo satellite.

For a detailed explanation, Figure 7 presents the statistics of ambiguity biases for each satellite. Their RMS values reach 0.07 to 0.31 cycles, which are normally 0.03 to 0.05 cycles for GNSS receivers such as u-blox modules. For the mean values, they vary from 0.02 to 0.20 cycles, which means almost 0 cycles compared with u-blox. In other words, the existence of smartphone ambiguity biases cannot be ignored before achieving AR using current methods.

In summary, this static experiment with Xiaomi Mi 8 proves that the proposed method demonstrates a significant improvement in terms of AR efficiency. The results show that the search-and-shrink procedure coupled with the majority test and the DD residual test is efficient in obtaining the correct ambiguity vectors from candidates. It also proves the existence of ambiguity biases in smartphone GNSS data, which further demonstrates the necessity of the proposed method.

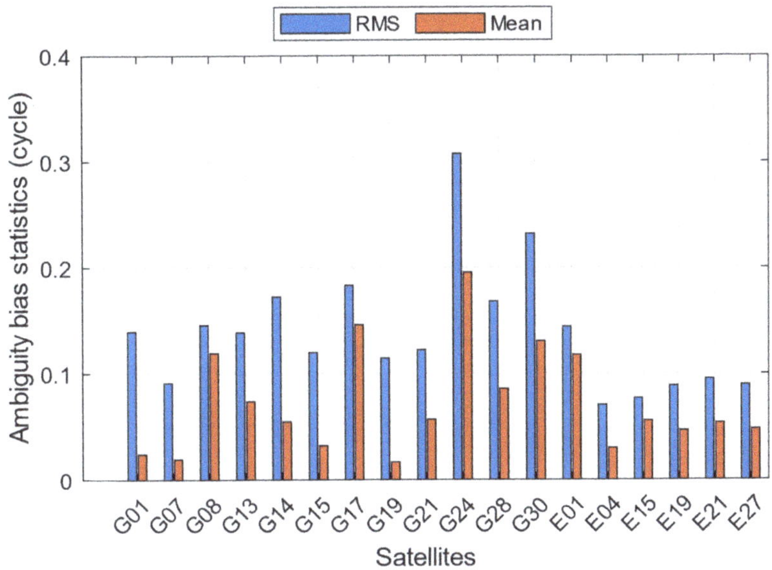

Figure 7. Ambiguity bias statistics for each GPS and Galileo satellite.

4. Kinematic Experiment—Smartphone Positioning Performance

The kinematic experiment uses the smartphone Google Pixel 5, whose ground trajectory and data collection platform are shown in Figure 8. Similar to the static experiment, we apply no external antennas or signal repeaters to perform data collection. As can be seen, an open area is selected, where dynamic ground multipath, antenna offsets and smartphone orientation variations are the primary causes of ambiguity biases. To provide positioning reference, two survey-grade GNSS antennas with u-blox ZED-F9P receivers are used to provide RTK fixed solutions with centimeter-level accuracy. In this way, the relative location of Google Pixel 5 can be described through two directions, that is, the along-track and the cross-track directions with respect to the antennas of GNSS receivers, which are denoted as Receiver #1 and #2, respectively. With a long-term calibration of the along-track and the cross-track offsets of Google Pixel 5, 36.51 cm and 9.90 cm can be obtained for their ground truth values, respectively. This has made us capable of conducting kinematic accuracy evaluations with high levels of confidence.

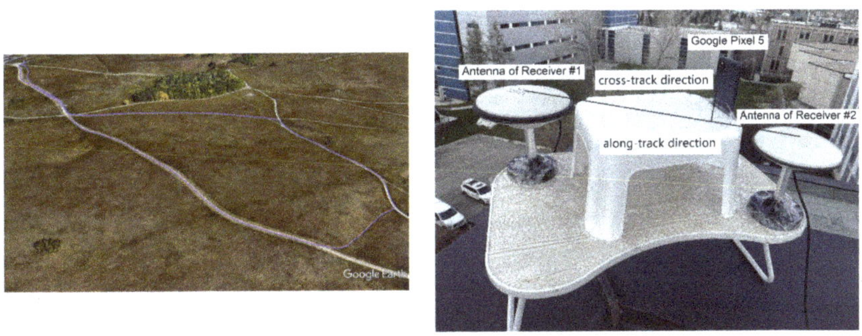

(**a**) (**b**)

Figure 8. Ground trajectory (**a**) and data collection platform (**b**) of the kinematic experiment using Google Pixel 5.

This experiment was conducted in Calgary, Canada, at the local time of 8 PM, 4 April 2022, where smartphone GNSS data are collected by Geo++ RINEX Logger, version 2.4.3. The base station is equipped with a Trimble NetR9, with a 5 km distance from the smartphone. The dataset includes a total of 1771 s. The GPS, Galileo, and GLONASS constellations are used, where GLONASS satellites are unavailable for AR but are used as an additional source for geometry-based cycle-slip detection [55]. In this experiment, GLONASS is needed because, in the kinematic applications, cycle-slips are more frequent than the previous static applications, which this study should consider and minimize by increasing accessible satellites. During the kinematic experiment, the first and last 200 epochs are static when verifying the performance of Google Pixel 5 compared with Xiaomi Mi 8. The rest of the configurations are the same as the static experiment. In Figure 9, the numbers of satellites involved and the frequency signals are plotted, which are, on average, 9.8, 2.7, 9.8, 8.4, and 7.7 for GPS L1, GPS L5, Galileo E1, Galileo E5a, and GLONASS L1, respectively.

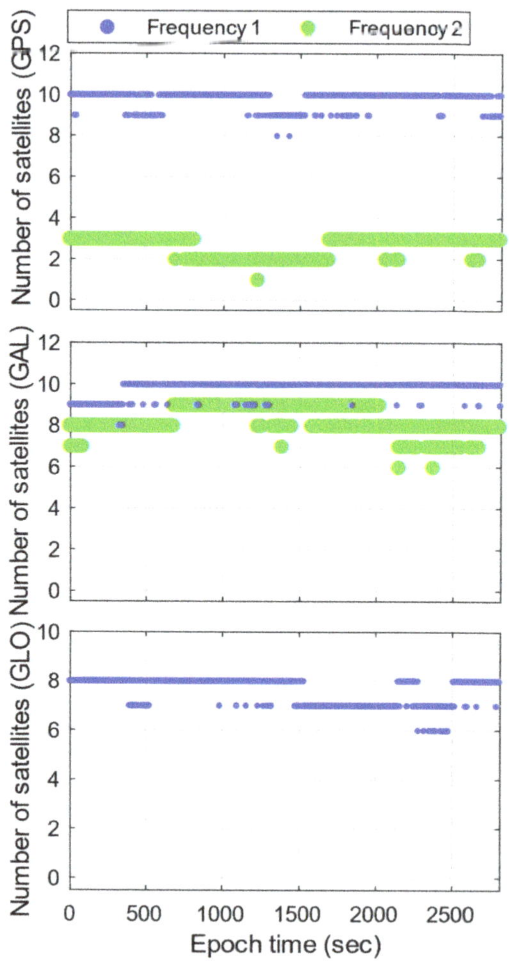

Figure 9. Numbers of GPS, Galileo (GAL), and GLONASS (GLO) satellites of Google Pixel 5 for dual frequency. For GPS, Frequency 1 is L1 C/A, and Frequency 2 is L5 (Q). For Galileo, Frequency 1 is E1 (C), and Frequency 2 is E5a (Q). For GLONASS, Frequency 1 is L1.

Figure 10 provides the 2D, along-track, and cross-track positioning errors during the experiment. In general, the solutions using the LAMBDA method are scattered, where the average error distance is 11.74 cm. For the proposed method, it is 4.6 cm, which indicates a significant improvement. For 95% of the data, the proposed method has an error distance within 10.2 cm versus 39.9 cm for the LAMBDA method. Therefore, it is concluded that the proposed method improves positioning accuracy by considering ambiguity biases in smartphones.

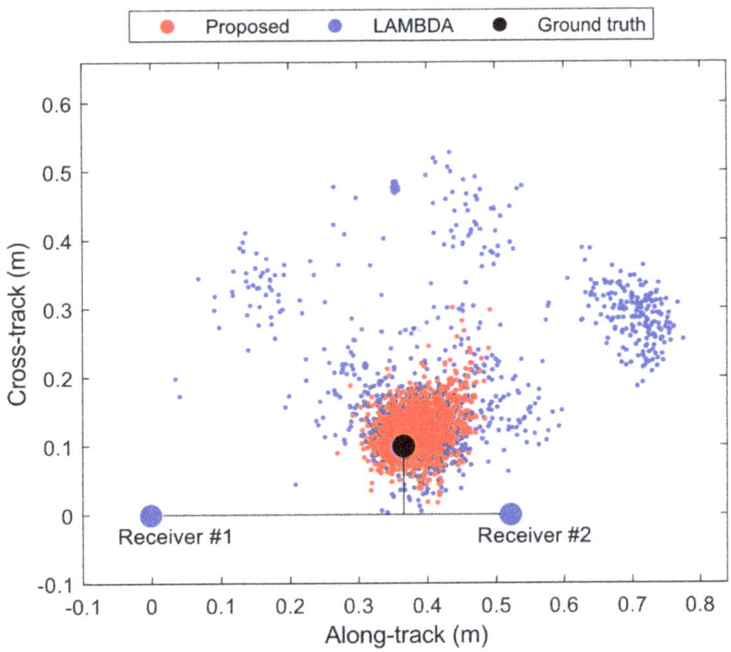

Figure 10. Along-track and cross-track positioning errors of Google Pixel 5 with respect to the ground truth (black dot) and antennas of two GNSS receivers (blue dots).

Figure 11 presents the time-series of the along-track and cross-track positioning errors, comparing the proposed method with LAMBDA and the float solutions. The RMS values of the float solutions are 9.8 cm and 8.1 cm for the along-track and the cross-track directions, in contrast to 12.8 cm and 11.7 cm for LAMBDA, respectively. Generally, LAMBDA shows lower performance than the float solutions due to the existence of ambiguity biases that frequently affect its search-and-shrink procedure; therefore, the resolved ambiguities are less trustworthy. For the proposed method, the optimized solution reaches accuracy values of 3.8 cm and 3.9 cm. In addition, it can be seen that the first and last 200 epochs have smoother solutions, and this is because, when static, the quality of smartphone GNSS measurements, including the noise levels of the carrier phases and pseudoranges [56], and satellite availability are relatively better [57]. Overall, it is evident that the proposed method outperforms the LAMBDA method and the float solutions in terms of AR and positioning accuracy.

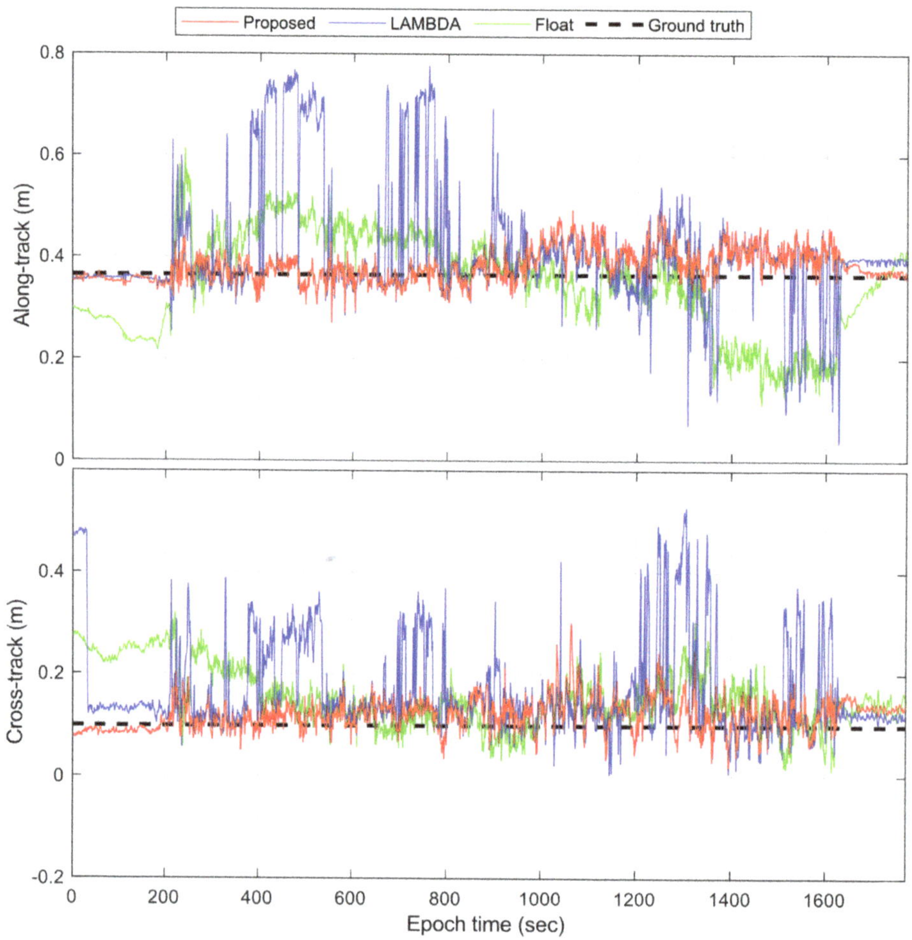

Figure 11. Along-track and cross-track positioning error time series of Google Pixel 5.

5. Conclusions

This study proposes an improved AR algorithm for smartphone positioning by considering ambiguity biases, where the search-and-shrink method is used with the testing methods, including a multi-epoch DD model residual test and majority tests for candidate ambiguities and vectors. The static dataset is first applied to evaluate smartphone AR efficiency. Secondly, the kinematic data verify the improvement in smartphone positioning performance. The key points are summarized as follows:

1. The existence of ambiguity biases is not negligible for AR based on smartphone devices. In the static experiment performed with Xiaomi Mi 8, the average level of ambiguity biases ranges from 0.07 to 0.31 cycles.
2. The proposed AR scheme using the search-and-shrink procedure coupled with the majority test and the multi-epoch DD residual test can overcome the problem of AR. The majority test can identify the actual ambiguity vector from the candidates with an accuracy of 71% for the first rank and 6% for the second rank versus 74% and 2% for the DD residual test.
3. The proposed method achieves AR to improve the positioning accuracy of smartphones. For the static test, the RMS values are 1.1 cm, 1.7 cm, and 2.1 cm for east,

north, and upward directions, in contrast to 0.2 m, 0.2 m, and 0.1 m for the float solutions, respectively. For the kinematic test, the RMS values are 3.8 cm and 3.9 cm for the along-track and the cross-track directions versus 9.8 cm and 8.1 cm for the float solutions.

Author Contributions: Y.J. and Y.G. (Yuting Gao) designed the research algorithm, performed the analysis, and wrote the paper; Y.J. and W.D. conducted the kinematic experiment and performed the analysis. F.L. conducted the static experiment and revised the manuscript. Y.G. (Yang Gao) supervised the project and revised the manuscript. All authors have read and agreed to the published version of the manuscript.

Funding: This research received no external funding.

Institutional Review Board Statement: Not applicable.

Informed Consent Statement: Not applicable.

Data Availability Statement: The GNSS dataset associated with this study is publicly accessible in the GitHub repository (https://github.com/Yuting1117/Smartphone-AR-Dataset, accessed on 15 May 2023).

Acknowledgments: The authors would like to acknowledge Profound Positioning Inc. for providing experiment devices and facilities.

Conflicts of Interest: The authors declare no conflict of interest.

References

1. Banville, S.; Van Diggelen, F. Precise positioning using raw GPS measurements from Android smartphones. *GPS World* **2016**, *27*, 43–48.
2. European Union Agency for the Space Programme. World's First Dual-Frequency GNSS Smartphone Hits the Market. 2018. Available online: https://www.euspa.europa.eu/newsroom/news/world-s-first-dual-frequency-gnss-smartphone-hits-market (accessed on 15 May 2023).
3. Cozzens, T. Qualcomm Launches 3 Dual-Frequency + NavIC Smartphone Modules. GPS WORLD. 2020. Available online: https://www.gpsworld.com/qualcomm-launches-3-dual-frequency-navic-smartphone-modules/ (accessed on 15 May 2023).
4. Håkansson, M. Characterization of GNSS observations from a Nexus 9 Android tablet. *GPS Solut.* **2019**, *23*, 21. [CrossRef]
5. Robustelli, U.; Baiocchi, V.; Pugliano, G. Assessment of Dual Frequency GNSS Observations from a Xiaomi Mi 8 Android Smartphone and Positioning Performance Analysis. *Electronics* **2019**, *8*, 91. [CrossRef]
6. Robustelli, U.; Paziewski, J.; Pugliano, G. Observation Quality Assessment and Performance of GNSS Standalone Positioning with Code Pseudoranges of Dual-Frequency Android Smartphones. *Sensors* **2021**, *21*, 2125. [CrossRef]
7. Teunissen, P.; Verhagen, S. GNSS carrier phase ambiguity resolution: Challenges and open problems. In *Observing Our Changing Earth*; Springer: Berlin/Heidelberg, Germany, 2009; pp. 785–792.
8. Li, G.; Geng, J. Characteristics of raw multi-GNSS measurement error from Google Android smart devices. *GPS Solut.* **2019**, *23*, 90. [CrossRef]
9. Geng, J.; Li, G. On the feasibility of resolving Android GNSS carrier-phase ambiguities. *J. Geod.* **2019**, *93*, 2621–2635. [CrossRef]
10. Riley, S.; Lentz, W.; Clare, A. On the Path to Precision—Observations with Android GNSS Observables. In Proceedings of the ION GNSS+ 2017, Portland, OR, USA, 25–29 September 2017.
11. Zangenehnejad, F.; Gao, Y. GNSS smartphones positioning: Advances, challenges, opportunities, and future perspectives. *Satell. Navig.* **2021**, *2*, 24. [CrossRef] [PubMed]
12. Darugna, F.; Wübbena, J.B.; Wübbena, G.; Schmitz, M.; Schön, S.; Warneke, A. Impact of robot antenna calibration on dual-frequency smartphone-based high-accuracy positioning: A case study using the Huawei Mate20X. *GPS Solut.* **2021**, *25*, 15. [CrossRef]
13. Dabove, P.; Di Pietra, V. Towards high accuracy GNSS real-time positioning with smartphones. *Adv. Space Res.* **2019**, *63*, 94–102. [CrossRef]
14. Paziewski, J.; Sieradzki, R.; Baryla, R. Signal characterization and assessment of code GNSS positioning with low-power consumption smartphones. *GPS Solut.* **2019**, *23*, 98. [CrossRef]
15. Gao, R.; Xu, L.; Zhang, B.; Liu, T. Raw GNSS observations from Android smartphones: Characteristics and short-baseline RTK positioning performance. *Meas. Sci. Technol.* **2021**, *32*, 084012. [CrossRef]
16. Wu, Q.; Sun, M.; Zhou, C.; Zhang, P. Precise Point Positioning Using Dual-Frequency GNSS Observations on Smartphone. *Sensors* **2019**, *19*, 2189. [CrossRef] [PubMed]
17. Benvenuto, L.; Cosso, T.; Delzanno, G. An Adaptive Algorithm for Multipath Mitigation in GNSS Positioning with Android Smartphones. *Sensors* **2022**, *22*, 5790. [CrossRef]

18. Wanninger, L.; Heßelbarth, A. GNSS code and carrier phase observations of a Huawei P30 smartphone: Quality assessment and centimeter-accurate positioning. *GPS Solut.* **2020**, *24*, 64. [CrossRef]
19. Heßelbarth, A.; Wanninger, L. Towards centimeter accurate positioning with smartphones. In Proceedings of the 2020 European Navigation Conference (ENC), Dresden, Germany, 23–24 November 2020; IEEE: Piscataway, NJ, USA, 2020; pp. 1–8.
20. Yong, C.Z.; Odolinski, R.; Zaminpardaz, S.; Moore, M.; Rubinov, E.; Er, J.; Denham, M. Instantaneous, Dual-Frequency, Multi-GNSS Precise RTK Positioning Using Google Pixel 4 and Samsung Galaxy S20 Smartphones for Zero and Short Baselines. *Sensors* **2021**, *21*, 8318. [CrossRef]
21. Teunissen, P.J.G. Success probability of integer GPS ambiguity rounding and bootstrapping. *J. Geod.* **1998**, *72*, 606–612. [CrossRef]
22. Parkins, A. Increasing GNSS RTK availability with a new single-epoch batch partial ambiguity resolution algorithm. *GPS Solut.* **2011**, *15*, 391–402. [CrossRef]
23. Hatch, R. The synergism of GPS code and carrier measurements. In Proceedings of the International Geodetic Symposium on Satellite Doppler Positioning, Las Cruces, NM, USA, 8–12 February 1982; Volume 2, pp. 1213–1231.
24. Wen, Q.; Geng, J.; Li, G.; Guo, J. Precise point positioning with ambiguity resolution using an external survey-grade antenna enhanced dual-frequency android GNSS data. *Measurement* **2020**, *157*, 107634. [CrossRef]
25. Zeng, S.; Kuang, C.; Yu, W. Evaluation of Real-Time Kinematic Positioning and Deformation Monitoring Using Xiaomi Mi 8 Smartphone. *Appl. Sci.* **2022**, *12*, 435. [CrossRef]
26. Teunissen, P. The least-squares ambiguity decorrelation adjustment: A method for fast GPS integer ambiguity estimation. *J. Geod.* **1995**, *70*, 65–82. [CrossRef]
27. Li, B.; Verhagen, S.; Teunissen, P.J. GNSS integer ambiguity estimation and evaluation: LAMBDA and Ps-LAMBDA. In *China Satellite Navigation Conference (CSNC) 2013 Proceedings*; Springer: Berlin/Heidelberg, Germany, 2013; pp. 291–301.
28. Teunissen, P.J.G. Theory of integer equivariant estimation with application to GNSS. *J. Geod.* **2003**, *77*, 402–410. [CrossRef]
29. Odolinski, R.; Teunissen, P.J. Best integer equivariant estimation: Performance analysis using real data collected by low-cost, single-and dual-frequency, multi-GNSS receivers for short-to long-baseline RTK positioning. *J. Geod.* **2020**, *94*, 91. [CrossRef]
30. Odolinski, R.; Teunissen, P.J.G. Best integer equivariant position estimation for multi-GNSS RTK: A multivariate normal and t-distributed performance comparison. *J. Geod.* **2022**, *96*, 3. [CrossRef]
31. Yong, C.Z.; Harima, K.; Rubinov, E.; McClusky, S.; Odolinski, R. Instantaneous Best Integer Equivariant Position Estimation Using Google Pixel 4 Smartphones for Single-and Dual-Frequency, Multi-GNSS Short-Baseline RTK. *Sensors* **2022**, *22*, 3772. [CrossRef]
32. Li, L.; Li, Z.; Yuan, H.; Wang, L.; Hou, Y. Integrity monitoring-based ratio test for GNSS integer ambiguity validation. *GPS Solut.* **2016**, *20*, 573–585. [CrossRef]
33. Khanafseh, S.; Pervan, B. New approach for calculating position domain integrity risk for cycle resolution in carrier phase navigation systems. *IEEE Trans. Aerosp. Electron. Syst.* **2010**, *46*, 296–307. [CrossRef]
34. Verhagen, S.; Li, B.; Geodesy, M. *LAMBDA Software Package: Matlab Implementation*; Delft University of Technology: Delft, The Netherlands, 2012.
35. Teunissen, P.J. Integer least-squares theory for the GNSS compass. *J. Geod.* **2010**, *84*, 433–447. [CrossRef]
36. Teunissen, P.J. Least-squares estimation of the integer GPS ambiguities. In *Invited Lecture, Section IV Theory and Methodology*; IAG General Meeting: Beijing, China, 1993; pp. 1–16.
37. Chen, D.; Lachapelle, G. A Comparison of the FASF and Least-Squares Search Algorithms for on-the-Fly Ambiguity Resolution. *Navigation* **1995**, *42*, 371–390. [CrossRef]
38. Giorgi, G.; Teunissen, P.; Buist, P. A search and shrink approach for the baseline constrained LAMBDA method: Experimental results. In Proceedings of the International GPS/GNSS Symposium, Tokyo, Japan, 11–14 November 2008.
39. De Jonge, P.; Tiberius, C. *The LAMBDA Method for Integer Ambiguity Estimation: Implementation Aspects*; Publications of the Delft Computing Centre, LGR-Series; Delft University of Technology: Delft, The Netherlands, 1996; Volume 12, pp. 1–47.
40. Chang, X.-W.; Yang, X.; Zhou, T. MLAMBDA: A modified LAMBDA method for integer least-squares estimation. *J. Geod.* **2005**, *79*, 552–565. [CrossRef]
41. Teunissen, P.; Massarweh, L.; Verhagen, S. Vectorial integer bootstrapping: Flexible integer estimation with application to GNSS. *J. Geod.* **2021**, *95*, 99. [CrossRef]
42. Koch, K.-R. *Parameter Estimation and Hypothesis Testing in Linear Models*; Springer Science & Business Media: Berlin/Heidelberg, Germany, 1999.
43. Teunissen, P.J. *Testing Theory: An Introduction*; D. C. Jones: New York, NY, USA, 2009; pp. 123–126.
44. Baarda, W. *A Testing Procedure for Use in Geodetic Networks*; Publication on Geodesy, New Series; Rijkscommissie Voor Geodesie: Delft, The Netherlands, 1968; Volume 2.
45. Chang, G. On kalman filter for linear system with colored measurement noise. *J. Geod.* **2014**, *88*, 1163–1170. [CrossRef]
46. Gao, Y.; Jiang, Y.; Gao, Y.; Huang, G. A linear Kalman filter-based integrity monitoring considering colored measurement noise. *GPS Solut.* **2021**, *25*, 59. [CrossRef]
47. Verhagen, S.; Li, B.; Teunissen, P.J. Ps-LAMBDA: Ambiguity success rate evaluation software for interferometric applications. *Comput. Geosci.* **2013**, *54*, 361–376. [CrossRef]
48. Verhagen, S.; Teunissen, P.J. New global navigation satellite system ambiguity resolution method compared to existing approaches. *J. Guid. Control Dyn.* **2006**, *29*, 981–991. [CrossRef]

49. Takasu, T.; Yasuda, A. Development of the low-cost RTK-GPS receiver with an open source program package RTKLIB. In Proceedings of the International Symposium on GPS/GNSS, International Convention Center, Jeju, Republic of Korea, 4–6 November 2009; Volume 1.
50. Saastamoinen, J. Atmospheric correction for the troposphere and stratosphere in radio ranging satellites. *Use Artif. Satell. Geod.* **1972**, *15*, 247–251.
51. O'Neill, M.; Ryan, C. Grammatical evolution. *IEEE Trans. Evol. Comput.* **2001**, *5*, 349–358. [CrossRef]
52. Klobuchar, J.A. Ionospheric time-delay algorithm for single-frequency GPS users. *IEEE Trans. Aerosp. Electron. Syst.* **1987**, *AES-23*, 325–331. [CrossRef]
53. Goad, C.C. Short distance GPS models. In *GPS for Geodesy*; Springer: Berlin/Heidelberg, Germany, 1996; pp. 239–262.
54. Li, B.; Lou, L.; Shen, Y. GNSS elevation-dependent stochastic modeling and its impacts on the statistic testing. *J. Surv. Eng.* **2016**, *142*, 04015012. [CrossRef]
55. Li, B.; Qin, Y.; Liu, T. Geometry-based cycle slip and data gap repair for multi-GNSS and multi-frequency observations. *J. Geod.* **2019**, *93*, 399–417. [CrossRef]
56. Shinghal, G.; Bisnath, S. Conditioning and PPP processing of smartphone GNSS measurements in realistic environments. *Satell. Navig.* **2021**, *2*, 10. [CrossRef]
57. Zhang, X.; Tao, X.; Zhu, F.; Shi, X.; Wang, F. Quality assessment of GNSS observations from an Android N smartphone and positioning performance analysis using time-differenced filtering approach. *GPS Solut.* **2018**, *22*, 70. [CrossRef]

MDPI
St. Alban-Anlage 66
4052 Basel
Switzerland
www.mdpi.com

Sensors Editorial Office
E-mail: sensors@mdpi.com
www.mdpi.com/journal/sensors

www.ingramcontent.com/pod-product-compliance
Lightning Source LLC
Chambersburg PA
CBHW061549210926
51850CB00009B/2045
* 9 7 8 3 0 3 6 5 8 8 1 8 6 *